ENVIRONMENTAL CHANGE

The ecological environment (or ecosphere) consists of life and life-support systems (air, water, soils and sediments, and topography). It is surrounded by the geological environment below and the cosmic environment above. Environmental change results from the interactions within ecosystems, from forces and events inside the Earth, and from forces and events in interplanetary and interstellar space. This book explores the nature, causes, rates, and directions of environmental change throughout Earth history. In short, it surveys the evolving ecosphere.

Since the Earth first formed, the environment has constantly changed. Details of the changing environment are complex and richly varied. Cutting through the morass of detail are three big questions. First, what causes environmental change? Is it mechanisms inside the ecosphere? Or is it external forces and events in the Cosmos at large and in the solid Earth? Second, what is the rate of environmental change? Is it a gradual and stately affair? Or is all the action crammed into short and hectic episodes? Third, where does environmental change lead? Does it maintain an approximate steady state? Or does it go in cycles? Or does it follow trends?

Environmental Change: The Evolving Ecosphere looks at change in life, climate (air and water), soils, sediments, and landforms. Descriptions of the cosmic, geological, and ecological environments should be valuable to student readers. Carefully selected examples are used to address the basic questions about environmental change – what causes it, how fast it occurs, and what directions it goes in. The chief conclusion is that environmental change is advantageously viewed in an evolutionary context – the ecosphere, biosphere, atmosphere, hydrosphere, pedosphere, and toposphere all unceasingly adapt to ever-changing circumstances.

Richard John Huggett is a Senior Lecturer in Geography at the University of Manchester.

ENVIRONMENTAL CHANGE

The evolving ecosphere

Richard John Huggett

London and New York

First published 1997
by Routledge
11 New Fetter Lane, London EC4P 4EE

Simultaneously published in the USA and Canada
by Routledge
29 West 35th Street, New York, NY 10001

Typeset in Garamond by LaserScript, Mitcham, Surrey
Printed and bound in Great Britain by
Butler & Tanner Ltd, Frome and London

British Library Cataloguing in Publication Data
A catalogue record for this book is available from the British Library

Library of Congress Cataloguing in Publication Data
A catalogue record for this book has been requested

ISBN 0–415–14520–1
0–415–14521–X (pbk)

For Jamie, Sarah, Edward, Daniel, Zoë, and Ben

CONTENTS

COLOUR PLATES

The following plates appear between pages 172 and 173

BLACK-AND-WHITE PLATES

FIGURES

TABLES

PREFACE

Books about the environment are immensely popular. A huge number are published each year. Most of the works deal with current concerns over human environmental impacts. Some discuss the issues in a historical context. Very few look back further than Holocene or Pleistocene times. However, past and future human environmental impacts are, arguably, best viewed from the full perspective of geological time. *Environmental Change* aims to provide that broadest of perspectives. It takes up, and enlarges upon, the combined ecological and evolutionary approaches established in my *Geoecology: An Evolutionary Approach* (1995). In brief, the ecosphere is defined as life plus life-support systems – biosphere, toposphere, atmosphere, pedosphere, and hydrosphere. It sits upon the geological environment (lithosphere and barysphere), and looks outwards to the cosmic environment. The ecological approach sees all parts of the ecosphere as interdependent and interacting with their cosmic and geological milieux. The evolutionary approach sees the ecosphere and its parts unendingly adapting to ever-changing circumstances.

Geoecology probed the present interdependencies within the ecosphere. *Environmental Change* focuses on the changing nature of these interdependencies, and the connections with cosmic and geological events and processes. It contemplates these changes and connections through Earth history. Indeed, an alternative title might be 'A History of the World in 8½ Chapters'. Much of the discussion revolves around environmental change over the last 500 million years, with a bias towards Pleistocene and Holocene changes. Taking such a long time-span inevitably poses problems of coverage. Some readers may regret the absence of several important topics, and the rather sketchy accounts of others. But I have had to be mightily selective in the examples used. All the chosen examples illustrate points about the big themes of environmental change – nature, causes, rates, and directions. A wide coverage also creates a problem with the definition of technical terms. To what extent will hard-rock geologists understand climatological terms, pedologists understand astronomical terms, or geomorphologists understand ecological terms? Many senior readers will probably be well versed in the several languages of Earth and life science. To smooth the way for those who are not yet scientific polyglots, I start each chapter with a simple account of the structure and functioning of the 'sphere' in question. Students, in particular, may find this a helpful stratagem.

I am grateful to many people involved with the production of this, my latest offering. Graham Bowden, 'King of CorelDRAW!', painstakingly adapted all the diagrams to my specifications. Sarah Lloyd at Routledge courageously took another Huggett book on board. All the scientists listed in the bibliography (and more) unknowingly provided me with hours of reading, and inspired me to write this book. Several people kindly provided me with photographs. Peter Dicken and other colleagues in the Geography Department at Manchester University cheerily continued to make work a pleasure. Derek Davenport willingly provided further beery discussions on life, the universe, and (this time) digital computers. And, last but not least (even numerically), my wife and family patiently stood by while 'Daddy wrote his story'.

Richard Huggett
Poynton
September 1996

1

INTRODUCING ENVIRONMENTAL CHANGE

The notion of change fascinates most people. All humans experience change during their lifetimes. They are all keenly aware of life cycles, of birth and death. They are all familiar with the passage of night and day and with the march of the seasons. Some will notice that the weather in one year is never quite the same as the weather in previous years, that severe winters or summer droughts will do damage to some plants but not to others. Some will have heard that winters in western Europe were harsher a couple of centuries ago, that the Earth has recently emerged from an Ice Age, and that very hot and humid climates prevailed in the remote past. Such popular notions of environmental change are derived from firsthand experience, hearsay, and media coverage of scientific investigations.

Scientists from environmental, geographical, and geological disciplines study environmental change. They explore changes in the ecosphere and its component systems – the atmosphere, biosphere, hydrosphere, pedosphere, and toposphere. Many of these scientists are interested in past environments. They flaunt their penchant for things and times past by adding the root 'palaeo' to their subject names – palaeoanthropology, palaeobiology, palaeoclimatology, palaeogeography, palaeoecology, and palaeoceanography. Archaeology, which studies the human historical past, is an exception.

The work of all scientists who delve into past environments confirms that environmental change is real, various, and complex. Some changes take an afternoon to accomplish, some take aeons. Some changes are unique, some are recurrent. Some changes are unidirectional, some are cyclical. Some changes are linear, some are non-linear. In short, environmental change comes in a splendid multiplicity of styles, none of which may be held as the norm. This finding is borne out by modellers seeking to predict future environmental change, and particularly the changes likely to occur in the next century as a result of human actions.

WHAT IS ENVIRONMENTAL CHANGE?

Most people have an idea what environmental change is, but they would be hard pressed to define it satisfactorily. The problem lies in the natural variability of environmental factors. Take the example of air temperature. This varies on a daily and a yearly basis. How then may sustained temperature changes be measured? A yardstick is required, and this is normally provided by taking average values for days, months, seasons, years, or longer periods. Like averages may then be compared. So, it might be found that the mean temperature for one year is greater than that in the previous year – a change has occurred.

There is nothing wrong with taking average values provided they are not treated as some sort of norm. Average values for all the main atmospheric variables over thirty-year periods are used to define the misleadingly named 'climatic normals'. The danger here lies in imagining that the climatic normals are somehow immovable. Current thinking suggests that climate is always changing, so that one generation's climatic norms are another generation's climatic extremes. The tenor of this argument applies to all environmental factors – change is the norm, constancy the exception.

Jumps, trends, and fluctuations

Environmental variables display three basic types of change – discontinuities, trends, and fluctuations (Figure 1.1; cf. R. G. Barry 1969). A discontinuity is an abrupt and permanent change in the average value. A trend is smooth increase or decrease, not necessarily a linear one, in the average. A fluctuation is regular or irregular change characterized by at least two maxima (or minima) and one minimum (or maximum). There are several kinds of fluctuation. An oscillation is smooth and gradual progress between maxima and minima. A periodicity is the recurrence of maxima and minima after a roughly constant time interval. Less than regular periodicity is quasi-periodicity. An episodicity is a sustained minimum (or 'norm') interrupted by an abrupt but temporary switch to a maximum. It marks an event, such as a large flood or a landslide. Repeated events may occur randomly or quasi-periodically. Another subtle possibility of change is that the average value is sustained for a time while the variability about the average increases or decreases. Subtler still is chaotic change. Chaos describes a cryptic pattern of change so irregular that it is easily mistaken for randomness.

All this sounds simple enough, but detecting discontinuities, trends, oscillations, and chaos in data sets is challenging. Commendably, mathematicians have proved equal to the task and devised a battery of sophisticated statistical procedures for analysing time series. Despite this sterling endeavour by the numerically gifted, many of the changes discussed

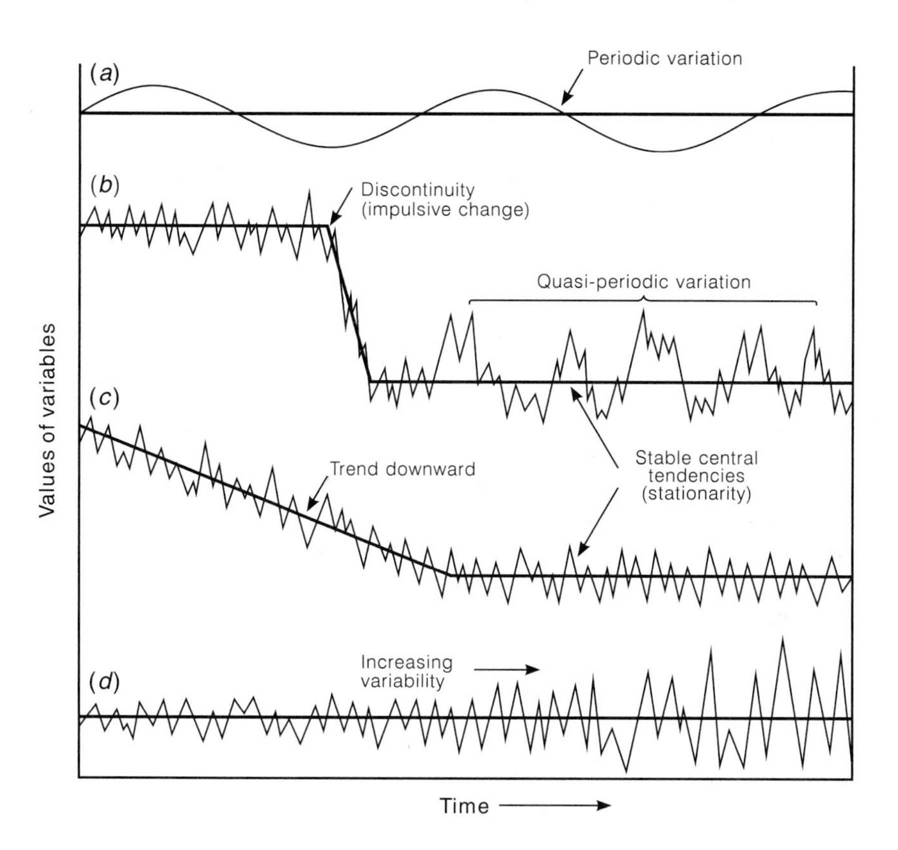

Figure 1.1 The chief styles of environmental change. The idealized time series apply to all variables that are continuous in time. Temperature and pressure are examples
Source: After Hare (1996)

in the context of Earth history are established by the arcane practice of 'eyeballing'. Only recently have time series suitable for rigorous statistical analysis been extracted from ice, rocks, sediments, and trees.

Trends, or directional changes, are an important aspect of environmental change. There are two alternatives – the environment may change in a definite direction, or else it may stay the same. In the first case there is a changing or transient state; in the second case there is a steady state. Oscillations complicate this simple picture. Steady and changing environmental variables may follow cycles. Indeed, it is probably safe to say that all environmental variables display some degree of cyclical variation. That being the case, non-cyclical change may be disregarded or viewed as very low amplitude periodic change.

Rates

How fast does mean annual temperature increase? How rapidly does a glacier retreat? How quickly did trees colonize newly exposed land after deglaciation? Questions such as these are relatively easy to answer for the instrumental period. Less sure answers can be given for historical and geological times. Indeed, before the coming of geochronometric dating methods, estimates of geological rates were at best a first approximation. Luckily, many geological rates can now be gauged using an absolute time-scale kept by geochronometers, which will be described in a later section.

Rates of environmental change range from slow to fast. Slow rates are usually referred to as gradual environmental change. Fast rates are sometimes described as catastrophic environmental change, though this term carries connotations of suddenness and violence and is shunned by many environmental, geographical, and geological scientists. Convulsive change is a euphemistic alternative. None the less, catastrophes are common in the natural world, catastrophic change does occur, and there is no good reason for dropping a perfectly good word. All rates between the slow and fast extremes occur in the environment.

Scales

Environmental change occurs across all spatial scales, from a cubic centimetre to the entire globe, and across all time-scales, from seconds to aeons. Space and time are continuous. However, it is normal practice to slice the Earth's spatial and temporal dimensions into convenient portions and to label them. No agreed labelling system exists. The prefixes micro, meso, macro, and mega are very popular. They produce the expressions microscale, mesoscale, macroscale, and megascale. It is far simpler, but far less impressive, to use the terms small-scale, medium-scale, large-scale, and very large-scale. The boundaries between the categories are somewhat arbitrary. How large does small have to be before it becomes medium? A consensus of sorts has emerged, though the exact divisions adopted depend to some extent upon what it is that changes – climate, soils, slopes, landscapes, and so forth. It is possible that the environment has a fractal structure. If this were the case, there would be natural divisions of time and space scales. This question is under investigation. Until a sure conclusion is reached, it seems best to create divisions using logical steps, say whole powers of ten. My current preferred scheme is shown in Figure 1.2. In this scheme, the spatial dimension has the following limits bestowed upon it: microscale is up to 1 km^2; mesoscale is 1–10,000 km^2; macroscale is 10,000–1,000,000 km^2; and megascale is from 1,000,000 km^2 to the entire surface area of the Earth.

The micro–meso–macro–mega designation may be applied to the time dimension. However, the terms short-term, medium-term, long-term, and

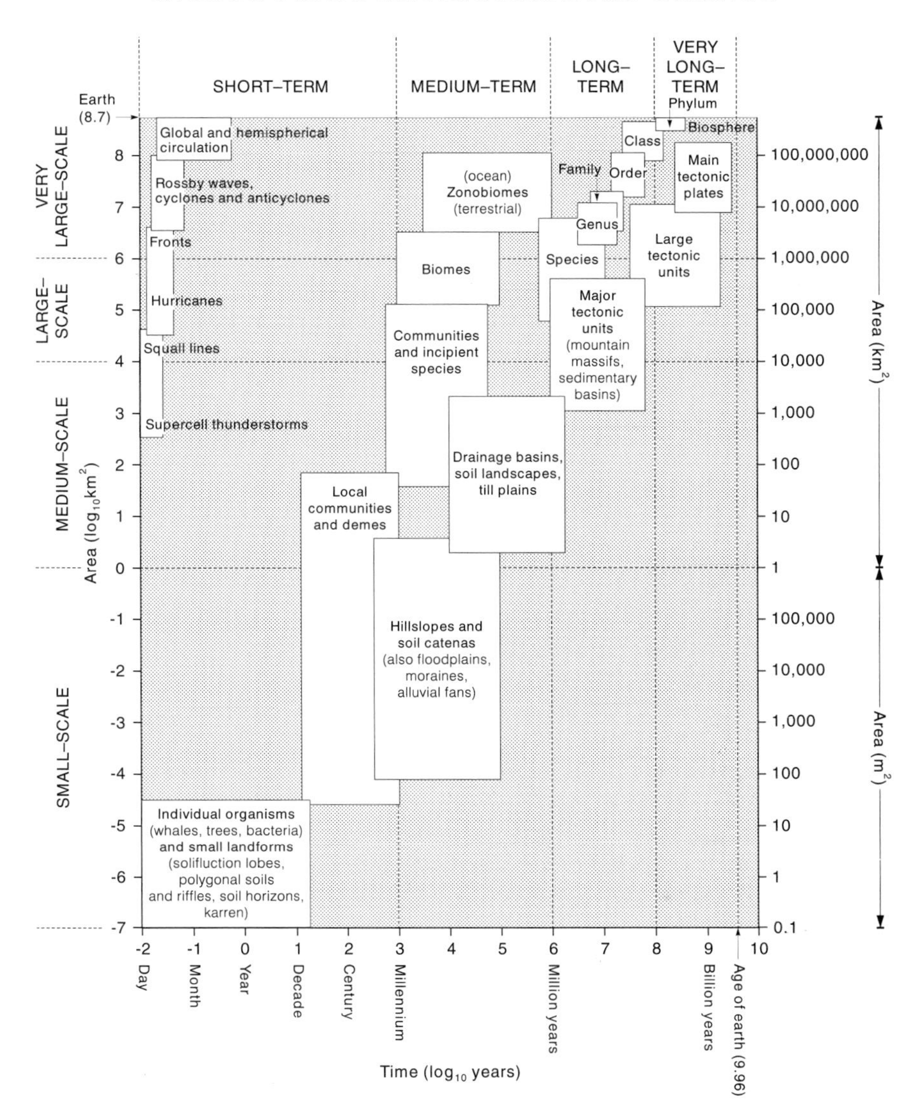

Figure 1.2 Scales of environmental change
Source: After Huggett (1991)

very long-term seem plainer. As with spatial divisions, so with temporal ones: it is a matter of personal preference where the limits between the time domains are set. My preference is to define short-term as up to 1,000 years, medium-term from 1,000 to 1,000,000 years, long-term from 1,000,000 to 100,000,000 years, and very long-term from 100,000,000 to 4,600,000,000 years. The terms 'megayear' (= 1 million years) and 'gigayear' (= 1 billion = 1,000 million years) are useful when talking of long-term and very long-term environmental changes.

Components of the environment relate to these spatial and temporal scales. Each component has a size and a 'life-span'. Some of the components discussed in this book (and others that are not) are plotted in Figure 1.2. Notice that atmospheric systems, be they local or global, survive for less than a year. This is due to the (literally) fluid nature of air. Tectonic units of the lithosphere may survive for billions of years. Soils, landscapes, and communities are intermediate between atmosphere and lithosphere, surviving for anything from centuries to hundreds of megayears.

WHY DOES THE ENVIRONMENT CHANGE?

Ecospheric influences

Environmental changes come from three sources (Figure 1.3). The first source is the ecosphere itself. The atmosphere, biosphere, hydrosphere, pedosphere, and toposphere interact amongst, and within, themselves. These interactions produce environmental cycles, steady states, and trends.

Directional changes (trends) often occur when a threshold within the ecosphere is crossed. This happened in Antarctica as the Eocene epoch closed. Late Eocene atmospheric cooling reached a critical point and Antarctic ice expanded. The Antarctic ice sheet has lasted ever since. The arrival of the modern ice age triggered profound changes in the biosphere. Some 2.5 million years ago in Africa, as the ice age began, savannahs (open grassy communities adapted to drier conditions) expanded at the expense of woodlands (closed forests adapted to wetter conditions). This event, which swept across Africa, led to the disappearance of forest-dwelling animal species and the emergence of grassland species. The human family was affected by this event. Gracile (slender) australopithecines (*Australopithecus afarensis* and *A. africanus*), which depended on forests for food and refuge, gave way to the first members of the human genus – *Homo* – who became adapted to life on open plains. Were it not for a trend towards cooling and aridification during Tertiary times, modern humans might not exist. Paradoxically, the sapient species whose origin was occasioned by a global environmental change is now changing the environment on a global scale.

Cosmic influences

The other two sources of environmental change lie outside the ecosphere (Figure 1.3). The ecosphere interacts with its surroundings on two fronts: it interacts 'below' with the solid Earth (the lithosphere and barysphere); and it interacts 'above' with the rest of the Cosmos (the cosmosphere).

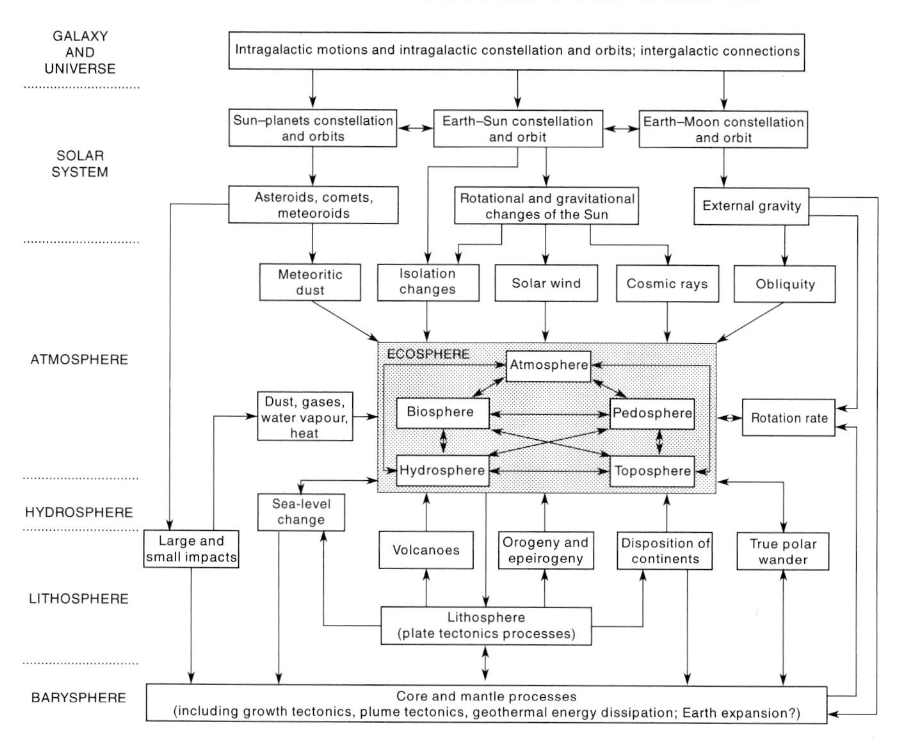

Figure 1.3 Interactions within and between the ecosphere, lithosphere, barysphere, and the rest of the cosmosphere
Source: After Huggett (1991, 1995)

Geological and cosmic forces cause cycles, steady states, and trends in the environment.

The ecosphere is powered chiefly by the Sun. Solar influences tend to promote steady-state changes about periodic and quasi-periodic fluctuations. Variations in solar output are commonly cyclical. The 11-year solar cycle is an example. Many ecospheric processes are finely adjusted to solar cycles. Other cycles are driven by orbital changes in the Sun and planets, by the Solar System's passage around the Milky Way galaxy, and possibly by the Milky Way's interaction with its closest neighbouring galaxies. Some ecospheric processes are attuned to these various astronomical pulses.

Bombardment by asteroids and comets may instigate and encourage trends in the biosphere. Somewhat ironically, it is possible that evolution depends on an occasional asteroid or comet strike causing widespread extinction, so opening up new opportunities for organic diversification.

Geological influences

Geological forces also produce cycles, steady states, and trends in the ecosphere. The ultimate seat of change is core and mantle processes. These processes act through the agency of the lithosphere. Plate tectonic processes drive sea-level change, volcanic activity, orogeny and epeirogeny, continental drift, and true polar wander.

Long-term biodiversity trends are an example of geological influences on the ecosphere. Biodiversity may depend on the relative fragmentation of land masses. During the Cambro-Ordovician explosion, biodiversity soared as available niches filled. It remained high for the rest of the Palaeozoic era because the continents were separated. As the continents came together to form the supercontinent Pangaea, biodiversity dropped, but rose again from the Triassic period as the supercontinent broke apart. An alternative explanation of long-term biodiversity trends is that climatic and geological processes endlessly increase the complexity of the physical environment. Biodiversity tracks these changes in the physical world, with occasional checks caused by mass extinctions, and never attains a steady state. This issue will be considered in detail in Chapter 8.

The solid Earth is not immune to cosmic influences. The Earth interacts with other bodies in the Solar System, and elsewhere, through the force of gravity. This interaction alters the Earth's rotation rate and the tilt of its rotatory axis. It also influences its orbital characteristics (the eccentricity of the orbit and the precession of the equinoxes). Space debris (asteroids, comets, and meteoroids) strikes the Earth. Large impacting bodies may induce changes in the core and mantle that will eventually be felt in the ecosphere. Large and medium impacting bodies stir up dust, gases, and water vapour and have a direct impact on the ecosphere.

EVIDENCE OF ENVIRONMENTAL CHANGE

It is fortunate indeed that the environment records its own history. Were this not the case, environmental reconstruction would be impossible. The record is incomplete, but it does hint at the way things were. All history – human and environmental – is a matter of imperfect record. Reading the record is difficult, even more difficult than reading a book in a foreign language with many pages torn out, some pages and paragraphs put in the wrong order, and large portions of text rendered illegible. Shreds of evidence must be gathered, their significance assessed, and then woven together. At best, the finished picture is an interpretation of the past, and there is no way of knowing whether the interpretation is correct. Accepted interpretations will explain indications of past conditions according to current theories of environmental evolution. New theories often call for new interpretations. It is doubtful whether this process of theorizing and evidence-gathering will ever lead to a grand theory of environmental evolution that explains everything in Earth history. An ability to understand the past is limited by lost evidence, and is probably constrained by the limits of human comprehension.

The evidence of environmental change is varied. It falls into these categories: geochemical, physical, biological, historical and instrumental.

Geochemical evidence

The conditions in air, water, and soil influence most chemical and physical processes. These processes exert some control over the mineralogy, elemental composition, and isotopic composition of sediments and soils. It is possible, therefore, to use geochemical data as a guide to past environmental states. Interpreting geochemical evidence is no mean task. It requires an understanding of the chemical and physical processes themselves, of their influence on chemical properties of sediments and soils, of their preservation in the sediments and soils, and of their association with climate. The main lines of geochemical evidence are listed in Table 1.1.

Evaporite minerals number more than eighty. Where evaporation exceeds precipitation, they are precipitated from concentrated brines in a definite order. In addition, their stability is circumscribed by temperature. For example, gypsum has an upper temperature limit of 58°C. Calcite is deposited in caves and mines as secondary precipitates in speleothems (stalagmites, stalactites, and flowstones). Annual luminescent growth-bands in the flowstones, caused by the presence of humic substances occluded in the calcite, reveal details of temperature and precipitation (e.g. A. Baker, Smart, Edwards, and Richards 1993; A. Baker, Smart, and Ford 1993; Shopov *et al.* 1994; A. Baker, Smart, Barnes, *et al.* 1995; A. Baker, Smart, and Edwards 1995). Carbonate sediments bespeak warm conditions. Calcium carbonate in marine shells dissolves in water containing carbon dioxide. The amount of calcium carbonate that can be dissolved increases as temperature decreases. So calcium carbonate precipitates as calcareous muds or oolites when waters become warmer, as in shallow parts of tropical seas. Coal suggests abundant vegetation and an adequate water supply. Fulgurites, which are natural glasses formed by lightning strikes, may be used as indicators of thunderstorms (Sponholz *et al.* 1993).

Mineral ratios often vary according to salinity and seawater composition. The magnesium/calcium ratio reflects calcium carbonate content of sea water (Tucker 1992). The strontium/calcium ratio in benthic ostracods (crustaceans) from cores in the Aral Sea, south-western Asia, reveals past salinity levels (Boomer 1993).

Clay minerals are sensitive to precipitation. Clays derived from basic igneous rocks and sampled along a moisture transect in California show a clear relationship with precipitation (Jenny 1980: 329–31; see also Singer

Table 1.1 Some geochemical evidence of environmental change

Type of evidence	*Examples of evidence*	*Environmental factors indicated*
Mineralogy		
Minerals and mineral ratios	Evaporites: anhydrite, gypsum, halite, and about eighty others	Temperature
	Calcite in flowstones (sinter, travertine)	Temperature, precipitation, acidity of precipitation
	Carbonates: limestone, dolomite, chalk	Temperature
	Coal	Abundant vegetation, moisture
Mineral ratios	Magnesium/calcium ratio	Salinity and seawater composition
Clay minerals	Kaolinite, goethite, smectite	Precipitation
Duricrusts	Alcretes, calcretes, ferricretes, gypcretes	Rainfall
Elemental composition		
Substitute elements	Strontium and magnesium substitution in marine shells	Temperature
Trace elements	Boron, gallium, rubidium, and strontium	Salinity
Isotopes		
Cosmogenic nuclides	Beryllium-10, carbon-14	Solar modulation of cosmic radiation
Oxygen isotope ratio ($^{18}O/^{16}O$ or $\delta^{18}O$)	Proxy temperature records from foraminifera in deep-sea cores	Temperature
Carbon isotope ratio ($^{13}C/^{12}C$ or $\delta^{13}C$)	Carbon isotope ratios in soil carbonates and in organisms	Continental biomass or variations in the extent and density of vegetation
Strontium isotope ratio ($^{87}Sr/^{86}Sr$ or $\delta^{87}Sr$)	Strontium isotope ratio in marine carbonates	Continental runoff and submarine volcanism

1980). The percentages of kaolinite, halloysite, gibbsite, and iron oxides all rise as mean annual precipitation increases. Conversely, the percentages of montmorillonite and illite decrease, and are virtually absent when the mean annual precipitation exceeds 80 cm. As a rule, large accumulations of kaolinite are formed under the extreme weathering regimes found in the tropics. Exceptionally, kaolin-rich soils are associated with swamps.

Duricrusts provide pointers to moisture regimes. Calcretes are presently formed chiefly in semiarid areas where the annual rainfall is about 200–500 mm (Watkins 1967; Goudie 1985). It is not unreasonable, therefore, to use their presence in ancient sediments to infer desert-like conditions with an annual water deficit (e.g. Wright 1990). Ferricretes and alcretes are widely accepted as being products of humid climates, ferricretes requiring alternations of wet and dry seasons. There is evidence that ferricrete may form under permanently moist tropical climates (McFarlane 1976). Silcretes seem to form under two distinct climatic regimes (Summerfield 1983).

'Non-weathering profile' silcretes are produced under a mainly arid to semiarid climate (e.g. Thiry and Simon-Coinçon 1996). Under these dry climatic conditions, the weathering environment tends to be alkaline. The alkalinity means that silica becomes locally mobile and builds up high concentrations. 'Weathering profile' silcretes are created under a humid climate. Under wet conditons, the weathering environment tends to be highly acidic with poor drainage. These circumstances, too, encourage silica accumulation.

Substitution of elements is common and may indicate temperature. Strontium and magnesium substitute for calcium in marine shells; minor and trace elements often show greater substitution at higher temperatures. Several trace elements, including boron, gallium, rubidium, sodium, and strontium, are sensitive to salinity, though their concentrations are also affected by sediment source, clay mineralogy, and grain size.

Isotopes are particularly valuable in environmental reconstruction. They are atoms with the same number of protons, a different number of neutrons, and a different atomic mass. The kinetic and thermodynamic properties of molecules depend on mass. This means that, during chemical and physical processes, different isotopes of the same element become partially separated. The separation, or fractionation, of isotopes leads to either an enrichment or depletion of individual isotopes. As isotopes of carbon, oxygen, and sulphur are abundant in the environment and partake in several important biochemical and geochemical reactions, their mass differences may be used in reconstructing past environmental conditions. Oxygen isotope ratios in foraminifers provide an invaluable palaeothermometer. They may also be used as a proxy for local surface temperature in tooth enamel. Oxygen isotope ratios for human teeth found in the Julianehaab Bay area of Greenland covering the last millennium pick out the Medieval Optimum, the Little Ice Age, and a return to warmer conditions in modern times (Fricke *et al.* 1995). Carbon isotope ratios in soil carbonates indicate the relative proportion of C_3 to C_4 plants in the local biomass. C_3 plants use the C_3 biochemical pathway for fixing carbon dioxide, which uses the enzyme ribulose-diphosphate carboxylase. They include nearly all trees, shrubs, herbs, and grasses that prefer a cool growing season. C_4 plants use the C_4 biochemical pathway for fixing carbon dioxide, which uses the enzyme phospho-enol-pyruvate carboxylase. They include a few shrubs in the families Euphorbiaceae and Chenopodiaceae and grasses that prefer a warm growing season. A carbon-isotope record from Devils Hole, Nevada, suggests fluctuations in the extent of vegetation (Coplen *et al.* 1994). Carbon isotope ratios in Devonian–Mississippian brachiopods from North America, Europe, Afghanistan, and Algeria probably reflect an expansion of terrestrial or marine biomass (or both), or burial of carbon in soils and sediments (Brand 1989). Strontium-isotope ratios reflect continental runoff and submarine volcanism.

Physical evidence

Physical evidence of environmental change is highly varied (Table 1.2). Some geomorphological features point to specific erosional environments. Glacial troughs (U-shaped valleys), rock bars and basins in trough floors, hanging valleys, cirques (corries, cwms), and fjords, for example, are uniquely glacial forms produced by ice erosion. Periglacial features, such as ice wedges and permafrost horizons, indicate cold climates (e.g. Ran *et al.* 1990). Tors and inselbergs are indicative of tropical weathering regimes (e.g. Büdel 1982: 252–98). Old river channels (palaeochannels) and river

Table 1.2 Some physical evidence of environmental change

Type of evidence	*Examples of evidence*	*Environmental factors indicated*
Erosional landforms	Glacial: glacial troughs, hanging valleys, fjords, striations, erratics, rock drumlins, rock basins, meltwater channels, cirques	Ice movement, snowline (cirques)
	Periglacial: permafrost horizons, ice wedges, pingos, rock glaciers, felsenmeer (block fields)	Frigid climate
	Tropical: tors and inselbergs	Torrid climate
	Fluvial: palaeochannels and river terraces	Stream discharge
Sediment textures, structures, and properties	Dunes, alluvial deposits (including river terraces), tempestites, tsunamites, till fabric	Air and water current directions, air current velocities, floods, storms, tsunamis
	Magnetic susceptibility in loess sequences and deep-sea cores	Mean wind intensity
Terrestrial soils and sediments	Till plains, fluvioglacial features, periglacial features	Frigid climates
	Tillites	Ancient glaciations
	Palaeosols (buried soil profiles)	Vegetation type, climate, atmospheric composition
Lacustrine sediments	Old shorelines, lake sediments	Lake levels – rainfall, evaporation, soil moisture
	Varves	Runoff, rainfall, wind intensity
Marine sediments	Deep-sea cores	Surface- and deep-water temperature, salinity
	Glaciomarine sediments	Ice-rafting
	Dropstones	Ice-rafting
Glaciers	Advances, retreats	Temperature, duration of melting season, sunshine and cloudiness, snowfall
Ice sheets	Annual layers	Snowfall
	Ice cores (e.g. Vostok ice core, Antarctica, Camp Century ice core, Greenland)	Atmospheric conditions, e.g. temperature, carbon dioxide content, aerosol content, nitrate content

terraces may suggest past stream discharge (e.g. Dury 1953; Leigh and Feeney 1995).

Sedimentary evidence is revealing. Sediment particles are sensitive to fluid motions. Sedimentary textures and structures may suggest current velocities and directions. Cross-bedding, which results from the lateral migration of ripples, and ripple marks furnish evidence of current direction. Sedimentary gradients (such as particle size gradients), unique sources of sediment constituents, and particle orientations (fossil orientations) allow direction of flow to be guessed. Magnetic susceptibility of sediments and soils, which reflects the concentration of magnetic minerals in a sample, is a guide to sediment source and transport intensity. Particle size data may also reveal fluid flow velocities. Specific examples of sedimentary indicators are numerous. Here are two. The taphonomy of benthic shell concentrations in siliciclastic sediments reflects the hydrodynamics of the sedimentary environment (Pickerill and Brenchley 1991). The form and composition of relict beach barriers around the margins of former pluvial lakes in the California desert indicate the waves and currents that formed them (Orme and Orme 1991). Lake waves are produced by wind, and the relict barriers allow palaeowind regimes to be inferred.

Some terrestrial sediments offer a general guide to past environments. Till plains, fluvioglacial landforms, and periglacial landforms are signs of frigid climates. The glacial origin of tills (and tillites, their lithified equivalent), for instance, is indicated by several sedimentary characteristics – a lack of sorting, a great range of grain size (including boulders), the preferred orientation of clasts, the presence of erratics, the angular shape of the stones, striations on the stones, and their extensive distribution. But the signs are not unequivocal – submarine mass-flow deposits possess similar features. Indeed, the climatic significance of most depositional landforms is debatable.

Palaeosols are found in sedimentary rocks and surface deposits. They are invaluable in reconstructing past environments. Older pedological indicators were somewhat crude. Reddish horizons, which result from iron staining, were generally thought to form in hot and seasonally humid climates. Red beds in the rock record were therefore taken as sure signs of desert conditions. The Devonian red beds are a classic example. However, the red pigmentation may result from post-depositional changes and bear no relation to climate. A recent interest in fossil soils has thrown up several interesting findings, not the least of which is that palaeosols may register atmospheric oxygen and carbon dioxide levels from the time of their formation. The carbonate component of goethite in a Late Ordovician ironstone palaeosol suggests that atmospheric carbon dioxide was about sixteen times higher than today (Yapp and Poths 1992).

Old lake shorelines and lacustrine sediments mark the former extent of lakes. Lake levels, especially those in closed basins (where there is no outlet for the lake water), are excellent indicators of regional water balances (precipitation, evaporation, and soil moisture). Lacustrine sediments often possess built-in chronometers. Varves in lakes normally chronicle annual deposition cycles and provide information on runoff, rainfall, and wind intensity.

Sediment cores furnish information on environmental conditions. Measurements from deep-sea cores, especially oxygen isotope ratios, have revolutionized ideas about global climatic change, providing detailed time series of surface-water and deep-water temperatures. Specific sediments give vital clues to the action of some processes. Particle size, composition, and mass flux of aeolian material in deep-sea sediments point to the intensity of past atmospheric circulations, the source of atmospheric dust, and the availability of sediments from source regions. Glaciomarine sediments are laid down by icebergs. Thick and extensive masses of unsorted deposits on the ocean floor are attributed to ice-rafted sedimentation. Similarly, dropstones (outsized stones that disrupt finely layered marine sediments) are often a sign of iceberg involvement in sediment deposition (see M. R. Bennett *et al.* 1996). For example, three deltaic features perched on the flanks of Mt Baldy, Thunder Bay, Ontario, Canada, lie above inferred postglacial lake levels (B. A. M. Phillips and Fralick 1994). They appear to be glaciolacustrine deltas containing dropstone units, which implicate icebergs in the depositional process. Tillites, which are generally thought to be lithified tills, are taken as evidence of old glaciations.

Ice is a good source of palaeoenvironmental information. Records of glacier advances and retreats, either historical records or dated moraines, suggest changes in temperature, the duration of the melting season, sunshine and cloudiness, and snowfall. Ice sheets, which have grown year by year, are huge repositories of historical data. Ice cores extracted from the Greenland and Antarctic ice sheets yield up a variety of palaeoclimatic and palaeoenvironmental signals. Snowfall is reflected in the thickness of annual layers. Atmospheric conditions are revealed by measurements of isotope ratios, entrapped carbon dioxide, aerosols, and nitrates. Beryllium-10 concentrations catalogue fluctuations in cosmogenic sources.

Biological evidence

Most animals and plants are sensitive to climatic conditions (Table 1.3). The more sensitive ones are admirably suited for enlistment as indicators of past climates.

Plants

Seeds, fruits, leaves, twigs, and wood fragments are the most frequently identifiable plant macrofossils. Most are found in sediments, but some survive as stomach contents of herbivores, and in fossil droppings. Finds of

Table 1.3 Some biological evidence of environmental change

Type of evidence	*Examples of evidence*	*Environmental factors indicated*
Botanical		
Macrofossil plants	Seeds, fruits, achenes, leaves, twigs, wood fragments, trunks	Flora, environment, temperature, rainfall
Microfossil plants	Pollen grains, spores, micro-sporangia, leaf hairs and epidermis, diatom and desmid skeletons (indicate nutrient supply and climatic conditions in lakes), opal phytoliths	Plant communities, temperature, rainfall
Floral lists	Compilations of plants present in particular places	Regional climate
Leaf physiognomy	Leaf size, leaf shape, leaf-margin form	Climate
Tree-growth rings	Ring width, ring-width variability, width of early wood, variations in cell size	Water availability, temperature, frosts, growing season
Zoological		
Mammals	Faunal composition, size of individuals, population structure	Climate, habitat, community structure
Amphibians and reptiles	Species distribution	Warmth, presence of water
Insects, especially beetles	Species distribution	Temperature, rainfall, and habitat
Molluscs	Species distribution	Climate, ecology, salinity (freshwater species)
Cladocerans and ostracods	Abundances in lake cores	Climate, water conditions, nutrients, and productivity in lakes
Corals	Species distribution	Temperature, salinity
Foraminifera	Species distribution, oxygen-isotope ratio	Temperature, salinity

stomach contents and droppings provide invaluable information about the diet and environment in which the animals lived. Plant macrofossils are extensively used in palaeoenvironmental reconstruction (e.g. Birks 1993; Kuhry 1994; Pellatt and Mathewes 1994; Jackson and Givens 1994). They are usually found near the place they grew.

Plant microfossils comprise the pollen grains of higher plants, the spores of lower plants, and less commonly leaf hairs and epidermis, and the skeletons of diatoms and desmids. Pollen analysis has been a mainstay of much Pleistocene and Holocene environmental reconstruction. Pollen grains and spores mostly fall into sediments from the air. This pollen rain is derived from local and regional vegetation. A change in the pollen spectrum at a site thus registers local and regional changes in vegetation.

This is why it is essential to consider macrofossil and microfossil evidence together. In western Norway, late glacial pollen diagrams from the islands of Utsira and Blomoy show birch (*Betula pubescens*) to occur at Blomoy and warmth-loving taxa at Utsira (Birks 1993). But the plant macrofossil evidence fails to reveal birch at either site. Reconstruction of the local vegetation points to willow shrub and crowberry (*Empetrum*) dwarf-shrub heath at Utsira, and a community dominated by least willow (*Salix herbacea*) at Blomoy. The birch pollen appears to have arrived by long-distance dispersal, aided by the strong late glacial winds.

Diatoms, desmids, and other members of the aquatic microflora originate in the water-body from which the sediment is deposited. These organisms are sensitive to nutrient supply and climatic conditions. Their abundance in sediments thus provides a guide to change in lake environments. Opal phytoliths form in plants from monosilicic acid (H_4SiO_4). The monosilicic acid is absorbed in solution through plant roots, accumulates in living tissues, and precipitates in the spaces between cells and along cell walls. The opaline casts so formed normally hold the size and shape of the cellular structures within which, or around which, they formed. Many phytoliths in grasses are distinct enough to identify the tribe or genus to which they belong. They are thus excellent indicators of past flora, especially in grassland (e.g. Fisher *et al.* 1995).

Floral lists are helpful in reconstructing environmental conditions. For instance, Triassic and Jurassic floral lists from Eurasia, when subjected to ordination, revealed a latitudinal gradient that probably represents a response of vegetation to climate. Microphyllous conifers (associated with evaporites) characterized floras at the dry, subtropical end of the gradient; deciduous ginkgophytes and broad-leaved conifers characterize the floras at the cool temperate end (Ziegler *et al.* 1993).

Leaf physiognomy is closely related to climate. Analysis of leaf physiognomy is highly sophisticated. CLAMP (climate analysis multivariate program) uses multiple descriptors of leaf size, shape, and margin forms to reconstruct several climatic variables (Wolfe 1993, 1995).

Tree-growth rings are excellent palaeoclimatic indicators. Water availability, temperature, and the length of the growing season are mirrored in ring width, width of early wood, variations in cell size, and year-to-year variability of ring width. Wide rings with little width variation imply that climate does not limit growth. Uninterrupted secondary wood implies that conditions are always favourable for growth – that the climate is uniform and not seasonal.

Where the fossil record permits, analyses of plant community structure, vegetational and leaf physiognomy, and growth rings and vascular systems in wood make it possible to define terrestrial palaeoclimatic parameters, such as mean annual temperature and mean annual range of temperature, with better resolution than most sedimentological techniques.

Animals

Many animal species are good indicators of past climates and habitats. Modern mammal species tend to have specific habitat preferences. Fossil forms belonging to the same genera as living species give a guide to past habitats. A Pleistocene fauna near Austin, Texas, contains six extant mammal taxa that now live to the north and east, in either cooler or more humid climates; they suggest a marsh-like habitat (Lundelius 1991). Amphibians and reptiles indicate warm conditions and the presence of water (e.g. Bailon and Rage 1992). Most large amphibians and reptiles are today exclusively tropical in distribution. Crocodiles, for example, are not found poleward of the 15°C mean annual isotherm. Reptiles and amphibians living in temperate climates also betoken relative warmth. European pond terrapin (*Emys orbicularis*) fossils, for instance, suggest warm summers.

Insects, and especially beetles, are a good guide to past environmental conditions (e.g. Lowe *et al.* 1995). Many have a narrow feeding range and are associated with specific plant communities. *Oodes gracilis*, a marshland beetle, today ranges over central and southern Europe, but not Britain. However, it is found in 120,000-year-old river terrace deposits at Trafalgar Square, London, that contain hippopotamus remains (Coope 1965). It points to the warmth of the last interglacial in southern England.

Terrestrial molluscs give a first-rate indication of climate and local ecology. The terrestrial gastropod *Columella columella*, which today has a high Arctic and alpine distribution, is associated with glacial gravels at Thriplow, Cambridgeshire (Sparks 1957). Freshwater molluscs register water conditions and salinity. The freshwater bivalve *Corbicula fluminalis* (now extinct in Europe, but occurring in the rivers of North Africa) is found in interglacial mammal-bearing deposits in southern Britain (Sparks 1964). A study of Holocene environmental history at Bátorliget marsh, north-east Hungary, used aquatic and terrestrial mollusc remains in conjunction with pollen analysis and other environmental indicators (Willis *et al.* 1995). Five malacological time-zones were distinguished. Starting with the oldest, they were characterized by cold-resistant molluscs, molluscs tolerant of periodic water supply and with a Palaeoarctic distribution, water bank and marshland species, warmth-loving species, and marshland species with fewer warmth-loving types.

Ostracods were used in palaeoecological reconstructions by Charles Lyell. They have proved highly successful in palaeoenvironmental analysis (e.g. Masurel 1989; J. A. Holmes 1992; Keen 1993). Cladocerans (a type of crustacean) are also useful palaeoenvironmental indicators. Cladocerans and ostracods are common in lake sediments. Cladoceran, diatom, and lithostratigraphical evidence pointed to Holocene closed-basin lake-level changes in western Finnish Lapland (Hyvarinen and Alhonen 1994).

Corals are indicants of sea temperatures. Coral reefs are thought to be reliable indicators of tropical climates, although some small accumulations occur in temperate latitudes. Modern coral reefs require a minimum temperature of 21°C, and clear, aerated, shallow water of normal salinity. These requirements mean that they are confined to the littoral zone between 30° N and 30° S. Exact temperature requirements vary with the type of coral. Hermatypic corals, for example, have narrow temperature tolerances, with an optimum in the range 23–29°C.

Foraminifera are unicellular or colonial protists. Most of them secrete a chambered calcium carbonate shell or test, and live as planktonic or benthic organisms. Planktonic foraminifera assemblages are closely linked to water-mass distribution in the oceans. Particular species indicate cold or warm waters and salinity. Foraminiferal tests in deep-sea sediments are used in oxygen-isotope determinations of palaeotemperature.

Historical and instrumental evidence

Historical and instrumental evidence covers the last three hundred years or so in fair detail, though only in some places. Records go back to about 3000 BC, but they are exceedingly fragmentary. Historical and instrumental evidence of past climates is shown in Table 1.4.

Documentary evidence comprises descriptive weather registers and weather diaries, ships' logs, a miscellany of accounts, annals, books, chronicles, and documents, and grain prices (see Lamb 1995). This type of evidence may be specific. For instance, in western Norway (from Ryfylke to Sunnmore), *avtaksforretning* reports detail successful tax relief petitions following physical damage to farmlands (Grove and Battagel 1989). The reports reveal abnormally heavy rains in December 1743. Ancient written sources have been used to chronicle ice appearances (freezing episodes) on the River Rhône (Jorda and Roditis 1993). Indeed, records of ice breakup in rivers (cryophenological evidence) are exceedingly useful climatic indicators (e.g. Kajander 1993). Some documentary evidence allows a more general picture of past climates to be sketched. Documentary evidence from the summer of 1783 records a widespread dry and smoky haze in western Europe that was probably associated with the Laki fissure eruption in Iceland (Grattan and Charman 1994). The haze was created by volcanic gases in the lower atmosphere. Records described severe damage to crops and trees in Britain and western Europe, which would be consistent with the impact of acid deposition. Another general climatic pattern is revealed by continental-scale comparisons of documentary evidence in western Europe and China during the Maunder Minimum (AD 1645–1715) (Pfister *et al.* 1994). Similarly, pre-instrumental records from western Europe, east Asia, North America, and Africa show many differences in regional weather, but all point to the uniqueness of twentieth-century warming when taken in the context of the last half millennium (Bradley 1991). (A word of caution is needed here: many kinds of documentary evidence tend to stress colder

Table 1.4 Historical and instrumental evidence of climate

Type of evidence	*Indications*	*Start of records and examples*
Documentary sources		
Descriptive weather registers and weather diaries (daily)	Wind, weather, rain and snow frequencies, and others	Earliest examples from Lincolnshire, AD 1337–44. Mainly parts of Europe. Scattered records from early expeditions to North America and elsewhere
Ships' logs	Wind, weather, rain and snow frequencies, and others	Records began 1670–1700. Fragments much earlier. Mainly European waters and some longer voyages (e.g. to the Indies and Far East)
Miscellaneous (accounts, annals, books, chronicles, documents)	Weather, especially extremes and long spells of weather, droughts, floods, frost, snow, great heat, great cold, and others	Records begin about AD 1100. Occasional reports much earlier (e.g. Italy 400 BC, Britain 55 BC, central Europe AD 500)
Grain prices	Crop yields and climate	Exeter from AD 1316–1820?
Meteorological instrument records		
Standard meteorological instrument observations	Surface pressure, temperature, precipitation, winds, humidity, and others	Records from 1650s in parts of Europe. Antarctica not covered until 1956. Southern Ocean still largely not covered except by satellite observations
Upper air measurements	Upper air temperature, humidity, pressure, winds	Records from 1930s in parts of Europe and North America (fragmentary records much earlier from mountain stations in Europe)
Ship-borne instruments	Sea temperatures, salinity, ocean currents	Records began in the 1850s, though fragments from 1780 are sufficient to deduce 40–50-year means. Mainly the Atlantic Ocean for the first 50–870 years
River and lake records		
River flood levels	Precipitation, evaporation, soil moisture	Earliest record for the River Nile (in fragmentary form from 3100 BC)
Lake levels	Precipitation, evaporation, soil moisture, subsoil moisture	Records start about AD 1650. Earliest reports are from Siberian lakes

Source: Partly after Lamb (1995)

intervals, and may overlook warmer periods!) Meteorological instrument records are usually more precise than documentary sources, but data are patchy for most places before about 1850.

Documentary information about environmental conditions other than climate also exists. Written sources on the forest in the upper Vicdessos Valley, southern France, contained ecological information to reassess the history of the region (Davasse and Galop 1990). The old view was that the forest, a former centre of an iron industry, suffered early and radical deforestation. It now seems that deforestation did not become widespread until the end of the eighteenth century. Documentary records of the cultivation of citrus trees and *Boehmeria nivea* (a perennial herb) in the eighth, twelfth, and thirteenth centuries AD in China indicate a medieval warm period (Zhang De'er 1994). In the thirteenth century, the northern boundaries of the citrus trees and of the herb lay further north than at present. The known climatic conditions for planting these species suggest that the annual mean temperature in south Henan province was, in the thirteenth century, 0.9–1.0°C higher than at present. Landscape changes in historical times could have a climatic or a human origin. Documentary evidence, in conjunction with palaeoecological information, can sometimes decide which origin applies. Aggradation in the upper Bowmont Water, an upland stream draining the northern Cheviot Hills, Scotland, has produced river terraces (Tipping 1994). The oldest terrace is undated, but the other five were formed in the last 250 years, the most recent in the eighteenth century. The documentary evidence rules out land-use changes having materially affected aggradation of the river, and increased precipitation and flooding during the Little Ice Age are invoked as likely causes. In Norway, documentary evidence fails to support a land-use hypothesis for lake acidification, as grazing intensity has increased in areas where lake waters are most strongly acidified (Patrick *et al.* 1990).

In some cases, biological evidence can be used to check documentary evidence. According to patchy documentary sources, the lowland landscape of eighteenth-century north-east Ireland consisted of enclosure by hedges, with extensive arable agriculture in some areas, and scrubby uncultivated land elsewhere. The modern pollen rain of a reconstructed nineteenth-century farm, characteristic of north-east Ireland, was used to interpret lowland lake deposits and peats for the last 500 years (V. A. Hall 1994). Scrub clearance in the eighteenth century produced an open landscape that was enclosed by hedges of hawthorn (*Crataegus*) and blackthorn (*Prunus spinosa*), taxa which are poorly represented in the modern pollen rain and the fossil pollen record. Arable agriculture, including the introduction of flax, increased in the eighteenth century.

River flood levels and lake levels are found in historical records. Records of lake levels for Lakes Erie and Michigan–Huron begin in 1819, and the modern data set starts in 1860 (C. T. Bishop 1990). Historical and archaeological evidence extends the record back 1,800 years and suggests that variations in maximum mean annual water levels have probably not exceeded those measured on the lakes since 1819. Historical evidence from Mexico shows that Lake Pátzcuaro has fluctuated over the last 600 years (O'Hara 1993). For rivers, the oldest year-by-year record is the flood levels of the Nile in Lower Egypt. Yearly readings at Cairo are available from the time of Muhammad, and some stone-inscribed records date from the First Dynasty of the Pharaohs, around 3100 BC.

DATING ENVIRONMENTAL CHANGE

Palaeoenvironmental indictors suggest what happened in the past, but this information is of limited use without knowing when it happened. A broad range of techniques is now available for dating events in Earth history (Table 1.5). Some are more precise than others. Relative dating simply puts events in the correct order. It assembles the 'pages of Earth history' in a numerical sequence. The technique relies on the principle of stratigraphic superposition. This states that, in undeformed sedimentary sequences, the lower strata are older than the upper strata. Some kind of marker must be used to match stratigraphic sequences from different places. Traditionally, fossils have been employed for this purpose. Distinctive fossils or fossil assemblages can be correlated between regions by identifying strata that were laid down contemporaneously. This was how the stratigraphic column was first erected by such celebrated geologists as William ('Strata') Smith (1769–1839). Although this technique was remarkably successful in establishing the broad development of Phanerozoic sedimentary rocks, and rested on the sound principle of superposition, it is beset by problems (see Vita-Finzi 1973: 5–15). It has been superseded by a battery of geochronometric dating techniques that has helped to establish the geological timetable (Figure 1.4).

There are several geochronometric dating techniques, and the number is growing. The main techniques are radiometric dating, fission-track dating, cosmogenic nuclide dating, palaeomagnetic dating, thermoluminescence dating, varve dating, tree-ring dating, and amino-acid racemization dating. These methods pinpoint when environmental change occurred. This information is crucial to a deep appreciation of environmental change. Without dates, nothing very useful can be said about rates.

Isotopic dating techniques

Radiometric dating

The environment contains a range of 'atomic clocks'. These tick precisely as a parent isotope decays radioactively into a daughter isotope. The ratio

Table 1.5 Some geochronometric dating techniques

Technique	*Dating range (years)*	*Datable material*
Isotopic techniques		
Radiometric		
Carbon-14	< 100,000	Animal tissue, bone, charcoal, peat, shells, wood, calcite in speleothems, tufa, soil carbonates, groundwater, sea water, ice
Thorium-230	< 200,000	Organic carbonate in deep-sea cores, corals, molluscs
Uranium-234	50–100,000	Marine carbonate, coral, molluscs
Potassium-40	> 100,000	Biotite, muscovite, hornblende, whole volcanic rock
Uranium-238	> 10,000,000	Zircon, uraninite
Rubidium-87	> 10,000,000	Biotite, muscovite, potassium-feldspar, whole igneous or metamorphic rock
Fission track	Few thousand to hundreds of millions	Uranium-bearing minerals – apatite, sphene, zircon
Cosmogenic nuclide		
Beryllium-10	Constrained by length of surface exposure	Quartz-bearing materials exposed at the land surface
Non-isotopic techniques		
Palaeomagnetic	> 1,000,000	Iron-bearing rocks
Thermoluminescence	~100,000	Quartz or zircon grains in loess, soils; pottery
Varve	~10,000	Mainly glacial lakes
Tree ring	~2,000	Trees; timbers from buildings, ships, etc.
Amino-acid racemization	~200,000	Organic remains

between parent and daughter isotopes allows age to be determined with a fair degree of accuracy, although there is always some margin of error, usually in the range $\pm$5–20 per cent. The decay rate of a radioactive isotope declines exponentially. The time taken for the number of atoms originally present to be reduced by half is called the half-life. Fortunately, the half-lives of suitable radioactive isotopes vary enormously. The more important isotopic transformations have the following half-lives: 5,730 years for carbon-14, 75,000 years for thorium-230, 250,000 years for uranium-234, 1.3 billion years for potassium-40, 4.5 billion years for uranium-238, and 47 billion years for rubidium-87. These isotopes are found in environmental materials.

Carbon-14 occurs in wood, charcoal, peat, bone, animal tissue, shells, speleothems, groundwater, sea water, and ice. It is a boon to archaeologists and Quaternary palaeoecologists, providing relatively reliable dates in late Pleistocene and Holocene times. Thorium-230 and uranium-234 are found mainly in marine carbonates of various kinds. They are used by

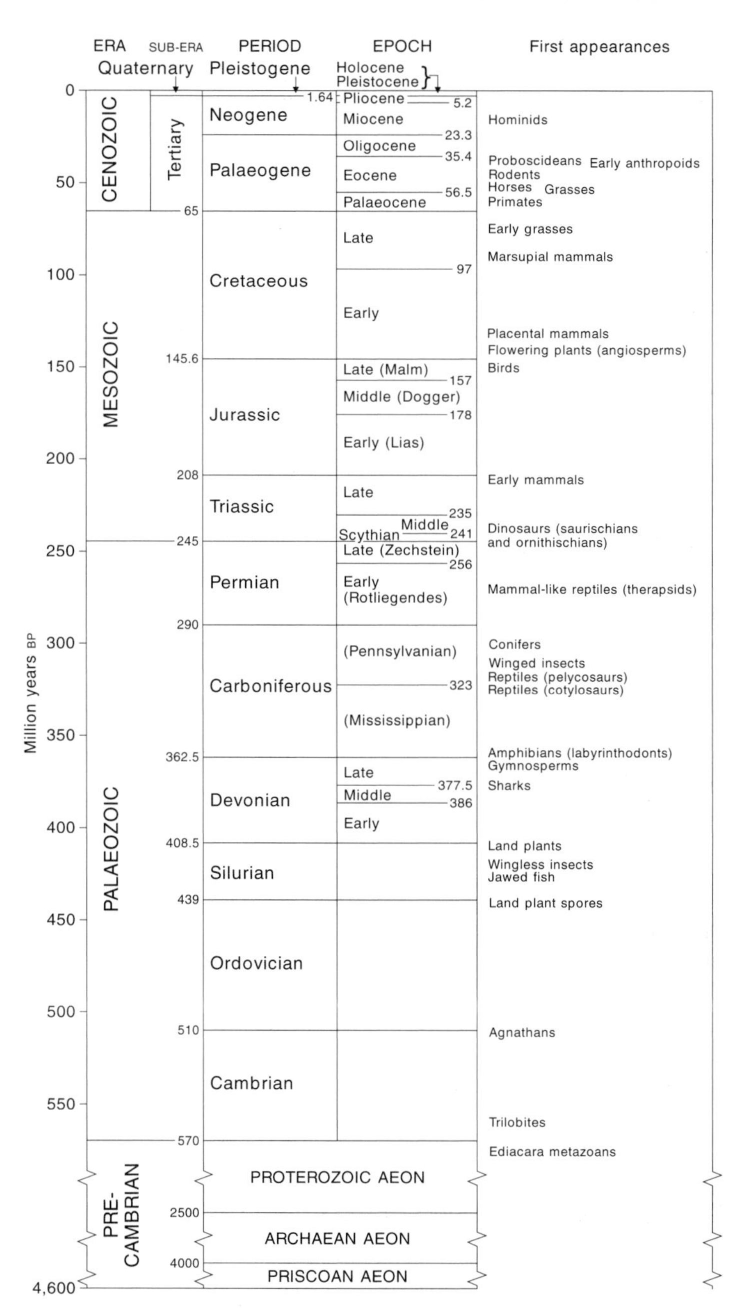

Figure 1.4 The geological time-scale
Source: After Harland *et al.* (1990)

geomorphologists and geologists to date marine terraces. Potassium-40, uranium-238, and rubidium-87 are found in some minerals and rocks. They have mainly geological applications.

Fission track dating

The spontaneous nuclear fission of uranium-238 damages uranium-bearing minerals such as apatite, zircon, and sphene. The damage is cumulative. Damaged areas can be etched out of the crystal lattice by acid, and the fission tracks counted under a microscope. The density of tracks depends upon the amount of parent isotope and the time elapsed since the tracks were first preserved, which only starts below a critical temperature that varies from mineral to mineral.

Cosmogenic nuclide dating

Radioactive beryllium-10 is produced in quartz grains by cosmic radiation. The concentration of beryllium-10 in surface materials containing quartz, in boulders for instance, is proportional to the length of exposure (e.g. Gosse *et al.* 1995). This technique gives a very precise age determination. Chlorine-36 may be used in a similar manner (Whitney and Harrington 1993).

Non-isotopic dating techniques

Palaeomagnetic dating

Some minerals or particles containing iron are susceptible to the Earth's magnetic field when heated above a critical level – the Curie temperature. Minerals or particles in rocks that have been heated above their critical level preserve the magnetic-field alignment prevailing at the time of their formation. Where the rocks can be dated by independent means, a palaeomagnetic time-scale may be constructed. This may be applied elsewhere using palaeomagnetic evidence alone.

Thermoluminescence dating

This involves measuring the glow-curve intensity in a sample of quartz or zircon grains (or both). The glow curve is the light emitted by a sample that has been irradiated and subsequently heated, plotted as a function of temperature. Suitable materials for thermoluminescence dating include loess (Pye *et al* 1995) and sandy soils (Lees *et al.* 1995).

Varve dating

The distinct layers of sediments (varves) found in many lakes, especially glacial lakes, are produced annually. In some lakes, the varve sequences run back thousands of years. Varves have also been detected in geological rock formations, even in Precambrian sediments.

Tree-ring dating

Tree rings grow each year. By taking a core from a tree (or suitable timbers from buildings, ships, and so on) and counting the rings, a highly accurate dendrochronological time-scale can be established and cross-referenced with carbon-14 dating. For example, an 8,000-year carbon-14 record has been built up from tree rings in bristlecone pine, *Pinus aristata* (Suess and Linick 1990).

Amino-acid racemization dating

Amino-acid racemization is based upon time-dependent chemical changes (called racemization) occurring in the proteins preserved in organic remains. The rate of racemization is influenced by temperature, so samples from sites of uniform temperature, such as deep caves, are needed.

PREDICTING AND 'POSTDICTING' ENVIRONMENTAL CHANGE

After having been reconstructed, past environmental conditions may be checked against predictions (or 'postdictions') made by computer simulations. The main areas of application are sedimentary basin evolution, biogeochemical cycles, and geological climates – Holocene, Pleistocene, and pre-Quaternary.

Sedimentary basin models

A sedimentary basin contains a source area (land) and a sink (ocean). Traditionally, different models have been applied to the subaerial and submarine sections of the basin. Landscape and landform models consider sediment production and sediment transport. Sediment production includes gains (by lowering of the weathering front) and transformations (by mechanical and chemical breakdown) of clastic sediments. Sediment transport is normally modelled using process laws. These 'laws' predict sediment transfer by rain splash, soil wash, and soil creep, landslides, rivers, and so forth, according to slope gradient and other environmental factors (see Huggett 1985: 161–97; Kirkby 1990). Sedimentary basin models consider the accumulation of sedimentary wedges in submarine environments. Like landscape models, they include sediment gains and losses, and sediment transfers defined by process laws. They commonly incorporate tectonic changes, sea-level changes, and compaction effects (e.g. Strobel *et al.* 1989; Tetzlaff and Harbaugh 1990; Waltham 1992).

Recently, more comprehensive models of sediment production, transfer, and deposition in complete (land–ocean) sedimentary basins have been constructed (e.g. Leeder 1991; Koltermann and Gorelick 1992; Rivenæs 1992). This promises to be a rewarding step, as the terrestrial sediment flux is a major control on sedimentary basin stratigraphy.

Biogeochemical models

Biogeochemical cycles are modelled by defining a set of geochemical reservoirs (stores) that are connected by geochemical fluxes. The models consist of a set of storage equations, one equation for each reservoir. Typically, reservoirs would include rocks, air, oceans, and organisms. Fluxes between reservoirs are either estimated using empirical evidence, or defined by process equations, which would define the rates as functions of temperature, reservoir size, and so on. The full model is a set of storage and process equations that is solved numerically using a computer. Solutions may be obtained for present-day conditions, where interest might focus on such problems as the effects of increasing atmospheric carbon dioxide on other parts of the ecosphere (see Huggett 1993 for a simple account). They may also be obtained for geological time-scales. GEOCARB II, for example, was devised to investigate the long-term carbon cycle (Berner 1994). It models the transfer of carbon between the atmosphere, oceans, biosphere, and lithosphere. Processes modelled include the weathering of calcium and magnesium silicates and carbonates, and sedimentary organic matter, on continents; the burial of organic matter and calcium–magnesium carbonates in sediments; the thermal breakdown of carbonates and organic matter at depth, with resultant carbon dioxide degassing (release into the atmosphere).

Climate models

Computer simulations using climate models run with appropriate boundary conditions provide a yardstick against which to test the palaeoclimates reconstructed from geochemical, physical, biological, and historical and instrumental evidence. Equally, the success of a general circulation model is gauged by considering the similitude between the predicted climate and the climate suggested by palaeoclimatic indicators. Thus theory and observation can be made to work in tandem, providing steps are taken to avoid circular argument.

The models used to simulate the chemical reactions and physical processes in the atmosphere are exceedingly sophisticated. Climate models range from simple zero-dimensional models to far more elaborate three-dimensional models. Four chief types of climate model exist:

1 Energy-balance models (EBMs). These predict the change in surface temperature resulting from a change in heating, with the proviso that the net flux of energy should remain unchanged.

2 Radiative–convective models (RCMs). These compute the vertical (one-dimensional) temperature profile of a single column of air. Normally, the predicted temperatures are global averages.

3 Statistical dynamical models (SDMs). These normally deal with a two-dimensional slice of air along a line, commonly a line of latitude. They thus combine the latitudinal dimension of energy-balance models with the vertical dimension of radiative–convective models.

4 General circulation models (GCMs). These are the most sophisticated climate models currently available. They tackle three-dimensional parcels of air as they move horizontally and vertically through the atmosphere. There are three types of general circulation model: atmospheric general circulation models (AGCMs), which predict the state of the atmosphere; oceanic general circulation models (OGCMs), which predict the state of the oceans and range from simple, mixed-layer oceans to full ocean general circulation models; and coupled atmosphere-ocean general circulation models (AOGCMs), which, as their name hints, consider the atmosphere and ocean simultaneously.

GCMs consist of a set of equations describing the physical and dynamical processes that determine climate. These equations are prognostic. This means that they enable the state of the atmosphere (or ocean) to be predicted. Additionally, many GCMs now include a heat- and water-balance model of the land surface, and a mixed-layer model of the ocean. Governing or primitive equations lie at the heart of a GCM. They are the equation of motion (conservation of momentum); the equation of continuity (conservation of mass or hydrodynamic equation); the equation of continuity for atmospheric water vapour (conservation of water vapour); and the equation of energy (thermodynamic equation derived from the first law of thermodynamics). To these storage equations are added the equation of state (hydrostatic equation), and, in some models, the surface-pressure tendency equation. To run a GCM, it is necessary to specify parameters, such as the solar constant and orbital parameters; and boundary conditions, such as the distribution of land and sea, topography, and total atmospheric mass and composition. GCMs also include diagnostic variables – clouds, surface albedo, vertical velocity, and the like. For prescribed boundary conditions and parameters, the full set of equations is solved to determine the rates of change in prognostic variables – temperature, surface pressure, horizontal velocity, water vapour, soil moisture, and so forth.

There is no doubt that the latest AGCMs and coupled AOGCMs are complicated, but they are powerful and sophisticated tools in modelling past and future climates. They are not, however, devoid of problems (e.g. Spicer 1993). First, they do not give consistent predictions. Run five different models with the same starting and boundary conditions and five different sets of predictions will emerge, though they will show some consistent results. The more the boundary conditions diverge from present conditions, the greater the disparities between different models become. Nevertheless, used wisely, climate models are helpful in identifying the importance of differing boundary conditions in the past and in the future. They have certainly met with considerable success in modelling past climates, as will be evident in this book.

WRITING ABOUT ENVIRONMENTAL CHANGE

This book has a rigid and unconventional structure. All chapters, save that on the cosmosphere and the concluding chapter, consist of five sections. In each case, the first section sets down the basic material about the structure and composition of the 'sphere' in question; and the second section describes the nature of change in the 'sphere'. The reason for including these two introductory sections is to enable non-specialists to equip themselves with sufficient definitions and ideas to understand the more discursive sections that follow. A problem in writing books that synthesize multifarious topics is to make the text intelligible to all readers. Some readers will have only a nodding acquaintance with some topics. Definitions of basic terms are thus helpful. The difficulty is knowing how many terms to define and where to define them. Glossaries can be helpful, but they do not discuss the subtleties and nuances of definitions, nor do they normally consider key ideas. Introductory sections outlining current thinking on the structure and dynamics of the cosmic and terrestrial spheres should be a more illuminating alternative. The other three sections in each chapter are more probing. The third section discusses the causes of change, focusing on external and internal causes. The fourth section considers the rate of change, paying special attention to fast changes. The fifth and final section considers directional aspects of change – cycles, steady states, and trends.

Chapters 2 and 3 look at the ecosphere's surroundings – the rest of the cosmosphere and the lithosphere and barysphere. The structure and function of these 'spheres' affects, directly or indirectly, all that happens in the ecosphere. It is necessary, therefore, to grasp their operations. The descriptions offered are meant to provide non-geologists and non-astronomers with a foundation of relevant material. Chapter 2 describes the nature of the Cosmos, selected aspects of cosmic change, the violent nature of the Solar System and Galaxy, and the evolution of the cosmic environment. Chapter 3 looks at the geological environment, paying close attention to geological factors that influence environmental change. After introducing the geological environment, it explores the nature of lithospheric change (plate tectonics), the possible causes of plate tectonics (including mantle plumes and cosmic influences), rates of lithospheric change (gradual changes and catastrophes), and steady-state, directional, and cyclical aspects of Earth history.

The next five chapters tackle in turn the components of the ecosphere – atmosphere, hydrosphere, pedosphere, toposphere, and biosphere and ecosphere.

Chapter 4 examines the atmosphere. It explains the nature of climate and climatic change. It looks into the mechanisms of climatic change – geological forcing (volcanoes, relief, geography, and true polar wander), cosmic forcing (the effects of solar activity, planetary interactions, and bombardment), and the internal dynamics of the world weather machine. It outlines some fast climatic changes, including stadial–interstadial climatic shifts (Dansgaard–Oeschger cycles), and rapid deglaciation. Finally, it notes some trends in geological climates, including secular changes caused by organisms, a slowly brightening Sun, a decelerating planetary rotation, and a possible long-term shift in axial tilt; and it discusses a 300-million-year cycle of hothouses and icehouses, and a 150-million-year cycle of warm and cool climatic modes.

Chapter 5 examines the hydrosphere. It delves into the nature of the hydrosphere and hydrospheric change, describing the global water cycle, ocean currents, rainfall and runoff, lake levels, and sea levels. It explores the mechanisms of change in water stores and water fluxes. It examines the somewhat controversial issue of ultra-fast hydrospheric changes – lake bursts, giant tsunamis, and impact superfloods. And it concludes by investigating selected trends in the world's waters – the origin of the hydrosphere, the late Precambrian water cycle, and cycles of sea-level change.

Chapter 6 looks into the pedosphere, which is defined to include soils (edaphosphere) and sediments (debrisphere). It investigates the nature of change in the soil–sediment system, focusing on soil formation, dynamic denudation, soil chronosequences, sediment sources and sinks, sediment budgets, and sedimentary sequences. It discusses autogenic (internal) and allogenic (external) causes of this change. It mentions rates of pedological change and the evidence for catastrophic sedimentation events. And to conclude, it examines secular trends in sediments, soils, and weathering features.

Chapter 7 looks at the Earth's continental and submarine relief features. It enquires into the nature of geomorphological change, including slope decline, slope recession, etching, equilibrium and non-equilibrium, and topographic chronosequences. It explains the geological influence on landscape change, the effect of exogenic processes (climatic and marine influences), and internal, self-organizing mechanisms. It contrasts superfast and superslow geomorphological changes. And to finish, it probes landscape cycles and episodes, and the possibility of evolutionary geomorphology.

Chapter 8 considers change in the biosphere from an ecological perspective. It covers community units and community properties, and the mechanisms of community change, including disharmonious communities and successional sequences. It investigates autogenic community change (new views on hydrosere succession and the peatland hummock–hollow cycle), and allogenic community change (the effects of climatic cycles and trends, and of geological changes). It scrutinizes the tempo of biotic change, looking at rates of species spread and biotic crises. And lastly, it ponders cycles, trends, and steady states (the Gaia hypothesis) in the organic world.

Chapter 9 looks at features of change shared by the terrestrial spheres, raises some disputatious issues, and makes a few suggestions for future work.

SUMMARY

Environmental change involves jumps, fluctuations, and trends. Fluctuations are regular and irregular. They oscillate with characteristic periods or quasi-periods, alter episodically, or follow a chaotic pattern. Some environmental changes occur quickly; most occur slowly. The environment changes through the operation of the ecosphere's internal machinery, and through the external agencies of cosmic and geological forces. Cosmic forces include the Sun, the Moon and planets, asteroids and comets, and supernovae. Geological forces include continental drift, volcanoes, and mountain building, all of which stem from plate tectonics. The environment keeps a faulty record of its history. Evidence of environmental change comes from geochemical, physical, biological, and historical and instrumental sources. Geochronometers (radiometric dating, fission-track dating, and so on) give vital clues about when events and changes occurred. They also permit rates of environmental change to be gauged. High-speed computers allow researchers to play with complicated, and tolerably realistic, models of environmental change. Modelling is particularly useful for studying change in sedimentary basins, biogeochemical cycles, and climate. General circulation models, run with appropriate boundary conditions, predict climates of the past. These predicted climates are then compared with palaeoclimatic indicators.

FURTHER READING

There is a veritable flood of books on environmental change. Most deal with the changes incurred by human activities. A few tackle the material covered here, though none covers it all. *Environmental Change* (Goudie 1992) and *Late Quaternary Environmental Change: Physical and Human Perspectives* (Bell and Walker 1992) are excellent introductions to Pleistocene and Holocene environmental changes. Further insights into the evidence of environmental change may be gleaned from *Reconstructing Quaternary Environments* (Lowe and Walker 1997), *Soils of the Past* (Retallack 1990), *Climate Modes of the Phanerozoic* (Frakes *et al.* 1992), *Quaternary Environments* (M. A. J. Williams *et al.* 1993), and *Climate, History and the Modern World* (Lamb 1995). A simple account of modelling is given in *Modelling the Human Impact on Nature* (Huggett 1993).

2

COSMOSPHERE

LIFE, THE UNIVERSE, AND EVERYTHING: THE COSMIC ENVIRONMENT

The cosmosphere is the domain of all non-living things and forces. Strictly speaking, it includes the Earth, but it is convenient to take it as all phenomena beyond the Earth's atmosphere. The Earth is intimately associated with the cosmosphere in at least three ways. First, it is 'connected' to extraterrestrial bodies by the force of gravity – it is a player in the music of the spheres. Second, it receives energy spewed out of stars, and primarily the Sun. Third, it collides with space debris.

Stretch, wobble, roll, and pitch

The Earth, and all other planets in the Solar System, turns about a rotatory or spin axis at the same time that it revolves around the Sun (Figure 2.1). The Earth's rotation is variable in the short term, producing minute fluctuations in the length of day (Hide and Dickey 1991). It has slowed down through Earth history, so the number of days per year decreases little by little.

The plane of the Earth's orbit is known as the ecliptic. The inclination of the ecliptic from the invariable plane of the Solar System varies within a 2° band owing to the gravitational attraction of other planets (Ward 1973; Muller and MacDonald 1995).

The Earth's orbit is a nearly circular ellipse that has the barycentre (centre of mass) of the Solar System at one focus. The divergence of an ellipse from a circle is measured as eccentricity. This is defined as the distance between the foci of the ellipse expressed as a percentage of the long-axis length. As the two foci move closer together, the ellipse becomes less eccentric until both foci meet and eccentricity is then zero. This parameter is known as orbital eccentricity or eccentricity of the elliptic. At present, the Earth's orbital eccentricity is 0.017.

The end-points of the long axis of the orbital ellipse (line *a* in Figure 2.1) define aphelion (the furthermost point of the Earth from the Sun) and

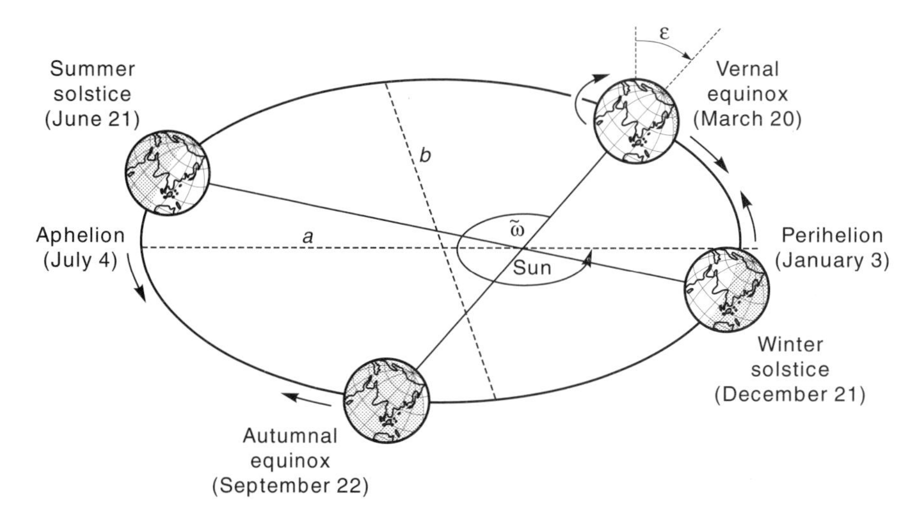

Figure 2.1 Elements of the Earth's orbit. The eccentricity, e, is given by $(a^2-b^2)^{1/2}/a$, where a is the semimajor axis and b is the semiminor axis. ω is the longitude of perigee relative to the moving vernal equinox. ε is the obliquity of the ecliptic (axial tilt). Aphelion, perihelion, the equinoxes, and the solstices currently occur on or about the dates shown

perihelion (the nearest point of the Earth to the Sun). Similar points for the Earth–Moon pair are called apogee and perigee. At present, the Earth is nearest the Sun, or in perihelion, on 3 January and is furthest from the Sun, or in aphelion, on 4 July. These dates change because the elliptical orbit itself rotates slowly. The Sun lies very close to a focus of the orbital ellipse, rather than at the orbital centre. Consequently, the distance between the Earth and the Sun is nearly 5 million km greater at aphelion than at perihelion. The mean distance from the Sun is called the semimajor axis. It is about 149.6 million km. It serves to define an astronomical unit, AU, which is in everyday use by astronomers. Very large distances in the Cosmos are normally measured in light-years, parsecs (pc), or multiples of parsecs. For the benefit of non-astronomers (like the author), 1 parsec is equal to 3.2616 light-years or 3.0857×10^{13} km.

The Earth's axis of rotation is tilted. If a line is drawn through the centre of the Earth and normal to the ecliptic, then the rotation axis stands at an angle, ε, from this 'vertical'. It is the same as the angle subtended by the Earth's orbital plane and the Earth's equatorial plane (technically called the equinoctial plane). It is commonly referred to as the obliquity of the ecliptic (or simply obliquity or tilt), since it measures the deviation of the ecliptic plane from equinoctial plane. At present, the angle of tilt is about 23.5°. If obliquity were 0°, which it very nearly is on Jupiter, then the equinoctial plane and ecliptic would coincide, and the rotation axis would stand bolt upright, at right-angles to the ecliptic. If obliquity were 90°, which it nearly is on Venus, then the equinoctial plane would lie at 90° to the ecliptic, and the rotation axis would lie on its side in the orbital plane.

The Earth's axial tilt causes the march of the seasons. The four seasons of the year start at the four points on the Earth's orbit corresponding to the two solstices and the two equinoxes. The solstices ('sun-stands') are the two points on the Earth's orbit where the Sun lies in the plane passing through the Earth's rotatory axis and normal to the ecliptic. They correspond to the times when the Sun reaches a position vertically above the Tropic of Cancer (21 June) and the Tropic of Capricorn (21 December). In the Northern Hemisphere, 21 December is called the winter solstice, and marks the start of winter because then the North Pole is tilted furthest away from the Sun. It is also the shortest day over the entire Northern Hemisphere. In the Southern Hemisphere, 21 December is called the summer solstice for it marks the start of summer and is the longest day of the year for the entire Southern Hemisphere. On 21 June, the North Pole is tilted towards the Sun: summer begins in the Northern Hemisphere, winter in the Southern Hemisphere; the Northern Hemisphere has its longest day, the Southern Hemisphere its shortest. The equinoxes are the points on the Earth's orbit where the Sun lies vertically above the equator and the two poles are equidistant from the Sun. This occurs at the cardinal points corresponding to 20 March and 22 September, which are referred to as the spring and autumn equinoxes because at these times all points on the Earth have equal hours of day and night.

The orientation of the rotatory axis slowly alters relative to the reference frame of the stars. In other words, the celestial poles (the points where the Earth's spin axis, when extended, pierces the celestial sphere) change. Currently, the North Pole points away from the Sun during the northern winter, when the Earth is near perihelion, and points towards the Sun during the northern summer, when the Earth is near aphelion. However, the North Pole slowly rotates or precesses in the opposite direction to the Earth's rotation. In doing so, it traces out a circle that, when joined to the Earth's centre of mass, describes a precessional cone. The South Pole moves in the same manner. This slow movement of the rotation axis is called axial precession (Figure 2.2). Axial precession is very slow, the North and South Poles circling the poles of the ecliptic once every 25,500 years (relative to a fixed perihelion). So, about 13,000 years in the future, the northern end of the rotation axis will point towards α-Lyrae, which then will be the Pole Star. However, at the same time that the Earth's axis of rotation is precessing, the elliptical orbit itself is rotating anticlockwise, much more slowly, in the orbital plane. The combined effect of axial precession and the rotation of the orbit leads to the precession of the equinoxes. This means that the four cardinal points on the Earth's orbit, corresponding to the

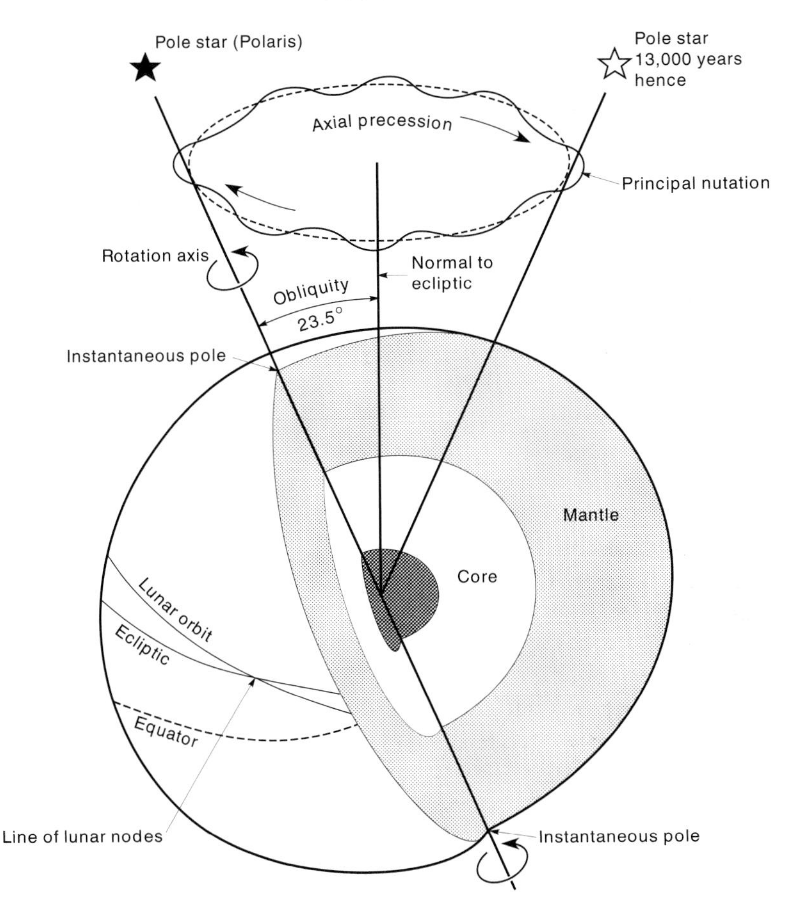

Figure 2.2 Axial precession, principal nutation, and the line of the lunar nodes

solstices and equinoxes, move slowly round the orbital path, and the dates of the solstices and equinoxes alter. To be precise, the time of the equinoxes occurs three seconds earlier each year.

The solar power house

The Sun, a middle-rate star, is the primary power source for the ecosphere. It is a nuclear fusion reactor, out of which pour electromagnetic radiation, cosmic rays, and the solar wind. Electromagnetic radiation is emitted at wavelengths ranging from very short (tens of nanometres) to long (metres). This spectrum spans extreme ultraviolet radiation, through ultraviolet, visible light, and infrared, to radio frequencies. The average power of the total solar irradiance, known as the solar constant, is about 1,370 $Watts/m^2$. (In my undergraduate days, it was 2.0 $cal/cm^2.min$, a much more satisfactory number.) The label 'constant' is, and is not, misleading. On the one hand, all portions of the Sun's radiative output vary over days and years according to the Sun's changing activity levels, but the variations hardly ever exceed 0.2 per cent. On the other hand, this is 0.2 per cent of a mighty energy stream, so even such modest changes in total solar output have significant effects on the Earth.

Cosmic rays are atomic nuclei, mainly protons with very few electrons, that stream through interplanetary and interstellar space, almost at the speed of light. They are high-energy particles with energies ranging from tens of millions to billions of electron volts. This compares with a photon of sunlight that carries about 2 electron volts of energy. Cosmic rays come from the Sun and from distant stars. Solar cosmic rays are generally of lower energy and tend to decay more quickly. Their flux increases during solar flares. Much of the cosmic ray flux is thought to be produced at the shock fronts created during supernovae, colossal stellar explosions that signalize the death of a massive star or the violent change from a white dwarf into a neutron star.

The solar wind is a ceaseless, electrified stream of low-energy particles, less than about 10,000 electron volts, issuing from the Sun at about 900 km per second. It is, effectively, the hot (100,000°C) solar corona expanding into interplanetary space as a rarefied plasma (that is, a completely ionized gas consisting of free electrons and free atomic nuclei in temperatures too high for atoms to exist).

Space debris

In orbiting the Sun, the Earth meets other bodies in the Solar System whose orbits it happens to cross. These bodies range in size from dust-like particles, which continuously rain into the atmosphere, to pieces of rock and ice the size of mountains, which strike the Earth on rare occasions. Three broad groups of extraterrestrial objects are known to collide with the Earth: asteroids, comets, and meteoroids.

Asteroids

Asteroids (meaning 'star-like' objects), when viewed through a telescope, appear as a point of light. Those that venture within the inner Solar System are placed in three groups: Amor asteroids, Aten asteroids, and Apollo asteroids (Table 2.1). They each have different collision probabilities and impact velocities (Shoemaker *et al.* 1990). Earth-crossing asteroids are generally thought to come from the Asteroid Belt, a region of space lying

Table 2.1 Earth-crossing asteroids

Type of asteroid	*Determinative orbital characteristic*	*Number observed (as at August 1993)*	*Collision probability (per year)*	*Impact velocity (km/s)*
Amor[a]	Distance from Sun at perihelion between 1.0167 and 1.3 AU	40	1.4×10^{-9}	17.5
Aten	Semimajor axis less than 1.0 AU; distance from Sun at aphelion more than 0.9833 AU	15	10.7×10^{-9}	16.2
Apollo	Semimajor axis less than, or equal to, 1.0 AU; distance from Sun at perihelion less than 1.0167 AU	125	4.1×10^{-9}	21.1

Note: [a] These are Earth-crossing Amors
Sources: Number observed from Rabinowitz *et al.* (1994); collision probabilities from Shoemaker *et al.* (1990); impact velocities (given as root mean square velocity for asteroids whose mean orbital elements are known) computed from data in Rabinowitz *et al.* (1994)

between the orbits of Mars and Jupiter. They may also come from comets in the Jupiter-family and Halley-family short-period system, which may themselves be the fragments of a single progenitor giant comet up to 180 km in diameter (M. E. Bailey *et al.* 1994).

As of August 1993, about 180 Earth-crossing asteroids had been identified (Rabinowitz *et al.* 1994). The largest discovered so far are the Earth-crossing Amors 1627 Ivar and 1580 Betulia, both with diameters of about 8 km, and the Apollo asteroids 1866 Sisyphus, with a diameter of about 10 km, 3200 Phaeton, with a diameter of about 6.9 km, and 2212 Hephaistos, with a diameter of about 5 km. The smallest detected Earth-crossing asteroid so far is probably 1993 KA2, with a diameter of 4–8 m, which passed within 0.001 AU (less than half the distance to the Moon) in May 1993. The full population of Earth-crossing asteroids larger than 100 m is estimated at over 136,000. The majority of these are small: there are about 20 larger than 5 km, 1,500 larger than 1 km, and 135,000 larger than 100 m (Figure 2.3). These estimates are more uncertain at the small end. All mountain-sized Earth-crossing asteroids (larger than about 6 km) are thought to have been discovered.

Comets

Comets (meaning 'long-haired' stars) are diffuse, unstable bodies of gas and solid particles that orbit the Sun. They have a dusty atmosphere, or coma, commonly with tails of plasma and dust during their active phases (which occur when their orbits bring them close to the Sun). Their orbits are highly elliptical, with a perihelion distance of less than 1 AU, and an average aphelion distance of about 10,000 AU. They are short-lived,

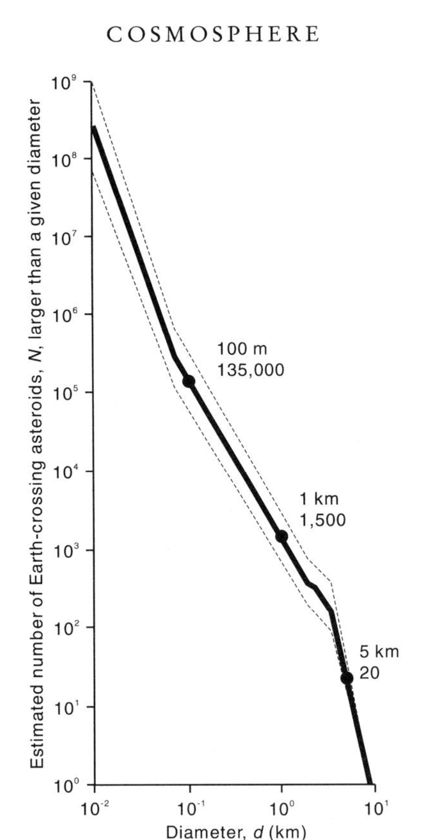

Figure 2.3 Estimated number of Earth-crossing asteroids larger than a given diameter
Source: After Rabinowitz *et al.* (1994)

surviving about a hundred perihelion passages. Comets that take more than 200 years to orbit the Sun are called 'long-period' comets (this includes comets that are not periodic at all, coming for the first time from the Oort cloud and being perturbed right out of the Solar System). Comets that take less than 20 years are called 'short-period' or Jupiter-family comets; those that take 20–200 years are called 'intermediate-period' or Halley-family comets.

By 1994, twenty-six short-period active Earth-crossing comets had been discovered, of which thirteen belong to the Jupiter family and thirteen to the Halley family; only two extinct short-period comets had been observed by the same date (Table 2.2). Some 411 long-period comets have been observed (Marsden and Steel 1994).

Table 2.2 Earth-crossing comets (1 km or more in diameter)

Type	*Activity*	*Number observed*	*Best estimated population*	*Collision probability (per year)*	*Impact velocity (km/s)*
Short- and intermediate-period comets[a]					
Jupiter family	Active	13	20 to 80	1.3×10^{-9}	22.9
	Extinct[b]	2	~780		
Halley family	Active	13	140 to 270	0.16×10^{-9}	45.4
	Extinct	0	2,100 to 4,050		
Long-period comets[c]					
		411	Large supply in Oort cloud	2.52×10^{-9}	52.8

Notes: [a] From data in Shoemaker *et al.* (1994)
[b] Extinct or dormant means that they do not, or cannot, display a detectable coma
[c] From data in Marsden and Steel (1994)

Meteoroids

A meteoroid is a natural solid object moving in interplanetary space that is smaller than about 100 m, but larger than a molecule. Meteoroids are too small to be observed by a telescope but are thought to be fragments of asteroids. A meteorite is a natural object of extraterrestrial origin that survives the brief journey through the atmosphere. It is a small asteroid or a meteoroid that has struck the Earth's surface. The annual accretion rate of small meteoroids, as found by examining hypervelocity impact craters on the space-facing end of the Long Duration Exposure Facility Satellite, is about 40,000 ± 20 tonnes (Love and Brownlee 1993).

Since 1975, 136 meteoroid impacts have been detected world-wide by infrared sensors in satellites (Tagliaferri *et al.* 1994). The sensors detect the heat emitted from the fireballs produced as meteoroids detonate in the atmosphere. The flux rate of meteoroids suggests that at least one 20 kilotonne airburst should occur every year. Such explosions are dangerous as they may be misinterpreted as nuclear explosions. On 1 October 1990, an explosion with over 1 kilotonne of TNT energy equivalent occurred 30 km over the central Pacific Ocean and was picked up by sensors on United States Department of Defense satellites. It took several months to decide that it had been caused by a 100-tonne stony asteroid striking the atmosphere (Tagliaferri *et al.* 1994).

The biggest and most well-documented encounter between a meteoroid and the Earth was the Tunguska event of the morning of 30 June 1908. A small asteroid, probably 60 m in diameter, travelling from south-east to north-west, hurtled over the Podkamennaya–Tunguska River region of Siberia. It exploded some 8 km above the ground, causing a great fireball about 60 km north-west of the remote trading post of Vanovara. The fireball could be seen 1,000 km away and the atmospheric shock heard at even greater distances. Trees within a 40 km radius of the blast were knocked flat, and there is some evidence that dry timber was ignited within an 8 km radius. The energy released by the explosion is estimated at 10–30 megatonnes of TNT equivalent energy (Adushkin and Nemchinov 1994).

Atomic and molecular clouds

In its passage around the Milky Way Galaxy, the Solar System passes through other kinds of matter. The interstellar medium consists of a hot, dilute, and very tenuous intercloud medium in which are diffuse atomic hydrogen clouds. The atomic clouds contain molecular clouds, which in turn contain denser clumps. The molecular clouds are widely scattered and are dense enough to be self-gravitating and exert gravitational forces on the Solar System. They may be very large, containing up to a million times more mass than the Sun, and are sometimes called giant molecular clouds. They tend to be sited relatively close to the galactic plane in spiral arms.

A CLOCKWORK MACHINE: COSMIC CHANGE

Newton constructed a clockwork Solar System that ticked with the precise regularity of a Swiss watch. His nice mechanical system could be represented, in simplified form, by a mechanical model – as an orrery. The discovery of non-linear dynamics has led to the demise of the comforting Newtonian world view. The cosmological clock is non-linear. It does not mark the regular beat once thought to exist in the cycles of the planets, in the processes of faraway galaxies, and in the calibration of the atomic clock (Shaw 1994: 317). Instead, the Cosmos displays the paradox of marvellous chaotic regularity. This does not mean that cosmic processes 'go crazy'. In the Solar System, for instance, it means that the mean orbital parameters of planets will probably progress stably for all time (Torbett 1989), but within as short a time as 5 million years, departures from these mean conditions may evolve unstable characteristics (see Shaw 1994: 133). From this 'stable chaos' (Milani and Nobili 1992), solar, lunar, planetary, and galactic beats ring out loud and clear.

The Sun

Solar activity varies over days and years. Several cycles of solar activity are known:

1 Sunspot or solar cycle. This has a long-term average cycle length of 11.14 years, but ranges from 6 to 17 years.

2 Hale, double sunspot, or heliomagnetic cycle. This has a period of about 22 years. It involves a reversal of the Sun's magnetic field and sunspot variations. It influences cycles of terrestrial magnetic activity and the atmospheric production of isotopes.
3 Gleissberg cycle. This is about 80 years long, and was first detected in records of auroral activity (Gleissberg 1955, 1965).
4 Sun's orbital or King-Hele cycle, or Sun–Jupiter repeat. This arises from the fact that the Sun does not sit plum centre in the Solar System, but orbits the barycentre. At maximum stretch (or apobac), its orbit extends a shade beyond two solar radii from the barycentre (Jose 1965). This motion follows a 178.73-year cycle that appears to influence solar activity (Jose 1965; Landscheidt 1983; Fairbridge and Shirley 1987).
5 Solar minima cycles. Prolonged periods of low sunspot activity seem to occur (but see Legrand *et al.* 1992). During the last millennium, there have been three such periods: the Wolf, Spörer, and Maunder Minima. In addition, there appears to be a 900-year cycle of solar activity (Henkel 1972). Both the prolonged solar minima and the 900-year cycle may be explained by two parameters, both of which are defined with reference to an inertial frame whose origin is the barycentre: the orientation of the solar orbit; and the change in the precessional amplitude of the solar orbit's orientation during successive solar orbits.

Gravity

The terrestrial environment is influenced by gravitational cycles. These cycles result chiefly from Earth–Moon interactions and from planetary orbital motions. The Sun's barycentric orbit is also influential, beating out a 178.73-year cycle of solar activity. All cycles in the Solar System are interlocked chronologically in a commensurable way. In other words, every fundamental cycle in the Solar System is related to every other cycle in small integer ratios. At any given time, the relationships between cycles will be approximate owing to the effects of ellipticity and other factors.

Moon

The Earth and the Moon orbit a common barycentre and have important gravitational influences upon one another. The chief lunar cycles are:

1 Diurnal and fortnightly tidal cycles. Everyone who has visited the seaside is familiar with these tidal cycles, except those travelling only to Mediterranean locations.
2 Lunar nodal or Metonic cycle. The Moon's orbit is inclined 5.1° to the Earth's orbit. The point where the lunar orbit intersects the ecliptic is called the lunar node (Figure 2.2). The lunar node turns backwards or

regresses (with respect to the sense of rotation of the Earth and revolution of the Moon), taking on average 18.6134 years to complete one revolution. Related to this cycle is the progression within this plane of the longitude of lunar perigee, which gives a mean apsides cycle of 8.849 years.

3 Other lunar cycles include the perigee–syzygy cycle of 31.008 years, the apsides–perihelion cycle of 62.013 years, the nodal–perihelion cycle of 93.020 years, the perigee progression cycle of 556.027 years, and the parallactic tidal cycle of 1,843.346 years (F. J. Wood 1985).

The lunar nodal cycle causes small variations in the lunar torque as the perigee and apogee come into the plane of the Earth's axial tilt. These variations are sufficient to make the spin axis nod up and down a little, a motion known as nutation. The effect of nutation is to produce 'waves' on the precessional cone (Figure 2.2). The Earth's principal nutation cycle entails a 9.18″ increase and decrease of axial tilt over an 18.6-year cycle. Minor nutations, sometimes called librations, involving less than a 1″ change in tilt, also occur. The main ones are a 9.3-year nutation cycle, which is half the length of the principal nutation cycle; annual and semiannual cycles, which are caused by variations of the solar couple at perihelion and aphelion; and fortnightly cycles, which are caused by variations in the lunar couple at perigee and apogee (Table 2.3).

In polar shifts, the Earth (or one or more of its solid shells) moves with the spin axis. Polar motions involve the Earth's wobbling about a fixed spin axis. Polar shifts leave latitude and longitude unaltered. Not so polar motions, in which the geographical poles move about a spin axis of fixed

Table 2.3 Motions and shifts of the Earth's spin axis

Motion	*Amplitude*	*Period*
Polar shift		
Steady precession	23.5°	25,700
Principal nutation	9.18″	18.6
Other periodic contribution to nutation in obliquity and latitude	< 1″	9.3 years, annual, semiannual, fortnightly
Polar motion		
Markovitz wobble	~.02″ (?)	20–40 years (?)
Chandler wobble	~.15″ (variable)	425–440 days
Annual wobble	~.09″	1 year
Semiannual wobble	~.01″	6 months
Secular motion of the pole	Irregular, ~0.2″ in 70 years	–

Source: After Rochester (1973)

posture. The chief polar motion is the Chandler wobble (Table 2.3). True polar wander is a large departure from the average geographical position of the spin axis. Polar motions result largely from the mass redistribution in the atmosphere, seas and oceans, ice sheets, seasonal changes of water tables, erosion and sedimentation, volcanism, and plate tectonics. They may also be forced by gravitational impulses.

Polar motions and polar shifts are linked with minor changes in the Earth's rotation rate. Lunar tidal friction brakes the Earth's rotation, causing a secular decrease in spin speed, which at present is about 40″ of arc per century. Tiny adjustments are made continuously to accommodate changes in the solar wind (which acts as a brake on the magnetosphere and thus Earth rotation), in gravitational tides, and in such geological processes as chemical differentiation and progressive heating that alter the Earth's moment of inertia.

Planetary cycles

The jostling of the planets, their satellites, and the Sun leads to variations in the Earth's orbit. Orbital readjustments are always taking place, almost imperceptibly. Large and significant changes take place over about 20,000–400,000 years. Orbital variations over this time-scale markedly alter the seasonal and latitudinal distribution of solar energy, without changing the annual total of solar radiation received by the Earth.

The chief short-term cycles are (Fairbridge 1984; Fairbridge and Sanders 1987):

1 Saturn–Jupiter lap. This is the principal cycle and involves Jupiter and Saturn coming into alignment on the same side of the Sun. It has a period of 19.857 years, and most other cycles are commensurable with it. For instance, the Sun's orbital cycle is 19.857 × 9 = 178.73 years. The half-cycle of 9.9295 years (Saturn and Jupiter in opposition) is found in spectral analyses of sunspot numbers.
2 Uranus–Saturn lap. This involves Uranus and Saturn coming into alignment every 45.4 years.
3 Outer planets repeat. The outer planets realign themselves in the same pattern every 4,448 years.
4 'All planets restart' cycle. This has a period of 93,408 years and is related to the orbital eccentricity cycle.

The four major medium-term orbital cycles are:

1 Tilt cycle. The tilt of the spin axis oscillates between 22° and 24° 30′. The oscillations have a major quasi-periodicity of 41,000 years, with minor components at 29,000 and 54,000 years.
2 Eccentricity cycle. The ellipticity of the orbit varies with average quasi-

periodicities of 100,000 and 410,000 years, with major periods at 95,000, 100,000, 120,000, and 410,000 years. The 100,000-year quasi-period is the short eccentricity cycle and the 410,000 quasi-period is the long eccentricity cycle. However, variations in the frequency of the obliquity cycle may produce also a strong 100,000 signal, too (Liu 1992, 1995).

3 Precessional cycle. It takes 25,800 years for the spin axis to precess once round the precessional cone relative to a fixed perihelion. The average quasi-periodicity of precession is 21,700 years, with major periods of 19,000 and 23,000 years, and with extreme periods of 14,000 and 28,000 years. From a climatic viewpoint, the precessional index is important. It is defined as $e \sin \omega$, where e is eccentricity and ω is the longitude of perihelion, and it describes the precession of the equinoxes.
4 Orbital-plane inclination cycle. The inclination of the orbital plane, compared to the invariable plane of the Solar System, varies by about 2° over a 100,000-year cycle (Muller and MacDonald 1995). This may account for changes in the accretion rate of interplanetary dust particles, as measured by helium-3 concentrations, in pelagic clays (Farley and Patterson 1995).

Cycles of orbital variations have slowed down during Earth history. For instance, the mean periodicities for tilt and precession were 60 per cent and 80 per cent lower during the Lower Silurian period, 440 million years ago (Berger *et al.* 1992; Bond *et al.* 1993; Berger and Loutre 1994). This change of orbital periodicity has implications for seeking Milankovitch rhythms in sediments – it would produce more sedimentary beds per unit thickness than in younger sequences.

Galactic cycles

Our Galaxy is a spiral star system. Its diameter is probably about 50 kiloparsecs, which is roughly 160,000 light-years. It has a massive 'spheroidal' centre (sometimes called 'the bulge'), and an extended and rather thin stellar 'disc' feature, in which hints of a spiral-arm structure are just discernible (M. E. Bailey *et al.* 1990: 26). The spiral arms contain the most massive molecular clouds and the formation sites of the most massive stars in the Galaxy. The Sun lies about 8.5 kiloparsecs from the galactic centre, around which it revolves at about 220 km/s. Its orbit has an eccentricity of 0.1. It takes very roughly 225 million years to complete one circuit. This, the galactic year (Parenago 1952), is poorly understood and possibly irregular (Shaw 1994: 245).

In moving round the Galaxy, the Sun moves in and out and up and down a little. The in-and-out epicycle, which moves in the same plane as the galactic plane, has a period of some 170 million years, and the up-and-down epicycle, which moves perpendicularly to the galactic plane, has a half-period of about

33 million years. The significance of these motions is that they lead to changes in the galactic-tide force experienced in the Solar System. These variations, along with stellar and molecular cloud perturbations, are capable of driving long-period comets into the inner Solar System. Stellar and molecular cloud perturbations are more likely to occur in spiral arms, through which the Sun passes every 100–200 million years.

It is possible that tidal interaction with the Large and Small Magellanic Clouds, companion galaxies to the Milky Way, produce effects on the Earth (G. E. Williams 1975a). The orbital period of the Large Magellanic Cloud is about 2 billion years and this may drive a very long-term cycle of asteroid and comet impacts (Kumazawa and Mizutani 1981).

THE CLOCK STRUCK ONE: RATES OF COSMIC CHANGE

The violent Universe

The verdict of modern astronomers is unanimous: space is a dangerous place. Several objects within the Milky Way Galaxy could enter solar space. Several latent dangers lurk in the Cosmos, waiting mindlessly for the laws of chance to call their number. As Isaac Asimov (1980) put it, humanity faces a choice of catastrophes: mini black holes, globs of antimatter, vagrant planets, supernovae, neutrinos, irregular solar activity, and bombardment all pose possible threats.

The first three phenomena, though a little fanciful, might exist. Supernovae do exist and have probably affected the Earth in the past (McCrea 1981; Hallam 1979; Russell 1979). Indeed, one theory holds that a nearby supernova triggered the formation of the Solar System. Galactic cosmic rays pour out of supernovae. The rate of supernovae occurrence in the Milky Way Galaxy suggests that the Solar System will be bathed in twice the background level of cosmic rays once every 100,000–1,000,000 years. And, during Earth history, some three or four supernovae have occurred within 10 parsecs of the Solar System, and some twenty within 20 parsecs (Shklovskii 1968). Massive explosions at such close quarters would have increased the cosmic ray flux by factors of 100. This would cause considerable ionization in the atmosphere and promote the catalytic destruction of about 90 per cent of the ozone layer, which would take a few years or perhaps a century to recover (Ruderman 1974). Such wild fluctuations in the cosmic ray flux might induce mutations in genetic information carried by organisms. Supernovae may emit vast quantities of high-energy photons, too. Every 2 million years or so a star explodes within about 100 parsecs of the Solar System, and every 2,000 million years within 10 parsecs. These stellar explosions increase the extreme ultraviolet and X-ray flux to, respectively, 100 and 10,000 times the solar norm, giving the

Earth the equivalent of an annual dose and a daily dose of solar ionizing radiation in a short burst (Torbett 1989). A nearby supernova would be a visual spectacle.

Space debris exists, too. It flies around the Solar System, some of it on a potential collision course with the Earth.

The violent Solar System

The Solar System, like the Universe at large, is now known to be a violent place. It is not the peaceful and secure clockwork machine envisaged by Isaac Newton. Space debris flying round the Solar System and solar flares are capable of modulating geological and ecological processes on Earth.

Asteroids and comets have struck the Earth. Nobody has witnessed a large strike in modern times, the Tunguska incident being the largest historical event for which there is direct evidence. But space probes have shown that impact craters are common on other planetary and satellitic bodies throughout the Solar System. The magnitude and frequency of impact events can be calculated from the crater-size distribution of other planets and satellites. The frequency of collision with asteroids, comets, and meteoroids is inversely proportional to the size of the impacting body. Meteoritic dust continuously rains in to the atmosphere. Small meteoroid strikes occur almost weekly. Asteroids with a diameter of about a kilometre strike about three times every million years. Mountain-sized asteroids or comets strike just once every 50 million years or thereabouts. A recent calculation gives 1-in-10,000 odds on a 2-km diameter asteroid smiting the Earth in the next century (Chapman and Morrison 1994).

Craters are found on the Earth. Some of them are unquestionably caused by volcanic activity. Many seem to be impact craters (Figure 2.4). Current estimates suggest that about 6 per cent of Phanerozoic terrestrial impact structures with diameters greater than 10 km, and 16 per cent of those with diameters greater than 20 km, have been discovered (Trefil and Raup 1990). Signs of an impact crater are shatter cones, shock-metamorphosed forms of silica, and shocked quartz crystals. All of these features are thought to be produced by the immense pressures created in the rocks around an impact site. Shatter cones are produced by an explosion above rock strata, rather than from a cryptovolcanic explosion from below them (e.g. Dietz 1947). Coesite and stishovite are forms of silica produced by immense pressures in laboratory experiments. Their natural occurrences in craters are signs of an impact origin. Crystalline quartz that has deformed and developed lamellar structures (shocked quartz) indicates an impact crater. Neither shatter cones, nor coesite and stishovite, nor shocked quartz provide incontrovertible evidence of an impact crater. The necessary shock and intense pressures might have been produced by other mechanisms. The only alternative geological mechanism is volcanism, but it is open to question

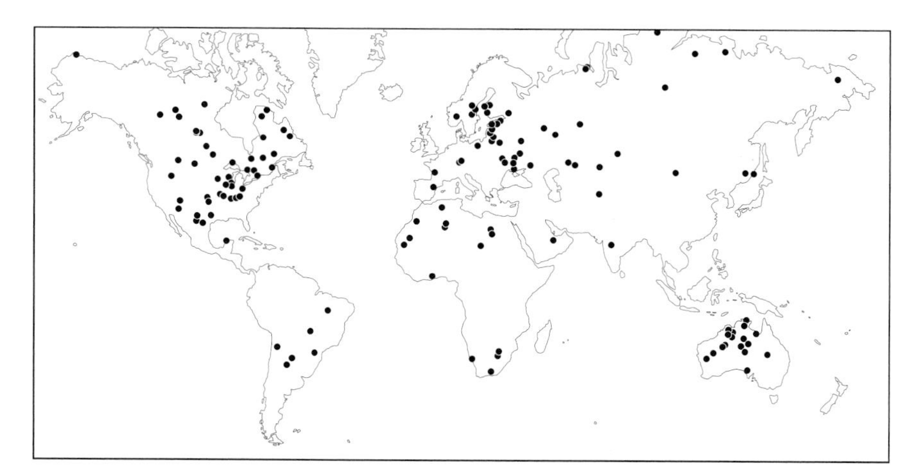

Figure 2.4 Sites of known terrestrial impact craters
Source: After Grieve and Shoemaker (1994)

whether even the biggest volcanic explosions are energetic enough to produce the shock-metamorphic features.

Violent times

A debate has raged over whether bombardment has occurred randomly or periodically. Originally, astronomers were inclined to the idea that asteroids and comets were stray bodies, taken out of the Asteroid Belt and Oort cloud (the vast cometary reservoir lying in the outer Solar System) by chance events. The latest thinking overturns this stray bolide hypothesis, or 'stochastic catastrophism' (Steel *et al.* 1994: 473), favouring instead a co-ordinated or coherent catastrophism.

Stochastic catastrophism, although demanding random strikes, does not preclude the possibility of bombardment episodes, which might also occur randomly. There are theoretical grounds, and some empirical evidence, for conjecturing that bombardment tends to occur as episodic showers ('storms' is a more apt description), roughly every 30 million years, each shower lasting a few million years (Clube and Napier 1982; Hut *et al.* 1987; M. E. Bailey *et al.* 1990: 412–15.). The empirical evidence is questionable. Time series analysis of terrestrial impact craters with diameters 5 km or more, 250 million years old or younger, and with a maximal age uncertainty of 20 million years produces a faint signal at 30 million years (Grieve and Shoemaker 1994).

Several mechanisms have been invoked to explain episodic storms of space debris. The Nemesis hypothesis argued that the Sun might have a companion star on a highly eccentric orbit that perturbs the Oort cloud at perihelion passage (M. Davis *et al.* 1984; Whitmire and Jackson 1984). The solar companion was named Nemesis after the Greek goddess who relentlessly persecutes the excessively rich. It was also dubbed the 'Death Star', presumably after the Empire's space station in *Star Wars* (Weissman 1984). The Planet X hypothesis proposes that an undiscovered tenth planet orbits in the region beyond Pluto, and produces comet showers near the Earth with a very stable frequency (Whitmire and Matese 1985; Matese and Whitmire 1986). Another hypothesis focuses on the up-and-down motion of the Solar System about the galactic plane. The period of oscillation about the galactic plane is roughly 67 million years, estimates of the period varying between 52 and 74 million years (see Innanen *et al.* 1978; Bahcall and Bahcall 1985). Because of this bobbing motion, the Solar System passes through the galactic plane, where interplanetary matter tends to be denser, every 33 million years, and reaches its maximum distance (about 80–100 parsecs) from the galactic plane every 33 million, too. Now, there is an approximate correspondence between galactic plane crossings by the Solar System and the boundaries between the geological periods (Innanen *et al.* 1978). The two phenomena could be related causally if the vertical motion of the Solar System about the galactic plane were to cause comet showers (Rampino and Stothers 1984a, 1984b). This is a reasonable scenario because most medium-sized molecular clouds are concentrated near the galactic plane. Thus, the vertical oscillation of the Sun through the galactic plane, which has a half-period of 33 million years, would modulate the rate at which the Sun would encounter stars and molecular clouds. The modulation may involve the perturbation of the Oort cloud and inner cometary reservoir leading to comet showers lasting several million years. It is also possible that the Solar System periodically passes through dense nebulae as it orbits the Milky Way, which would again perturb comets in the Oort cloud (Clube 1978; Clube and Napier 1982, 1984; Napier and Clube 1979).

Recent work proposes what might be termed a 'harmonized catastrophism' to supersede the older, stochastic catastrophism. Two main schools advocate this new view. One school, named coherent catastrophism by its creators (Steel 1991; Steel *et al.* 1994), contends that large comets disintegrate to produce clusters of fragments, ranging in size from microns, metres, tens and hundreds of metres, to kilometres. Such clusters will form a train of debris with a characteristic orbit. If the node of the orbit (the point at which it crosses the ecliptic) is near 1 AU, and if the cluster passes its node when the Earth is near, then it repeatedly crosses the Earth's orbit. The outcome is cluster-object impacts at certain times of the year, every few years, depending on the relationship between the Earth's and the cluster's orbital periods. But an impact occurs only when precession has brought the node to 1 AU, so only on time-scales of every few thousand years. One cluster – the Taurid complex – is presently active, and has been for the last 20,000 years. It has produced episodes of atmospheric detonations. These may have had material consequences for the ecosphere and for civilization.

The second school advocating non-random bombardment might be called co-ordinated catastrophism. It sees the Earth, Sun, and Solar System as coupled non-linear systems (Shaw 1994). This idea is immensely powerful. It leads to a new picture of Earth history that outlaws happenstance and instates chaotic dynamics as its centrepiece; a picture that shows a grand co-ordinated theme played out over aeons, and that portrays gradual and catastrophic change in the living and non-living worlds as different expressions of the same non-linear processes. A vital ingredient of this new view is that comets, asteroids, and meteoroids fly around the Solar System in critically self-organized, as opposed to random, regimes. This view is supported by evidence suggesting that cratering on planets and satellites has distinct patterns in space and in time. On Earth, this pattern is accounted for by the early stage of heavy bombardment (more than 4 billion years ago), during which colossal impacts, such as the one responsible for creating the Moon, led to an uneven distribution of terrestrial mass in the form of a meridional 'keel' of high-density rock. This 'keel' has subsequently influenced the incoming flight paths of space debris. The result is that impacts have tended to centre around three geographical nodes (one in north-central North America, one in north-eastern Europe, and one in Australia), at least during the Phanerozoic aeon. These three spatially invariant crater clusters are called Phanerozoic cratering nodes. Their significance will become clear in the next chapter.

A COSMIC CAROUSEL: DIRECTIONS OF COSMIC CHANGE

Several cosmic processes and events impinge upon the Earth and might well force secular trends in geological and ecospheric processes. The Sun has grown steadily brighter. Gravitational interactions have produced accelerations and decelerations of the Earth's spin rate superimposed on a slowing spin. They might also have caused protracted changes of obliquity. Impact events might have caused the Earth to tumble about its spin axis.

A faint young Sun

When the Earth was born, 4.6 billion years ago, the Sun was less luminous than it is now. Solar luminosity has steadily increased during geological time. The reason for this is that the solar core has become denser, and therefore hotter, as more and more hydrogen has been converted to helium. The luminosity is about 25–30 per cent greater than that at the dawn of Earth history (Hart 1978; Kasting 1987), and will go on increasing. Reduced sunlight in Precambrian times gives rise to the 'faint young Sun paradox'. With a fainter Sun, all the Earth's surface should have been frozen solid until about 2 billion years ago, but the oldest known sedimentary rocks, which demand the presence of oceans and a water cycle, are 3.8 billion years old. The answer to this paradox probably lies in atmospheric composition, and in particular the abundance of carbon dioxide in the early air.

A winding-down Earth

In theory, lunar tidal friction transfers angular momentum from the Earth to the Moon, and so retards the Earth's rotation rate. The current deceleration is some 5×10^{-22} radians per second. If the Earth's rotation rate should have decelerated, then the number of solar days per year, the number of solar days in a synodic month, and the number of synodic months in a year will have changed in a systematic manner (Figure 2.5).

The first evidence for spin deceleration was detected by John W. Wells in the skeletal growth rhythms of Middle Devonian corals living some 390 million years ago. At this time there appear to have been about 400 days in

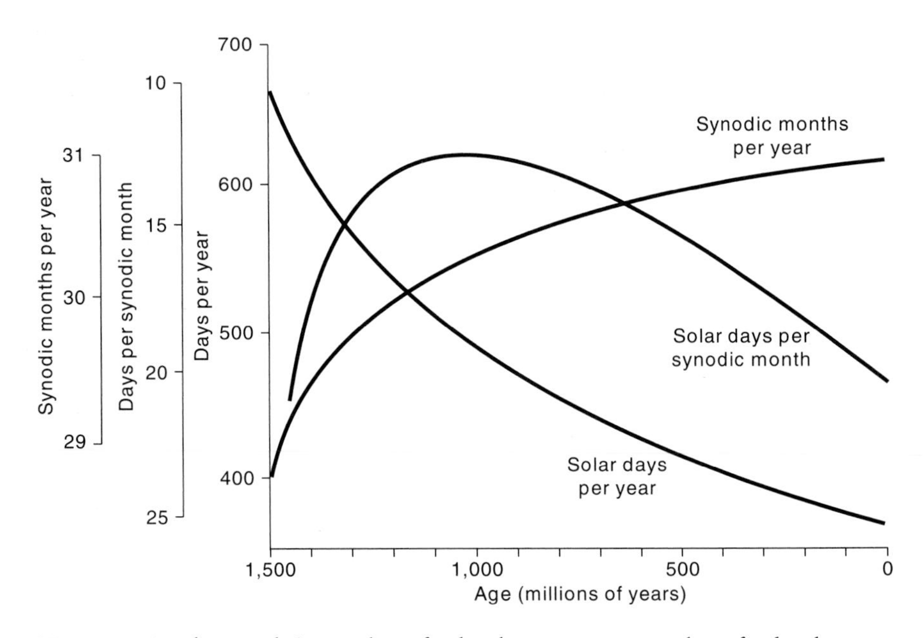

Figure 2.5 Secular trends in number of solar days per year, number of solar days per synodic month, and number of synodic months per year. The computations assume a 6° phase-lag of semidiurnal tides; a constant length of year and Earth mass; and a small tidal evolution of the Earth's orbit
Source: After Lambeck (1980)

the year (Wells 1963). Further work on fossil 'palaeontological clocks' (Pannella 1972), notably corals and bivalves, has confirmed Wells's findings. Moreover, it has shown that the rate of decrease of spin speed has not been uniform. Tidal friction was, it would seem, smaller during Proterozoic time and the rate of deceleration was commensurately less, as is indicated by fossil evidence (Piper 1990). During the Phanerozoic aeon, alternating periods of accelerating and decelerating spin velocity are separated by two types of turning point (Whyte 1977). First, spin maxima occur at transitions between acceleration and deceleration, which occurred 65 million and 375 million years ago. Second, spin minima occur at transitions between deceleration and acceleration, which occurred 235 million and 445 million years ago. These fluctuations about a general deceleration of spin velocity suggest mechanisms other than tidal friction are involved in the system, and a link with galactic processes seems possible (Fairbridge 1978).

A tumbling Earth

The Earth is stabilized by its equatorial bulge. Any small change in the distribution of its mass will cause a slight (probably imperceptible) adjustment of the position of the geographical poles as the entire mass of the Earth (or more likely the mantle and crust) tumbles about the spin axis, which remains fixed in the stellar reference frame. This is true polar wander. Large and sudden gains or losses of mass at particular places might cause substantial tumbles of the planet. Large tumbles do appear to have occurred on the Moon (Runcorn 1982, 1987). The Moon was originally orbited by several satellites, one of which, being drawn inwards by tidal forces (as Phobos is now being drawn in towards Mars), broke into fragments that smote the lunar surface leaving a trail of basins – regions of low mass – along the then equator. The equatorial position of the craters suggests that they were produced by a fragmented satellite, rather than by an asteroid or comet. The basins, being low-mass regions, would have destabilized the Moon, and a consequent redistribution of mass was effected by a tumble through 90°. This tumble brought the more massive regions to lie near to the equator, and the less massive basin regions to lie as near as they could to the poles. Later in the Moon's early history, another satellite was drawn in and fragmented, leaving a fresh trail of basins near the new equator. The new distribution of mass was again compensated for by a tumble. A similar sequence of events probably occurred a third time.

On Mars, too, there is evidence suggesting that true polar wander has taken place. The entire lithosphere of Mars has shifted, relative to its rotatory axis, in response to instabilities produced by mass redistribution. The cause of the mass redistribution may be impacts (P. H. Schultz 1985) or 'postglacial rebound' following the melting of giant polar ice caps (Rubincam 1993). The time distribution of microcraters on the Martian surface supports the notion of dramatic climatic fluctuations resulting from large changes of obliquity (Vasavada *et al.* 1993).

True polar wander has occurred on Earth, apparently owing to mass redistribution by geological and atmospheric process, including subduction (Spada *et al.* 1992). The entire lithosphere has moved about 5° during the last 65 million years (Courtillot and Besse 1987), and 75° in 75 million years during the Middle Palaeozoic era (Van der Voo 1994), relative to the spin axis. The mantle below the asthenosphere also wanders slowly (e.g. Ricard *et al.* 1993). A mantle roll of 17–19° over the last 90 million years has been detected (Livermore *et al.* 1984).

A tilting Earth

The present axial tilt appears stable, largely owing to the Moon. With the Moon present, axial tilt is stable between 0° and 30° and chaotic above 30°; without the Moon, the stable zone would be confined to 0–10° (Laskar *et al.* 1993). However, the Earth's axial tilt, like that of many other planets, might have changed radically during Earth history (see G. E. Williams 1993). The angle of axial tilt probably depends on the balance between the tidal friction effect, which reduces the angle, and internal energy dissipation, which increases the angle (Gold 1966). The conventional view is that solar and lunar tidal friction has led to a gradual increase in obliquity, the original axis standing at some 10–15° (e.g. MacDonald 1964). However, this view overlooked the possible effects of core–mantle coupling, and the history of the Earth's rotation and lunar orbit, some important details of which have become known only recently. Piecing together evidence from palaeoclimatology, geochronology, and geophysics has led to a model of evolving obliquity. The axis has slowly stood more upright from a primitively large value of somewhere between 54° and 90°, through a relatively rapid rise between about 650 and 430 million years ago, to the current value that has stayed roughly the same for most of the Phanerozoic aeon (Figure 2.6). The point of maximum obliquity change occurred 550 million years ago, coinciding with the Precambrian–Cambrian boundary. This model has enormous implications for interpretations of Earth's climatic and biological evolution and will be discussed more fully in later chapters.

Beginnings: solar and planetary formation

The initial conditions in the newly formed Earth, together with a few large planetesimal impacts in its very early history, placed severe restraints upon subsequent geological evolution. For this reason, understanding the origin of the Earth and Solar System is relevant to understanding many aspects of geology. Ancient Precambrian rocks, the surface of the Moon and other planetary bodies, and the asteroidal sources of meteorites record events in

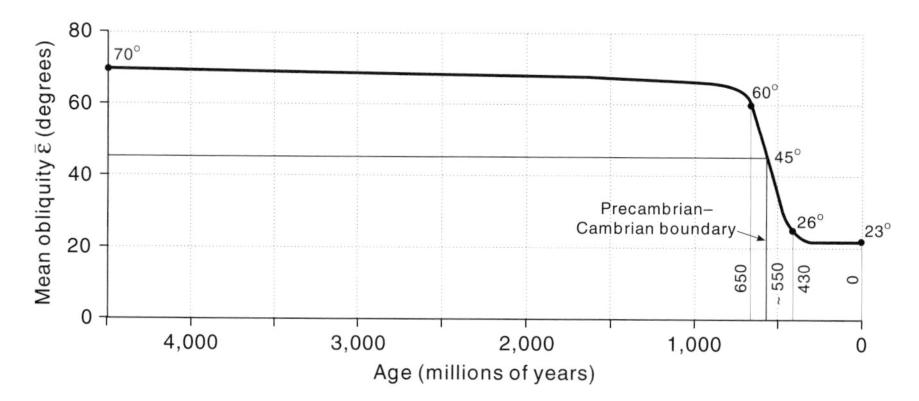

Figure 2.6 Proposed secular change of mean obliquity. This curve is consistent with a single giant impact hypothesis for the origin of the Moon, and with an interpretation of the geological record
Source: After G. E. Williams (1993)

the early history of the Earth and Solar System. Astronomical theory and observations of inchoate nebulae provide an insight into how the Solar System might have formed. Planetological, geological, and geochemical evidence largely accords with the astronomical observation and theory, though some conflicts arise. It is difficult, for example, to explain how an extremely hot and melted planet, as predicted by astronomical theory, could fail to produce a globally layered terrestrial mantle within several hundred million years by cooling, crystallization, and associated chemical fractionation (e.g. Wetherill 1990: 250–2). Such discrepancies between theory and evidence require further investigation.

It is now generally agreed that the Earth and other planets were created as a by-product as the Sun formed from a dense interstellar molecular cloud (e.g. Cameron 1988). This idea is expressed in the nebula hypothesis. It was first suggested by René Descartes and taken up by Immanuel Kant. It has the Sun and planets being formed together as a single process. Other models of planetary formation have been proposed. Some theorists claimed that planetary matter was torn from the Sun or a passing star, or both. Others have proposed that planetary matter was captured by the Sun while passing through an interstellar gas cloud.

The nebula hypothesis is currently fashionable. It has several variants, though a standard model has emerged since about 1970 (see Wetherill 1990; Torbett 1989). The standard account runs like this. In the beginning was a cloud of interstellar dust and gas. Compression eventually caused a gravitational instability. The upshot was that the cloud collapsed and fragmented under its own gravity. Initially, the collapse occurred in the core of the cloud. Density increased in the central regions producing a central object that was to become the proto-Sun. (Gravitational collapse in rotating clouds normally produces binary and multiple-star systems. Only where the specific angular momentum is relatively low will a single central star evolve. A companion star, smaller than the Sun and almost touching it, may have existed at an early stage in the solar nebula (Shaw 1994: 137). This star, named Genesis in its early stages and Siva in its final stage, was destroyed, or self-destructed, leaving its signature in the dynamical discontinuity between the inner (terrestrial) and outer (gaseous) planets. Alternatively, it is possible that Siva survives to this day as a small and weakly bound, highly eccentric, low-luminosity, distant solar companion that might, conceivably, correspond to the putative solar companion called Nemesis (or the Death Star) that has been invoked to explain periodic comet storms.) At a critical density, radiation, which was released as gravitational potential energy was converted to heat, became trapped. This radiational trapping generated an internal pressure within the cloud that counterbalanced the gravitational collapse. The outcome was a transition to a steady-state contraction in which dissipation could take place. Dissipational collapse then produced the solar nebula – a flattened rotating cloud or disc surrounding the proto-Sun (or Suns). At this point, the dust and gas, previously well mixed, started to separate. The dust settled towards the middle plane of the disc and coagulated to centimetre-sized particles in as little as 10–1,000 years. Frictional forces of gas drag continually dissipated the energy of solid-particle motion perpendicular to the disc. Consequently, the solid disc thinned. When the density of the solids was about a thousand times that of the gas, gravitational interactions between the solids led to gravitational fragmentation. The solid disc broke up into large regions, thousands of kilometres in extent, bound together by strong gravitational forces. These regions then collapsed into solid bodies called planetesimals.

The planetesimals were 1–10 km in diameter in the inner, terrestrial planetary zone, and 10–100 km in diameter in the outer planetary zone. In the outer planetary zone, rocky and icy planetesimals formed. In the inner planetary zone, only rocky planetesimals formed as it was too hot for icy planetesimal formation. Ultimately, planets formed from these planetesimals, which moved in roughly circular orbits, by a process called planetesimal accumulation. First, the planetesimals collided and merged to form planetary embryos the size of the Moon or Mercury. This took about 100,000 years. Planetesimal accumulation then slowed down because the greater distances between planetary embryos reduced the collision frequencies. The Earth took up to 100 million years to form in this way. The final stages of its formation involved impacts with residual planet-sized embryos. These giant impacts led to special events in early Earth history, including the formation of the Moon and the loss of the original atmosphere.

SUMMARY

The terrestrial environment is at the mercy of several cosmic forces. It is bathed in the stream of energy issuing from the Sun (electromagnetic radiation, cosmic rays, and the solar wind), and to a much lesser degree from other stars. It feels the results of gravitational jostling between the Earth and other objects in the Solar System, Galaxy, and, to a very small degree, other galaxies. And, it is sometimes at the receiving end of asteroids, comets, and meteoroids rocketing though near-Earth space. The character of these cosmic forces is strongly oscillatory, displaying periodic and quasi-periodic fluctuations. Solar, lunar, planetary, and galactic beats echo throughout the terrestrial environment. The Universe is a violent place. Several cosmic hazards threaten the Earth, of which bombardment and supernovae are perhaps the most worrying. Terrestrial impact craters betray past collisions with asteroids and comets. The impact record shows slight signs of periodicity. Bombardment is now seen to result, not from rare and random strikes by stray space debris, but from a harmonized pattern of activity played out over aeons. Secular trends are commonplace in the cosmic environment. Trends of significance to environmental change include the increasing brightness of the Sun, the frictional slowing-down of the Earth's rotation rate, large-scale true polar wander (possibly), and large changes in axial tilt (possibly). Events during the Earth's formation set tight constraints on subsequent planetary evolution. The Earth still bears the scars of spectacularly catastrophic events occurring during its early history, including the ejection of the Moon.

FURTHER READING

A readable and beautifully produced introduction to the Sun, planets, and their satellites is *The New Solar System* (Beatty and Chaikin 1990). An equally accessible and well-illustrated account of the Sun is *Sun, Earth and Sky* (Lang 1995). The Earth's orbital cycles are plainly explained in *Ice Ages: Solving the Mystery* (Imbrie and Imbrie 1986). *The Origin of Comets* (Bailey *et al.* 1990) contains a wealth of ideas about bombardment.

3

LITHOSPHERE AND BARYSPHERE

SHAKY FOUNDATIONS: THE GEOLOGICAL ENVIRONMENT

The layered Earth

During the nineteenth century, the nature of the Earth's interior was a matter of fierce and fascinating debate. All theories were hampered by a lack of evidence – the nature of rocks deep below the surface was unknown. In 1906, Richard D. Oldham observed that compressional seismic waves (P waves) slow abruptly deep within the Earth and can penetrate no further. This was strong evidence in favour of a liquid core. Three years later, Andrija Mohorovičić noticed that the velocity of seismic waves leaps from about 7.2 to 8.0 km/s at around 60 km deep. He had discovered the 'Moho' seismic discontinuity that marks the crust–mantle boundary. In 1926, Beno Gutenberg obtained evidence for a seismic discontinuity at the core–mantle boundary. This, the Gutenberg discontinuity, was confirmed during the 1950s when world-wide records of blasts from underground nuclear detonations were scrutinized. Subsequent studies of the Earth's seismic properties, using seismic waves propagated by earthquakes and by controlled explosions to 'X-ray' the planet (a technique called seismic tomography), have revealed a series of somewhat distinct layers or concentric shells in the solid Earth (Figure 3.1). Each shell has different chemical and physical properties.

The crust lies above the Moho. Its thickness ranges from 3 km in parts of ocean ridges to 80 km in collisional orogenic mountain belts. Continental crust is, on average, 39 km thick. The lithosphere is the outer shell of the solid Earth where the rocks are reasonably similar to those exposed at the surface. It includes the crust and the solid part of the upper mantle. It is the coldest part of the solid Earth. Cold rocks deform slowly, so the lithosphere is relatively rigid, it can support large loads, and it deforms by brittle fracture. On average, the lithosphere is about 100 km thick. Below continents it is up to 200 km thick, and beneath the oceans it is some 50 km

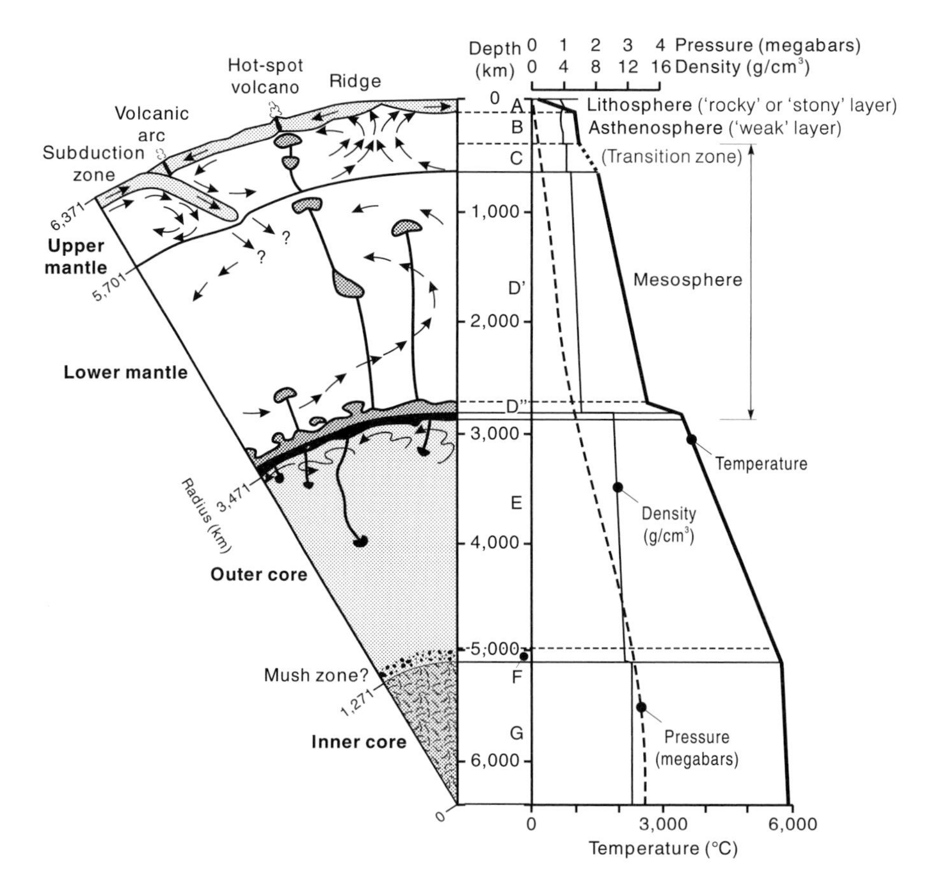

Figure 3.1 Layers of the solid Earth. The capital letters, A–G, are seismic regions

thick. Continental lithosphere is sometimes called the tectosphere. The differences in lithospheric thickness arise from temperature, and therefore viscosity, differences. The lithosphere under mid-ocean ridges is warm and thin; that under subduction zones is cold and thick; that under continents is cold, buoyant, and strong.

The mantle and core constitute the barysphere. Processes in the barysphere influence processes in the lithosphere and thus, indirectly, cause changes in the ecosphere, particularly those occurring over millions of years. For this reason, it is necessary to appreciate the basic structure and dynamics of the core and mantle.

The mantle is divided into upper and lower portions. The upper mantle consists of two shells. The asthenosphere (or rheosphere) lies immediately below the lithosphere. Temperatures increase with depth through the lithosphere. At around 100 km below the surface, lithospheric rocks are hot enough to melt partially, to weaken structurally, and behave rheidly (that is, like very slow-moving fluids). The asthenosphere, being relatively weak and ductile, is more readily deformed than the lithosphere. Its base sits at about 400 km below the Earth's surface. In most places, the top part of the asthenosphere is a 50–100-km-thick low-velocity zone.

Beneath the asthenosphere is the mesosphere. The uppermost part, which extends down to about 650 km, is a transition zone into the lower mantle. Rocks become more rigid again in the mesosphere because the solidifying effects of high pressures increasingly outweigh the effects of rising temperatures. The mesosphere continues as the lower mantle. Extending down to a depth of 2,890 km, the lower mantle accounts for nearly one-half of the Earth's mass.

The mantle rests upon the Earth's core, into which it merges through a fairly sharp and discontinuous transition zone known as the D″ (D double prime) layer. The core consists of an outer shell of mobile and molten iron, some 2,260 km thick, with a mush zone at its base. It sits on a solid inner ball, 1,228 km in radius, close to melting point, and composed of iron with, perhaps, some nickel (see Crossley 1993). The outer core drives the geodynamo that generates the Earth's magnetic field. This geomagnetic field changes its polarity from time to time (the magnetic north and south poles swap positions). Luckily for geologists, intervals of normal and reversed polarity leave traces as remnant palaeomagnetism in the rock record. Palaeomagnetic data enable the building of a geomagnetic time-scale. This is valuable in reconstructing rates of sea-floor spreading, the past positions of continents, and the degrees of true polar wander.

The sphere of rock

The outer shell of the solid Earth – the lithosphere – is composed of rocks and minerals. The average composition by weight of chemical elements is oxygen 47 per cent, silicon 28 per cent, aluminium 8.1 per cent, iron 5.0 per cent, calcium 3.6 per cent, sodium 2.8 per cent, potassium 2.6 per cent, magnesium 2.1 per cent, and the remaining eighty-three elements 0.8 per cent. These elements combine to form minerals. The chief minerals in the lithosphere are feldspars (aluminium silicates with potassium, sodium, or calcium), quartz (a form of silicon dioxide), clay minerals (complex aluminium silicates), iron minerals such as limonite and haematite, and ferromagnesian minerals (complex iron, magnesium, and calcium silicates). Ore deposits consist of common minerals precipitated from hot fluids. They include pyrite (iron sulphide), galena (lead sulphide), blende or sphalerite (zinc sulphide), and cinnabar (mercury sulphide).

Rocks are mixtures of crystalline forms of minerals. There are three main types:

1 Igneous rocks. These form by solidification of molten rock (magma). Most igneous rocks consist of silicate minerals, especially the felsic mineral group, which consists of quartz and feldspars (potash and plagioclase). Felsic minerals have silicon, aluminium, potassium, calcium, and sodium as the dominant elements. Other important mineral groups are the micas, amphiboles, and pyroxenes. All three groups contain aluminium, magnesium, iron, and potassium or calcium as major elements. Olivine is a magnesium and iron silicate. The micas, amphiboles (mainly hornblende), pyroxenes, and olivine constitute the mafic minerals, which are darker in colour and denser than the felsic minerals. Felsic rocks include diorite, tonalite, granodiorite, rhyolite, andesite, dacite, and granite. Mafic igneous rocks include gabbro and basalt. Ultramafic rocks, which are denser still than mafic rocks, include peridotite and serpentine. Much of the lithosphere below the crust is made of peridotite. Eclogite is an ultramafic rock that forms deep in the crust, nodules of which are sometimes carried to the surface by volcanic action. At about 400 km below the surface, olivine undergoes a phase change (it fits into a more tightly packed crystal lattice whilst keeping the same chemical composition) to spinel, a denser silicate mineral. In turn, at about 670 km depth, spinel undergoes a phase change into perovskite, which is probably the chief mantle constituent and the most abundant mineral in the Earth. For life and for soils, the distinction between acid igneous rocks such as granite, rich in quartz and potash feldspars, and basic igneous rocks such as basalt, rich in ferromagnesian minerals, is enormously important.
2 Sedimentary rocks. These are layered accumulations of mineral particles that are derived mostly from weathering and erosion of pre-existing rocks. They are clastic, organic, or chemical in origin. Clastic sedimentary rocks are unconsolidated or consolidated sediments (boulders, gravel, sand, silt, clay) derived from exogenic processes. Conglomerate, breccia, sandstone, mudstone, claystone, and shale are examples. Organic sedimentary rocks and mineral fuels form from organic materials. Examples are coal, petroleum, and natural gas. Chemical sedimentary rocks form by chemical precipitation in oceans, seas, lakes, caves, and, less commonly, rivers. Limestone, dolomite, chert, tufa, and evaporites are examples.
3 Metamorphic rocks. These form through physical and chemical changes in igneous and sedimentary rocks. The changes are caused by temperatures or pressures high enough to cause recrystallization of the component minerals. Slate, schist, quartzite, marble, and gneiss are examples.

Oceanic crust is 5–15 km thick and covers 50 per cent of the planet's surface. It is made largely of basalt, a dark, fine-grained igneous rock. It is called sima, as it has silica and magnesium as important constituents (Table 3.1). Continental crust is an admixture of igneous, sedimentary, and metamorphic rocks. It is called sial, as its dominant constituents are silica and aluminium (Table 3.1; see Rudnick and Fountain 1995). It consists of a thick base of igneous and metamorphic rocks, which has an andesitic bulk composition (Rudnick 1995), and a relatively thin cover of consolidated and unconsolidated sedimentary rocks. Its thickness ranges from 30 to 80 km. Transitional crust, with a thickness of 15–30 km, is found in islands, island arcs, and continental margins. The crust may be further divided into twelve crustal types (Table 3.2).

The open sea-floor has a thin cover of pelagic sediments. Continental surfaces are more varied in composition. They are underlain by shale (52 per cent), sandstone (15 per cent), granitic rocks (15 per cent), limestone and dolomite (7 per cent), basaltic rocks (3 per cent), and other rocks (8 per cent). All these crustal rocks provide much of the raw material from which soils evolve and on which life depends.

The lithosphere contains liquids and gases, as well as solid rocks. These are exchanged with gases and liquids in the ecosphere. The exchanges form part of the geological or rock cycle, which consists of three interrelated components – the water or hydrological cycle, the tectonic cycle, and the geochemical cycle (e.g. Wyllie 1971: 48):

1 Geochemical cycle. This is the circulation of chemical elements (carbon, oxygen, sodium, calcium, and so on) through the mesosphere, asthenosphere, lithosphere, and ecosphere (Chapter 8).
2 Water cycle. This is the circulation of meteoric water through the

Table 3.1 Average composition of continental and oceanic crust

	Continental			*Oceanic*
Elements expressed as oxides	*Upper (Weight %)*	*Lower (mafic) (Weight %)*	*Lower (felsic) (Weight %)*	*(ocean ridge basalt) (Weight %)*
Silicon (SiO_2)	65.5	49.2	61.0	49.6
Aluminium (Al_2O_3)	15.0	15.0	15.6	16.8
Iron (FeO)[a]	4.3	13.0	5.3	8.8
Calcium (CaO)	4.2	10.4	5.6	11.8
Sodium (Na_2O)	3.6	2.2	4.4	2.7
Potassium (K_2O)	3.3	0.5	1.0	0.2
Magnesium (MgO)	2.2	7.8	3.4	7.2
Titanium (TiO_2)	0.5	1.5	0.5	1.5
Phosphorus (P_2O_5)	0.2	0.2	0.2	0.2
Manganese (MnO)	0.1	0.2	0.1	0.2

Note: [a] Total iron as FeO
Source: After Condie (1989a)

Table 3.2 Crustal types and their physical properties

Major crustal types	*Area (%)*	*Volume (%)*	*Thickness (km)*	*Characteristics*
Continental shield	6	12	35	Stable parts of continents composed of Precambrian rocks (mainly metamorphic and plutonic) with little or no Precambrian sedimentary cover
Platform	18	35	41	Stable parts of crust consisting of a Precambrian basement with a 1–3 km blanket of relatively undeformed sedimentary rocks. Shield and platforms are both referred to as cratons
Palaeozoic orogenic belt	8	14	43	Long, curved belts of deformed Palaeozoic rocks bearing signs of igneous activity. They contain plateaux. Tectonic stability is intermediate to high
Mesozoic–Cenozoic orogenic belt	6	13	40	Similar to Palaeozoic orogenic belts but characteristically unstable tectonically
Continental rift system	<1	<1	28	Fault-bounded valleys involving grabens, half-grabens, and complex grabens. Found in unstable, tensional tectonic settings. Aulacogens are rift systems that open into ocean basins and peter out towards the interior of continents. They may be 'failed arms' formed at triple junctions
Volcanic island	⩽1	⩽1	14	Occur on or near ocean ridges (e.g. Ascension Island) and in ocean basins (e.g. the Hawaiian Islands). Tectonic stability highly variable
Island arc	3	3	22	Occur above subduction zones. Two types: continental-margin arcs (such as Japan and New Zealand) and island-arcs (collections of mainly andesitic volcanic peaks). Earthquakes and intense volcanism
Oceanic trench	3	2	8	From 5 to 8 km deep. Mark the start of subduction zones. Intense earthquake activity
Ocean basin	41	11	7	Stable with thin cover of deep-sea sediments. Basaltic volcanoes produce seamounts and guyots
Ocean ridge	10	5	5	Linear rift systems in the oceanic crust. Sites of sea-floor spreading. Unstable with basaltic volcanism that may produce islands
Marginal-sea basin	4	3	9	Portions of oceanic crust between island arcs (e.g. the Philippine Sea). Unstable, intermediate
Inland-sea basin	1	2	22	Partially or wholly surrounded by tectonically stable continental crust. Examples are Caspian and Black Seas. Thick accumulations of clastic sediments. Ordinarily stable
Average continental crust	41	79	36	–
Average oceanic crust	59	21	7	–

Sources: After Brune (1969) and Condie (1989a)

hydrosphere, atmosphere, and upper parts of the crust. It is linked to the circulation of deep-seated, juvenile water associated with magma production and the tectonic cycle (Chapter 5). Juvenile water may issue into the meteoric zone; meteoric water held in hydrous minerals and pore spaces in sediments, known as connate water, may be removed from the meteoric cycle at subduction sites.

3 Tectonic cycle. This is the repeated formation and destruction of crustal material. Volcanoes, folding, faulting, and uplift all bring rocks, water, and gases to the base of the atmosphere and hydrosphere. Once exposed to the air and meteoric water, these rocks begin to decompose and disintegrate by the action of weathering. The weathering products are transported by wind and by water to oceans. Deposition occurs. Burial of the loose sediments leads to compaction, cementation, and recrystallization and the formation of sedimentary rocks. Deep burial may convert sedimentary rocks into metamorphic rocks. Further alteration may produce granite. If uplifted, intruded or extruded, and exposed at the land surface, the loose sediments, consolidated sediments, metamorphic rocks, and granite, may join in the next round of the rock cycle.

The Earth's cracked shell

The lithosphere is not a single, unbroken shell of rock; it is a set of snugly tailored plates (Figure 3.2). At present there are seven large plates, all with an area over 100 million km^2. They are the African, North American, South American, Antarctic, Australian–Indian, Eurasian, and Pacific plates. Two dozen or so smaller plates have areas in the range 1–10 million km^2. They include the Nazca, Cocos, Philippine, Caribbean, Arabian, Somali, Juan de Fuca, Caroline, Bismarck, and Scotia plates, and a host of microplates or platelets.

In places, continental margins coincide with plate boundaries. Where they do, as along the western edge of the American continents, they are called active margins. Where they do not, but rather lie inside plates, they are called passive margins. The breakup of Pangaea created many passive margins, including the east coast of South America and west coast of Africa. Passive margins are sometimes designated rifted margins where plate motion has been divergent, and sheared margins where plate motion has been transformed. The distinction between active and passive margins is crucial to interpreting some large-scale features of the toposphere (Chapter 7).

The lithospheric plates and their subdivisions appear to form a set of nested polygons (Carey 1963). The main plates bounded by mid-ocean ridges – the Pacific, North America, South America, Africa, Europe, Siberia, India, Antarctica, and Australia – describe the first-order polygons. (As a non-believer in subduction, Carey declines to take subduction sites as

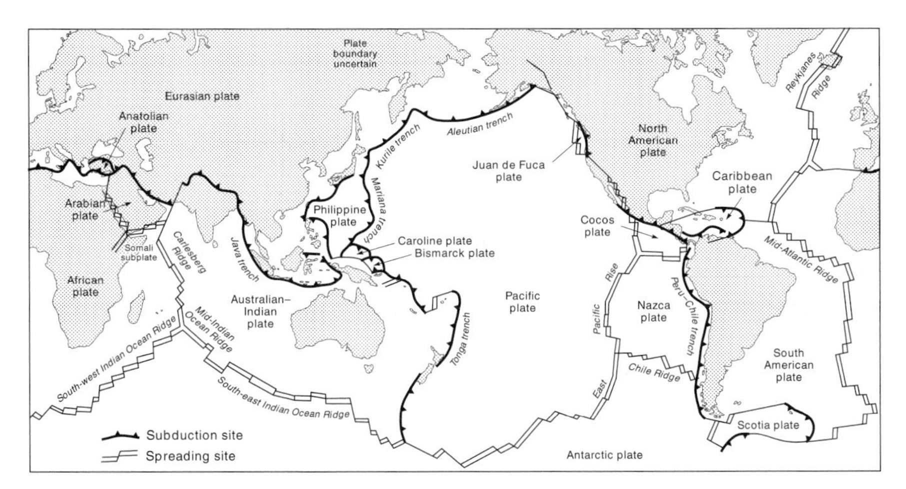

Figure 3.2 Tectonic plates, spreading sites, and subduction sites
Source: Partly after Ollier (1996)

plate boundaries: if the first-order polygons were split at subduction sites, then they would correspond to lithospheric plates as conventionally defined.) Second-order polygons are basin and swell features, such as cratonic basins, commonly bounded by raised rims and major faults and having a diameter of about 1,000 km. Third-order polygons are blocks some 10 km across. Examples are the fault blocks found in Japan (and most other places) that may move independently. Fourth-order polygons are created by master joints in rocks that lie about 10 m apart, and fifth-order polygons are ordinary joint blocks with a spacing normally less than a metre.

EVERYTHING ON A PLATE: LITHOSPHERIC CHANGE

The idea of lithospheric plates emerged with the acceptance of continental drift. If the continents have drifted, as Alfred Lothar Wegener had claimed, then large chunks of crust (including continental cratons and deep-ocean basins) have travelled several thousand kilometres without having suffered any appreciable lateral distortion. This lack of distortion is reflected in two features. First, it is seen in the excellent 'fit' of the opposing South American and African coastlines, which have taken 200 million years to drift 4,000 km apart. Second, it is manifested in the broad magnetic bands and faults of the deep-sea floor that have held their shape for tens of millions of years. This, and other evidence, suggests that the lithosphere is dynamic, that it changes.

Plate tectonics

Lithospheric change is currently explained by the plate tectonic (also called the geotectonic) model. This model is thought satisfactorily to explain geological structures, the distribution and variation of igneous and metamorphic activity, and sedimentary facies; in fact, it explains all major aspects of the Earth's long-term tectonic evolution (e.g. Kearey and Vine 1990).

The plate tectonic model comprises two tectonic styles. The first style involves the cooling and recycling system comprising the mesosphere, asthenosphere, and lithosphere lying under the oceans. The chief cooling mechanism is subduction. New oceanic lithosphere is formed by volcanic eruptions along mid-ocean ridges. The newly formed material moves away from the ridges. In doing so, it cools, contracts, and thickens. Eventually, the oceanic lithosphere becomes denser than the underlying mantle and sinks. The sinking takes place along subduction zones. These are associated with earthquakes and volcanicity. Cold oceanic slabs may sink well into the mesosphere, perhaps as much as 670 km or below the surface. Indeed, subducted material may accumulate to form 'lithospheric graveyards' (Engebretson *et al.* 1992).

It is not certain why plates should move. Several driving mechanisms are plausible. Basaltic lava upwelling at a mid-ocean ridge may push adjacent lithospheric plates to either side. Or, as elevation tends to decrease and slab thickness to increase away from construction sites, the plate may move by gravity sliding. Another possibility, currently thought to be the primary driving mechanism, is that the cold, sinking slab at subduction sites pulls the rest of the plate behind it. In this scenario, mid-ocean ridges stem from passive spreading – the oceanic lithosphere is stretched and thinned by the tectonic pull of older and denser lithosphere sinking into the mantle at a subduction site; this would explain why sea-floor tends to spread more rapidly in plates attached to long subduction zones. As well as these three mechanisms, or perhaps instead of them, mantle convection may be the number one motive force, though this now seems unlikely as many spreading sites do not sit over upwelling mantle convection cells. If the mantle-convection model were correct, mid-ocean ridges should display a consistent pattern of gravity anomalies, which they do not, and would probably not develop giant fractures (transform faults). But, although convection is perhaps not the master driver of plate motions, it does occur. There is some disagreement about the depth of the convective cell. It could be confined to the asthenosphere, the upper mantle, or the entire mantle (upper and lower). Whole mantle convection (Davies 1977, 1992) has gained much support, although it now seems that whole mantle convection and a shallower circulation may both operate.

The lithosphere may be regarded as the cool surface layer of the Earth's

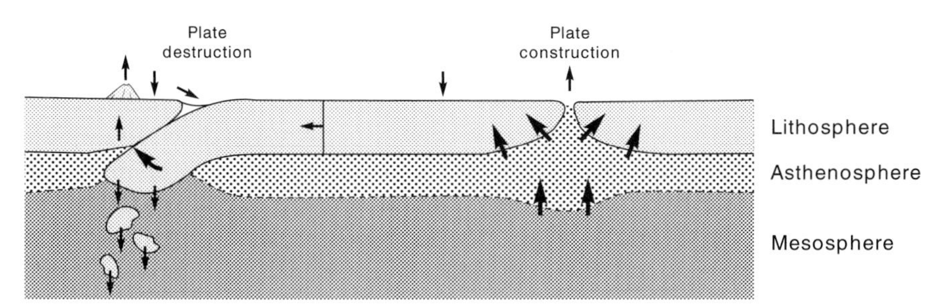

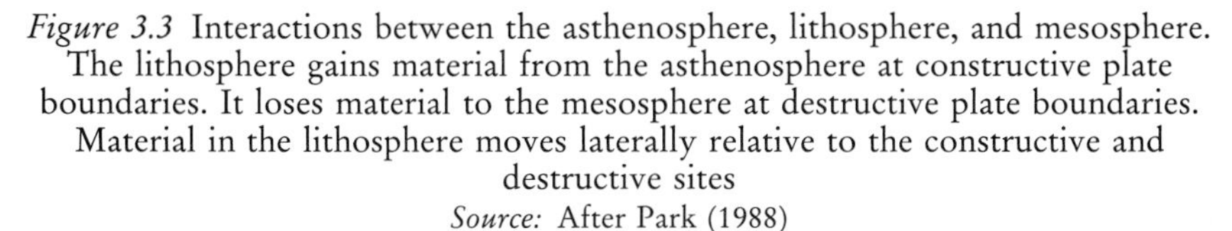

Figure 3.3 Interactions between the asthenosphere, lithosphere, and mesosphere. The lithosphere gains material from the asthenosphere at constructive plate boundaries. It loses material to the mesosphere at destructive plate boundaries. Material in the lithosphere moves laterally relative to the constructive and destructive sites
Source: After Park (1988)

convective system (Park 1988: 5). As part of a convective system, it cannot be considered in isolation (Figure 3.3). It gains material from the asthenosphere, which in turn is fed by uprising material from the underlying mesosphere, at constructive plate boundaries. It migrates laterally from mid-ocean ridge axes as cool, relatively rigid, rock. Then, at destructive plate boundaries, it loses material to the asthenosphere and mesosphere. The fate of the subducted material is not clear. It meets with resistance in penetrating the lower mantle, but is driven on by its thermal inertia and continues to sink, though more slowly than in the upper mantle, causing accumulations of slab material (Lay 1994; Fukao *et al.* 1994; Maruyama 1994). Some slab material may eventually be recycled to create new lithosphere, although the basalt erupted at mid-ocean ridges shows a few signs of being new material that has not passed through a rock cycle before: it has a remarkably consistent composition, which is difficult to account for by recycling, and emits such gases as helium that seem to be arriving at the surface for the first time; on the other hand, it is not 'primitive' and formed in a single step by melting of mantle materials – its manufacture requires several stages (Francis 1993: 49). It is worth noting that the transformation of rock from mesosphere, through the asthenosphere, to the lithosphere chiefly entails temperature and viscosity (rheidity) changes. Material changes do occur: partial melting in the asthenosphere generates magmas that rise into the lithosphere, and volatiles enter and leave the system.

The second tectonic style involves the continents. The continental lithosphere does not take part in the mantle-convection process. It is 150 km thick and consists of buoyant low-density crust (the tectosphere) and relatively buoyant upper mantle. It therefore floats on the underlying asthenosphere. Continents break up and reassemble, but they remain floating at the surface. They move in response to lateral mantle movements, gliding serenely over the Earth's surface. In breaking up, small fragments of continent sometimes shear off; these are called terranes. They drift around till they meet another continent, to which they become attached (rather than being subducted) or possibly are sheared along it. As they may come from a different continent than the one they are attached to, they are called exotic or suspect terranes. Much of the western seaboard of North America appears to consist of these exotic terranes. In moving, continents have a tendency to drift away from mantle hot zones, some of which they may have produced: stationary continents insulate the underlying mantle, causing it to warm. This warming may eventually lead to a large continent breaking into several smaller ones. Most continents are now sitting on, or moving towards, cold parts of the mantle. An exception is Africa, which was the core of Pangaea. Continental drift leads to collisions between continental blocks and to the overriding of oceanic lithosphere by continental lithosphere along subduction zones.

The continents, then, are affected by, and affect, underlying mantle and adjacent plates. They are maintained against erosion (rejuvenated in a sense) by the welding of sedimentary prisms to continental margins through metamorphism, by the stacking of thrust sheets, by the sweeping up of microcontinents and island arcs at their leading edges, and by the addition of magma through intrusions and extrusions (Condie 1989b: 62). The relative movement of continents over the Phanerozoic aeon is now fairly well established, though pre-Pangaean reconstructions are less reliable than post-Pangaean reconstructions. Figure 3.4 charts the probable breakup of Pangaea.

A third tectonic style is possible on the Earth (D. L. Anderson 1990). The lithospheric plates could remain at the surface and interact through 'ice-pack' underthrusting, but only if deep subduction did not occur. This would happen if there were a shallow temperature gradient between the lithosphere and mesosphere, or if lithospheric plates were created very quickly, or if the crust–lithosphere system were completely buoyant. Evidence for this tectonic style early in Earth history is found in North America, parts of western South America, and Tibet.

Surface tectonics

Plate motions explain virtually all tectonic forces that affect the lithosphere and thus the Earth's surface. Traditionally, tectonic (or geotectonic) forces are divided into two groups: diastrophic forces and volcanic (or vulcanic) forces. Diastrophic forces involve processes of folding, faulting, uplift, and subsidence that deform the lithosphere. They are responsible for the major features of the Earth's surface. Two categories are recognized: orogeny and epeirogeny. Orogeny is mountain building and uplift as a solely structural

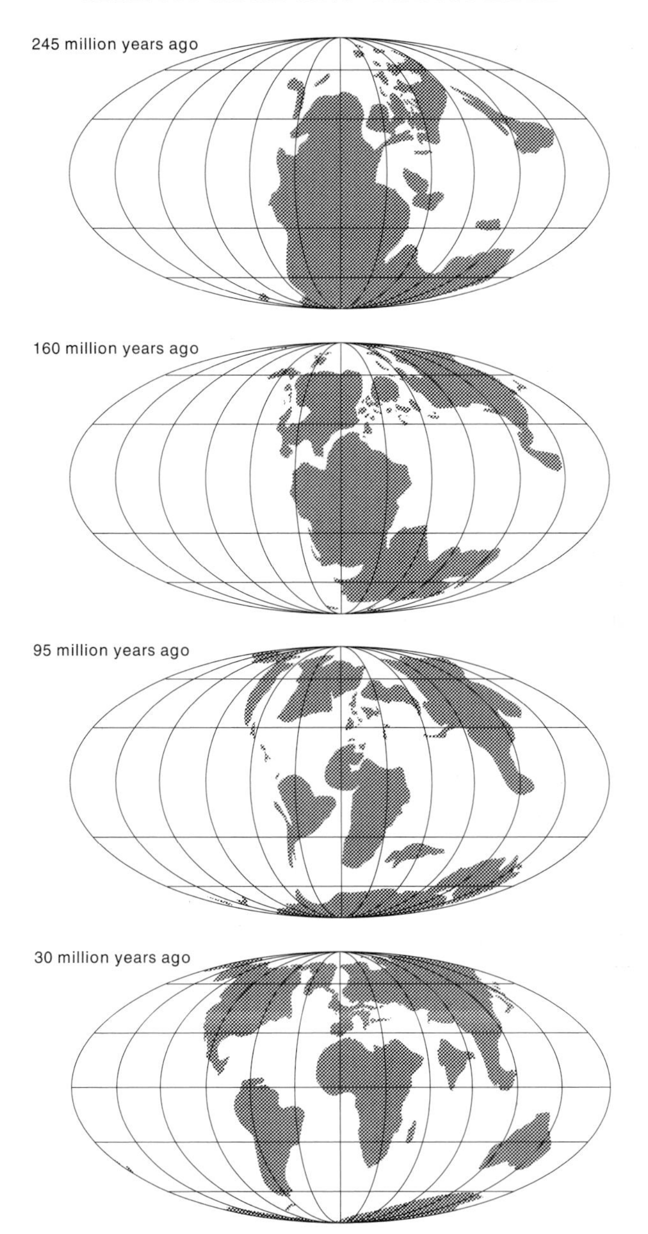

Figure 3.4 The changing arrangement of continents over the last 245 million years, showing the breakup of Pangaea, during the Early Triassic period; during the Callovian age (Middle Jurassic); during the Cenomanian age (Late Cretaceous); and during the Oligocene epoch. All maps use Mollweide's equal-area projection
Source: After maps in A. G. Smith *et al.* (1994)

or tectonic idea. It excludes the erosion that produces high peaks and deep valleys. Epeirogenesis is the upheaval or depression of large areas. It does not entail folding save for the broadest of undulations. It includes isostatic movements, such as the rebound of land after an ice sheet has melted, and cymatogeny (the arching, and sometimes doming, of rocks with little deformation over 10–1,000 km). Volcanic forces are either intrusive or extrusive forces. Intrusive forces are found within the lithosphere and produce such features as batholiths, dykes, and sills. Extrusive forces occur at the very top of the lithosphere and lead to exhalations, eruptions, and explosions of materials through volcanic vents.

The many tectonic forces in the lithosphere are primarily created by the relative motion of adjacent plates. Indeed, relative plate motions underlie almost all surface tectonic processes. Plate boundaries are particularly important for understanding geotectonics. They are sites of strain and associated with faulting, earthquakes, and, in some instances, mountain building (Figure 3.5). Most boundaries sit between two adjacent plates, but, in places, three plates come into contact. This happens where the North American, South American, and Eurasian plates meet (Figure 3.2). Such Y-shaped boundaries are known as triple junctions. Three plate boundary types produce distinctive tectonic regimes:

1 Divergent plate boundaries at construction sites, which lie along mid-ocean ridges, are associated with divergent tectonic regimes involving shallow, low-magnitude earthquakes. The ridge height depends primarily on the spreading rate. Incipient divergence occurs within continents,

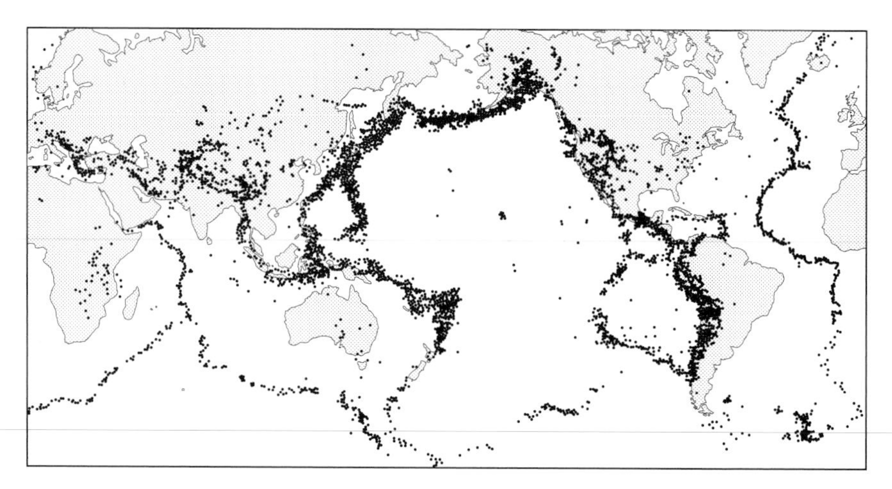

Figure 3.5 The global distribution of earthquakes
Source: After Ollier (1996)

including Africa, and creates rift valleys (which are linear fault systems and, like mid-ocean ridges, are prone to shallow earthquakes and volcanism). Volcanoes at divergent boundaries produce basalt.

2 Convergent plate boundaries vary according to the nature of the converging plates. Convergent tectonic regimes are equally varied; they normally lead to partial melting and the production of granite and the eruption of andesite and rhyolite. A collision between two slabs of oceanic lithosphere is marked by an oceanic trench, a volcanic island arc, and a dipping planar region of seismic activity (a Benioff zone) with earthquakes of varying magnitude. An example is the Scotia arc, lying at the junctions of the Scotia and South American plates. Subduction of oceanic lithosphere beneath continental lithosphere is manifest in two ways: first, as an oceanic trench, a dipping zone of seismic activity, and volcanicity in an orogenic mountain belt (or orogen) lying on the continental lithosphere next to the oceanic trench (as in western South America); and second, as intra-oceanic arcs of volcanic islands (as in parts of the western Pacific Ocean). In a few cases of continent–ocean collision, a slab of ocean floor has overridden, rather than underridden, the continent. This process, called obduction, has produced the Troödos Mountain region of Cyprus. Collisions of continental lithosphere result in crustal thickening and the production of a mountain belt, but little subduction. A fine example is the Himalaya, produced by India's colliding with Asia. Divergence and convergence may occur obliquely. Oblique divergence is normally accommodated by transform offsets along a mid-oceanic ridge crest, and oblique convergence by the complex microplate adjustments along plate boundaries. An example is found in the Betic cordillera, Spain, where the African and Iberian plates slipped by one another from the Jurassic to Tertiary periods (Tubia 1994).

3 Conservative or transform plate boundaries occur where adjoining plates move sideways past each other along a transform fault without any convergent or divergent motion. They are associated with strike-slip tectonic regimes and with shallow earthquakes of variable magnitude. They occur as fracture zones along mid-ocean ridges and as strike-slip fault zones within continental lithosphere. A prime example of the latter is the San Andreas fault system in California.

4 Tectonic activity also occurs within lithospheric plates, and not just at plate edges. This is called within-plate tectonics to distinguish it from plate-boundary tectonics.

Hot-spots and volcanoes

Most volcanoes are sited at plate boundaries (Colour plates 1 and 3). A few, including the Cape Verde volcano group in the southern Atlantic Ocean and the Tibesti Mountains in Saharan Africa, occur within plates. These 'hot-spot' volcanoes are surface expressions of thermal mantle plumes that may originate at the core–mantle boundary (p. 72). Hot-spots are characterized by topographic bumps (typically 500–1,200 m high and 1,000–1,500 km wide), volcanoes, high-gravity anomalies, and high heat flow (Crough 1983). Commonly, a mantle plume stays in the same position while a plate slowly slips over it (Wilson 1963). In the ocean, this produces a chain of volcanic islands, or a hot-spot trace, as in the Hawaiian Islands. On continents, it produces a string of volcanoes. Such a volcanic string is found in the Snake River Plain province of North America where a hot-spot, currently sitting below Yellowstone National Park, Wyoming, has created an 80-km-wide band across 450 km of continental crust, producing prodigious quantities of basalt in the process (Greeley 1982). Even more voluminous are continental flood basalts. These occupy large tracts of land in far-flung places. The Siberian province covers more than 340,000 km^2 (Renne and Basu 1991). India's Deccan Traps once covered about 1,500,000 km^2; erosion has left about 500,000 million km^2.

Hydrothermal vents

In 1977, a vent was discovered at the Galápagos spreading centre, out of which gushed hot water richly laden with solutes. Subsequent exploration has discovered many hydrothermally active sites – hydrothermal venting is common along the length of the 55,000 km ridge-crest system. Volcanism is intense along mid-ocean ridges. About half the thermal energy emanating from within the Earth is channelled into ridges. Of this excess thermal energy, about half is used in heating sea water that circulates through ridge systems. On average, about 1,000 km^3 of sea water annually are warmed in this way. This is equivalent to about 2.4 per cent of the annual water flux from land to sea. The difference between the two fluxes lies in the water temperature, pressure, and composition. The water circulating through ridges is warm (discharging at temperatures up to 450°C), salty, and at relatively high pressure. Hot, saline water at high pressure is a far better solvent than river water. Hot springs at ocean ridges carry about 1,000 ppm silica, whereas river waters carry about 10 ppm. The circulatory process appears to run like this (Figure 3.6): sea water penetrating the crust is heated to 350–400°C. It reacts with basaltic rocks and many chemical species are leached and transported as solutes. The highly reduced hydrothermal fluid rises and reaches the sea-floor, either directly through hot vents or indirectly after mixing with cold and oxygenated sea water before emission through warm vents. On mixing with sea water, polymetal sulphides and calcium sulphate (anhydrite) precipitate, either within subsurface lava conduits or else as 'chimneys' and the suspended particulate matter of the 'black smokers' (Jannasch and Mottl 1985). Owing to the highly efficacious

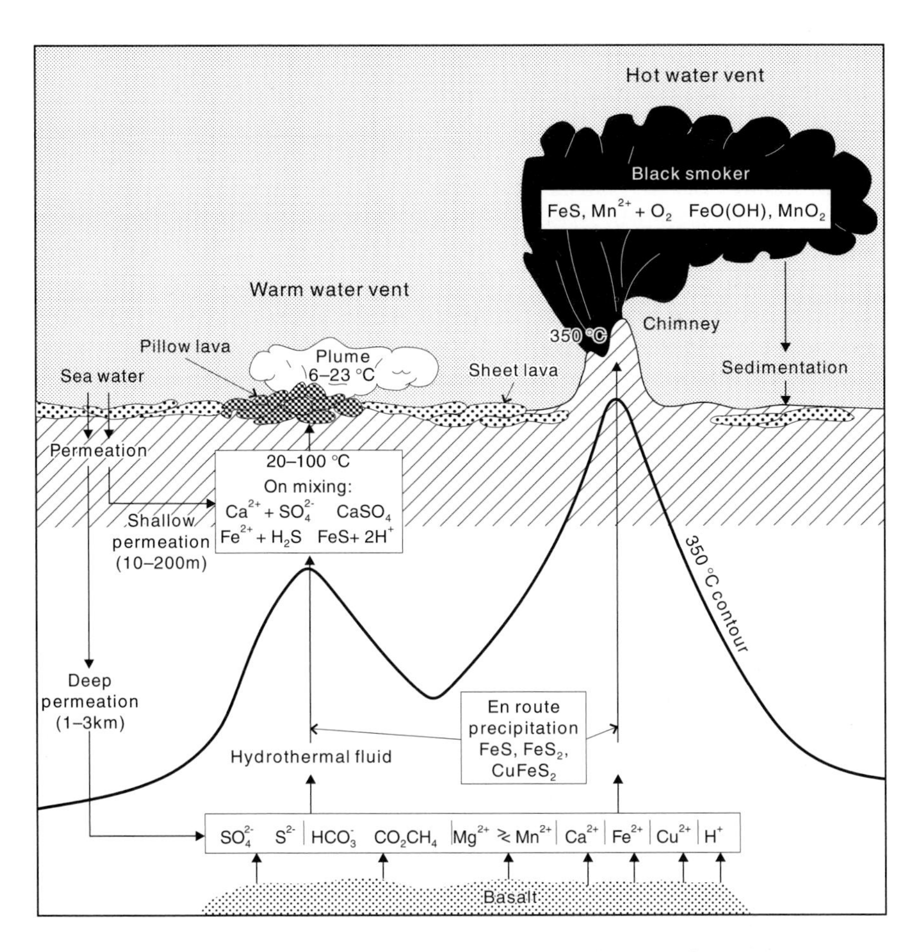

Figure 3.6 Inorganic processes within and on the deep-sea floor at oceanic spreading centres
Source: After Jannasch and Mottl (1985)

solvent action of hot water, the circulation of warm water through mid-ocean ridges delivers annually about one-half of the silica load entering the oceans (the other half being carried by rivers), and dominates the transport of several elements including iron, manganese, cobalt, nickel, copper, zinc, and lithium. Many of these trace elements are essential to the functioning of biospheric systems. Hot springs along mid-ocean ridges are therefore of immense significance to the biosphere (Rona 1988). On the other hand, the ocean crust is probably a net sink for alkalis.

ROCK BOTTOM: CAUSES OF LITHOSPHERIC CHANGE

The geological power house

The structure and dynamics of the solid Earth are, largely, a result of internal (geological) processes. The Earth is a huge heat engine. Most of the heat comes from the radioactive decay of potassium, uranium, and thorium isotopes, from heat of crystallization at the inner-core–outer-core boundary, and as a remnant of the Earth's primordial heat that was generated during planetary accretion and core formation. It may be released steadily, though pulses of more rapid release have been suggested.

Part of the energy released at the inner-core–outer-core boundary is used to drive the Earth's 'dynamo', which creates a geomagnetic field. Thermal instabilities develop at the all-important core–mantle junction. The reasons for their development are poorly understood, but might involve phase changes, segregation of chemical constituents, and positive and negative heat balances of physicochemical reactions. The thermal instabilities trigger the growth of mantle plumes. These may be hundreds of kilometres in diameter and rise towards the Earth's surface. Mantle plumes may also originate at the boundary between the upper and lower mantle. A plume consists of a leading 'glob' of hot material that is followed by a 'stalk'. On approaching the lithosphere, the plume head is forced to mushroom beneath the lithosphere, spreading sideways and downwards a little. The plume temperature is 250–300°C hotter than the surrounding upper mantle, so that 10–20 per cent of the surrounding rock is melted. This melted rock may then run onto the Earth's surface as flood basalt.

Superplumes may form. One appears to have done so beneath the Pacific Ocean during the middle of the Cretaceous period (Larson 1991). It rose rapidly from the core–mantle boundary about 125 million years ago. Production tailed off by 80 million years ago, but it did not stop until 50 million years later. It is possible that superplumes are caused by cold, subducted oceanic crust on both edges of a tectonic plate accumulating at the top of the lower mantle. These two cold pools of rock then sink to the hot layer just above the core and a giant plume is squeezed out between them (Penvenne 1995; see also Lenardic and Kaula 1994; A. B. Thompson 1991). Plume tectonics may be the dominant style of convection in the major part of the mantle: two super-upwellings (the South Pacific and African superplumes) and one super-downwelling (the Asian cold plume) appear to prevail (Figure 3.7). It influences, but is also influenced by, plate tectonics. Indeed, crust, mantle, and core processes may act in concert to create 'whole Earth tectonics' (Maruyama *et al.* 1994; Kumazawa and Maruyama 1994). Whole Earth tectonics integrates plate tectonic processes in the lithosphere and upper mantle, plume tectonics in the lower mantle,

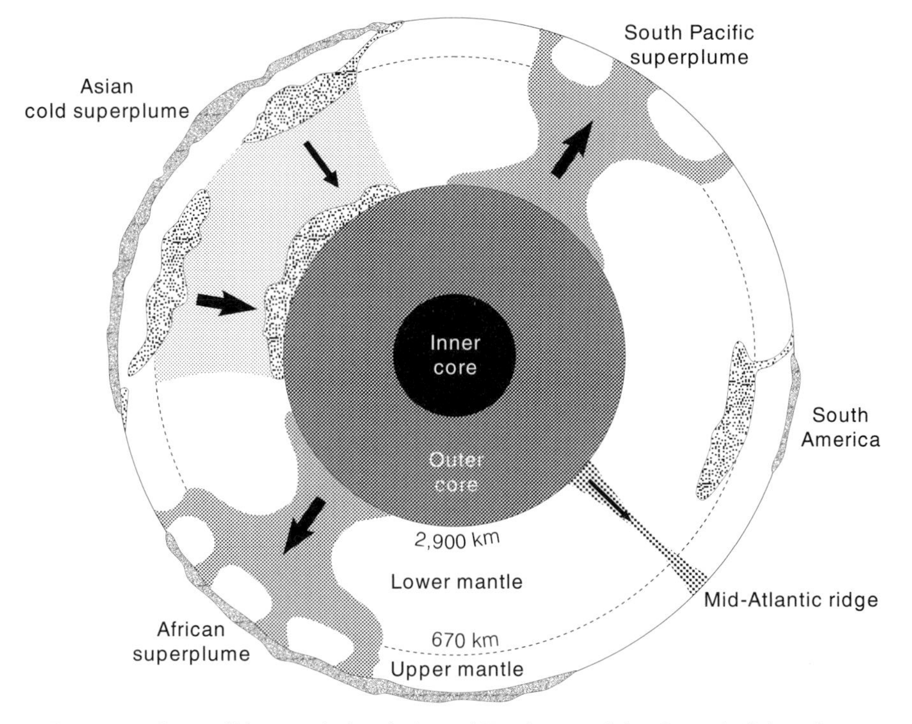

Figure 3.7 A possible grand circulation of Earth materials. Oceanic lithosphere, created at mid-ocean ridges, is subducted into the deeper mantle. It stagnates at around 670 km and accumulates for 100–400 million years. Eventually, gravitational collapse forms a cold downwelling onto the outer core, as in the Asian cold superplume, which leads to mantle upwelling occurring elsewhere, as in the South Pacific and Afrian hot superplumes
Source: After Fukao *et al.* (1994)

and growth tectonics in the core, where the inner core slowly grows at the expense of the outer core (Figure 3.8). Plate tectonics supplies cold materials for plume tectonics. Sinking slabs of stagnant lithospheric material drop through the lower mantle. In sinking, they create super-upwellings that influence plate tectonics, and they modify convection pattern in the outer core, which in turn determines the growth of the inner core.

Ultimately, the Earth's internal heat is liberated by conduction from rock to air and water, through volcanoes, through hydrothermal transport (Colour plate 4), and through plate recycling. About 70 per cent of this geothermal energy dissipation is associated with the formation of ocean crust, the remainder being lost by conduction to the atmosphere and hydrosphere (20 per cent), and radioactive decay of heat-producing elements in the continental crust (10 per cent). Current losses world-wide

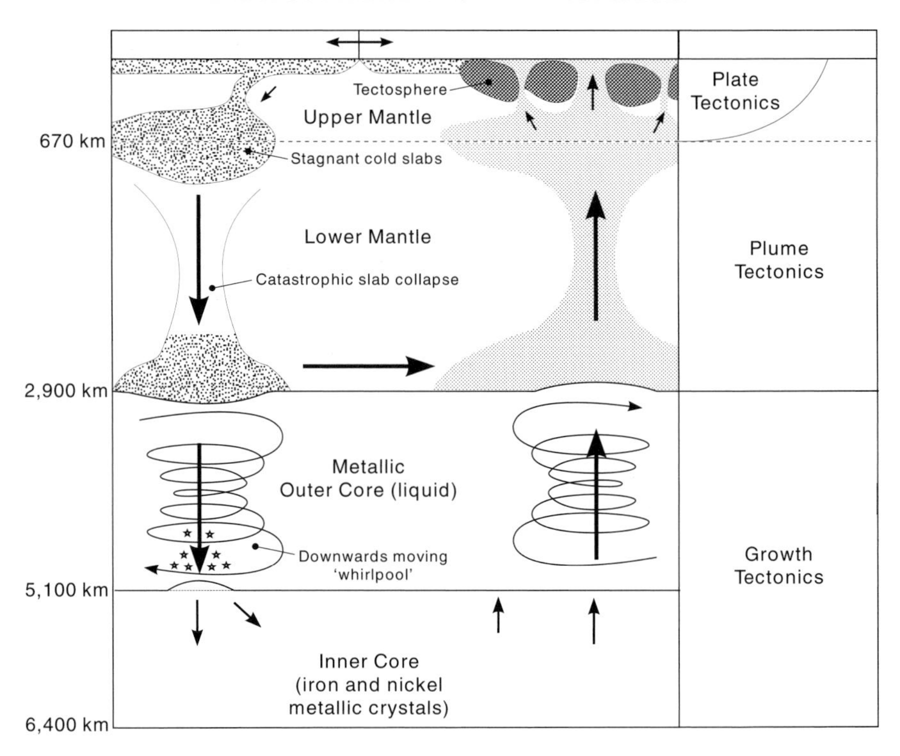

Figure 3.8 Whole Earth tectonics. Plate tectonics supplies cold lithospheric materials to the domain of plume tectonics. Stagnant slabs at around 670 km depth collapse catastrophically to produce super-upwellings, which influence plate tectonics, and modify convention patterns in the outer core, which control growth tectonics
Source: After Maruyama *et al.* (1994)

are 42 trillion Watts, which breaks down to 12 trillion Watts for continents and 30 trillion Watts for oceans (Sclater *et al.* 1980). The global figure averages about 81 milliWatts on each square metre of the Earth's surface. This is sufficient heat energy to power a 100 Watt light bulb from 1 km^2 (Francis 1993: 3). Some heat is lost by hot-spot volcanism and by the eruption of flood basalts.

Cosmic connections

The lithosphere is the product of internal Earth processes. However, cosmic processes and events appear to exert an influence on changes in the core, mantle, and lithosphere. There is mounting evidence that geological changes are locked into solar, planetary, and galactic rhythms. This is especially true of the galactic megacycles, the medium-term orbital cycles, and the host of short-term solar and lunar cycles. These pulses of geological change are, in some measure, forced by gravitational changes, by changes in electromagnetic radiation, and by the impact of asteroids, comets, and meteoroids.

Gravitational effects

Some geological changes occurring over days, years, decades, and centuries are tightly bound to the cyclical jostling of the Sun, Moon, and planets. This gravitational bonding is seen in the association of seismic activity with lunar cycles, in minute adjustments of rotation rate and axial tilt, and in the shape of the geoid. For instance, in seismically active regions of the world, there is a strong correlation between large earthquakes in the period 1897–1990 and the lunar nodal cycle (Du 1994). The seismic cycle in all regions consists of a 12.4-year active period and a 6.2-year quiet period, and may result from magma tides in the asthenosphere. Earthquakes of magnitude 4 and above for the period 1900–86 at Cairo, Egypt, and Alger, Algeria, were linked to the 11-year solar cycle (Mosalam Shaltout *et al.* 1992).

Many multi-millennial geological changes are firmly linked to orbital cycles in the Croll–Milankovitch frequency band. Geological megacycles may also be connected with gravitational cycles played out in the Milky Way Galaxy, and possibly even beyond.

Electromagnetic effects

Electromagnetic connections between the Sun and the Earth may well have a wider range of impacts than is commonly supposed. Sunspot cycles may produce pulses of energy that reach the Earth's magnetosphere and then cascade down to the core (Shaw 1994: 252). From the magnetosphere, the energy cascade runs through the ionosphere, atmosphere, glacier–ocean–atmosphere, tectosphere, magmasphere, and ends in the core. The Earth's dynamo is then perturbed and causes changes in the magnetosphere.

Impacts

Asteroids, comets, and even middle-rate meteoroids possess impressive kinetic energies. This is due much more to the speed at which they fly through space than to their size. Kinetic energy increases linearly with mass and with the square of the velocity. The kinetic energy of space debris is expressed as megatonnes (Mt) of TNT equivalent energy; in other words, as the kinetic energy equivalent to exploding so many megatonnes of TNT. The atomic bomb dropped on Nagasaki had a kinetic energy equivalent of 0.02 megatonnes of TNT. An asteroid with a diameter of 1 km has the energy equivalent of about 50,000 Mt, and one with a diameter of 10 km has the energy equivalent of about 50,000,000 Mt. By comparison, terrestrial energy is small beer: the explosion of Krakatau in 1883 was equivalent to about 50 Mt; and a major earthquake is equivalent to 100–500 Mt. A 100,000,000-Mt asteroid striking the Earth would release the same amount of energy, albeit at a lesser intensity, as the explosion of ten atomic bombs, roughly the size of the one dropped on Nagasaki in 1945, on every square kilometre of the Earth's surface (McCrea 1981).

Asteroids, comets, and meteoroids that lack the mass or velocity to survive the short but stressful journey through the atmosphere explode as airbursts. Those that reach the ground have several effects, one of the most immediate of which is the excavation of a crater. Crater size depends largely on the kinetic energy of the bolide and the density of the target rocks. Typical crater diameters are about 20 km for a 1-km-diameter asteroid, and 10 km for a 0.5-km-diameter asteroid. Because weathering, erosion, and transport are so active on the Earth's surface, craters eventually fade away until all that remains is an impact structure, or astrobleme ('star-wound'), recorded by morphological, gravity, and magnetic anomalies in the underlying rocks (Dietz 1961). Craters are formed in the ocean floor, too. On being struck, the sea water near the impact site is highly compressed by the shock and vaporizes upon decompression, spraying out of an expanding, though transient, water cavity. A system of giant waves radiates from the point of impact. After having passed almost instantaneously through the sea, the impactor pierces the oceanic crust and upper mantle. It leaves a clear morphological, gravity, and magnetic signature called a hydrobleme.

LIKE THE LIFE OF A SOLDIER: RATES OF LITHOSPHERIC CHANGE

Sudden or gradual?

What is the rate of lithospheric change? Is it a gentle, little-by-little process? Or does it occur in terse and turbulent bouts? These vexing questions have received much attention in the last two decades. They pose a dichotomy, each extreme of which suggests a radically different geological framework within which to interpret long-term environmental change.

Charles Lyell's uniformitarian dogma guided, some would now say blinkered, geological theory and practice for about 150 years. In a nutshell, Lyell claimed that geological processes had always run at current sedate rates (gradualism), and that the Earth's geological features had always been in roughly the same state – there had always been, for instance, the same approximate proportion of land and sea, even though the positions of the continents might have changed due to uplift and subsidence (steady-statism). By the end of the nineteenth century, most geologists had silently rejected the steady-state view of Earth history; all but a heretical handful still accepted unquestioningly the assumption of gradualism. During the present century, two important developments have occurred in the rate debate that have eventually led to rejection of strict gradualism. The first is the observation that many geological processes have an episodic character. The second is the acknowledgement that some episodic events in Earth history were triggered by cosmic processes.

Episodes in geological history

Derek V. Ager (1973) summed up geological history by comparing it to the life of a soldier – long periods of boredom and short periods of terror. This oft-quoted phrase captured in nine words the long-held view that periods of rapid geological change alternate with periods of slow geological change; that times of quiet are interrupted by times of paroxysmal action. The evidence for episodes in well-nigh all Earth history is now overwhelming. The list includes orogenic events, sea-floor spreading, rates of continental drift, continental flood-basalt eruptions, boundaries in sedimentary sequences, geomagnetic pole reversals, direction and rates of apparent polar wander, stable isotope geochemistry, anoxic events in the oceans, aridity, ice ages, biological evolution, growth rhythms, and extinctions. This evidence goes against the old gradualist dogma: gradual change does occur, but so do short, sharp changes, some of which have distinctly catastrophic characteristics.

Episodic change in the lithosphere is suggested by continental flood-basalt eruptions, continental drift rates, sea-floor spreading rates, and orogenic events. Over the last 250 million years, and probably well before, these geological processes and others (anoxic events, evaporite deposits, and sequence boundaries) show relatively short peaks of activity separated by long spells of inactivity (Figure 3.9a and b), the peaks showing a dominant period of about 26 million years (Figure 3.9b). These episodes appear to be correlated. They have been explained by a model of 'pulsation tectonics' (Sheridan 1987). Moreover, some of the geological changes during the activity peaks have a decidedly catastrophic character. Continental flood basalts are spewed out within geologically short time-spans – the Deccan flood basalts were erupted in, at most, 500,000 years (Courtillot 1990). Sea-floor spreading rates appear to display abrupt 'jumps' over the last 180 million years (Rich *et al.* 1986; see also Gaffin and O'Neill 1994). Continental drift rates are episodic. Over the last 300 million years, continental plates have drifted faster than the present average rate of 1–2 cm/yr (Figure 3.10). The faster drift rates came soon after the splitting of a continental plate (Condie 1989a: 154).

On a longer time-scale, the growth of continents appears to have been episodic (Condie 1989a, 1989b; Taylor and McLennan 1996). Spurts of growth occurred at the following times: 2.5 billion, 2–1.7 billion, 1.3–1.1 billion, and 0.5–0.3 billion years ago. The first of these is a period of rapid

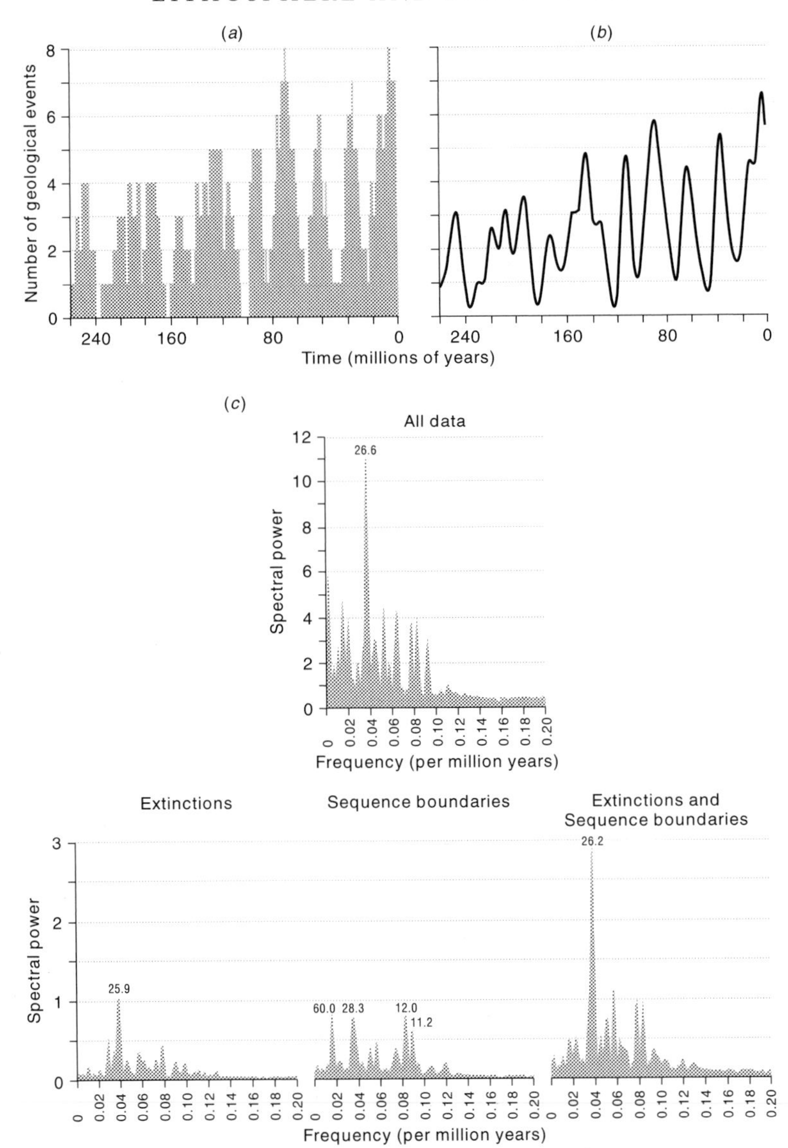

Figure 3.9 Number and spectral density of geological events (flood-basalt eruptions, fast sea-floor spreading episodes, orogenic events, anoxic events, evaporite deposition, sequence boundaries, and mass extinctions) over the last 260 million years. (a) Number of geological events within a 10-million-year moving window. (b) Number of geological events weighted by a Gaussian function with a scale length of 5 million years. (c) Spectral power for all data sets and selected subsets. A dominant period of about 26 million years is present in all cases, though the sequence boundary data set has dominant periods at around 60, 28, 12, and 11.2 million years

Source: Rampino and Caldeira (1993)

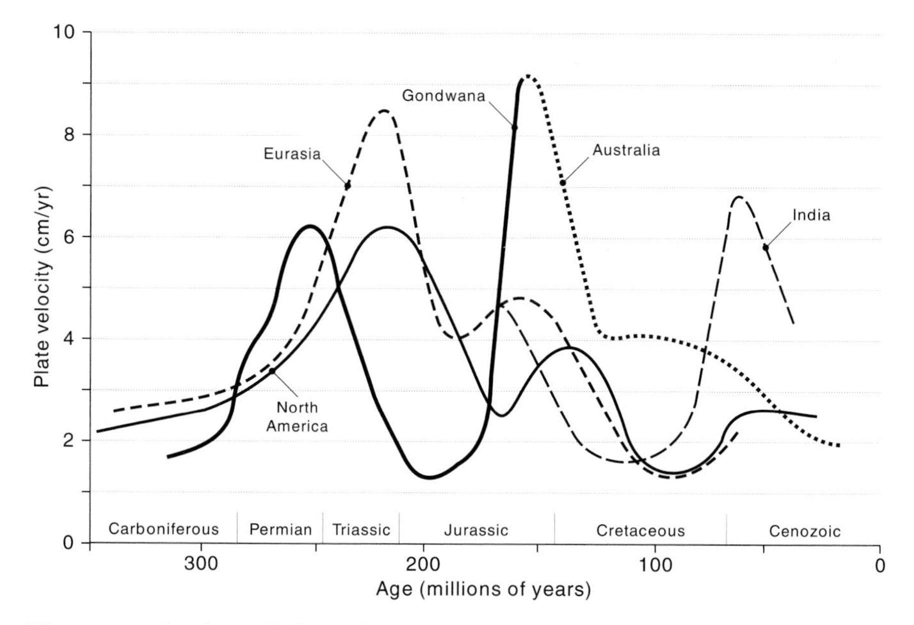

Figure 3.10 Continental plate velocities (root mean square) calculated from apparent polar wander paths for the last 300 million years
Source: After Piper (1987)

continent production in the later Archaean, an episode that appears to have been unique and perhaps catastrophic (Condie 1989b: 65). This episode might be linked to a catastrophic collapse of a primordial stable density-stratification in the core (Kumazawa *et al.* 1994). It is unclear why crust should grow intermittently when the power behind crust formation – the generation of internal heat and its liberation through crustal recycling – is always 'switched on'. The answer may lie in the supercontinent cycle, of which more will be said later in the chapter.

Cosmic catastrophes and geological history

In July 1994, the comet Shoemaker–Levy 9 smashed into Jupiter. Catastrophism was come of age in a spectacular release of raw energy that proved beyond any reasonable doubt that the Solar System is a violent and yet beautiful place. Nine blows in quick succession vindicated such illustrious authors as William Whiston, who had dared to speculate on the disasters that might befall a cometary encounter with the Earth. The inescapable conclusion was simple: the catastrophists were right, Lyell was wrong. Of course, the possibility of catastrophic events was raised by the heretical few while Lyellian gradualism was the ruling theory, but the establishment many would have no truck with such wild conjectures and dismissed catastrophes as very rare events of local importance that certainly had not guided Earth history. In 1980, a layer rich in iridium was found in Cretaceous–Tertiary boundary clays at sites around the world. Geologists sat up and took notice of bombardment as a possible geological process. The rest, as it were, is geological history.

Asteroids and comets have struck the Earth. The geological consequences of these strikes are open to question, some geologists preferring terrestrial causes to explain ultra-energetic events. None the less, bombardment is a highly plausible explanation for many sudden and violent events in Earth's past. Large asteroid and comet impacts would drastically alter the rate of some geological processes and stir others into action. The impacts would be felt throughout the whole Earth. Early speculations on this matter saw large impact events as triggers of substantial geological changes, including sharp surges of tectonic and volcanic activity (e.g. Nininger 1942; McLaren 1970). More recent investigations suggest that bombardment might play a basic role in plate tectonics (Clube and Napier 1982), in geomagnetic reversals (Clube and Napier 1982; Muller and Morris 1986; Raup 1985), in true polar wander (Runcorn 1982, 1987; P.H. Schultz 1985), in earthquakes (Clube and Napier 1982), in volcanism (Clube and Napier 1982; Rampino 1989; but see Loper and McCartney 1990), and in mantle hot-spot initiation (Rampino 1989). If these connections should exist, then they would represent a significant development in understanding the connection between terrestrial and cosmic processes. The notion that extraterrestrial impacts trigger geological changes unifies big events in Earth's history with astrophysical processes that operate on a galactic scale (Rampino 1989: 58).

A new theory viewing terrestrial dynamics in a cosmic setting revolves around the three Phanerozoic cratering nodes (p. 49). These three spatially invariant crater clusters have remained fixed relative to the principal axial moments of inertia that are determined by the Earth's mass distribution. They have mediated the internal Earth dynamics by influencing the inception and timing of magmatic processes. In turn, magmatic processes have mediated the inception and timing of continental drift, plate tectonics, core motions, geodynamic polarities, and their reciprocal relationships with mantle convection.

ROCK OF AGES: DIRECTIONS OF LITHOSPHERIC CHANGE

Is there any direction to change in the lithosphere? Three basic possibilities arise, for each of which there is evidence: steady states, secular trends, and long-running cycles.

A steady-state world

Since the Earth first formed, the lithosphere has been timebound – it has evolved. Over the full run of geological time, therefore, it has not displayed steady-state characteristics. But some aspects of lithospheric change do lend themselves to a steady-state interpretation. Plate tectonics stresses the timeless aspects of crustal change and the maintenance of a steady state. Crustal plates are continuously being created at mid-ocean ridges and destroyed in subduction zones. There is some evidence showing that sea-floor spreading does not occur at a uniform rate, but the variations appear not to be large enough to disrupt the overall steady condition for long periods. The pattern of crustal blocks is known to alter with time, but the net result of plate creation and plate destruction is a slowly changing pattern of oceanic and continental crust, of oceans and continents, of mountains and plains. The details of this pattern vary through time, but the general components remain much the same from one period to another. An approximate balance between internal geological processes of construction and destruction has obtained since the Proterozoic aeon. Thus, the unchanging world of the plate tectonicist seems at first sight not so very different from the timeless world envisaged by Lyell, the world whose state has remained unaltered through the ages. None the less, geological history is ultimately timebound, forced to new states by secular change.

A directional world

Earth history involves many secular changes, many of them irreversible. Some of these occur suddenly, but most are sluggish and result from secular Earth cooling, from rock recycling through the crust or mantle (which tends to produce exponential increases in such things as sedimentary volume and area), or from changes in atmospheric composition (Condie 1989a: 360). Crustal evolution and the controversial topic of Earth expansion are good examples.

The evolving crust

The study of Precambrian geology was vastly assisted by the discovery and wide application of isotopic dating methods, and the potassium–argon (K–Ar) method in particular. A reliable chronology of Precambrian events has helped to clarify geological change in igneous and metamorphic rocks during the first seven-eighths of Earth history. The evidence is not easy to interpret, but it points to an evolution of tectonic and petrologic styles (Figure 3.11; see also Miyashiro 1981; Kröner 1985; Rudnick 1995). This planetary evolution has been in part a response to the secular decrease in geothermal energy, and has led to a progressive evolution of the crust (Figure 3.12).

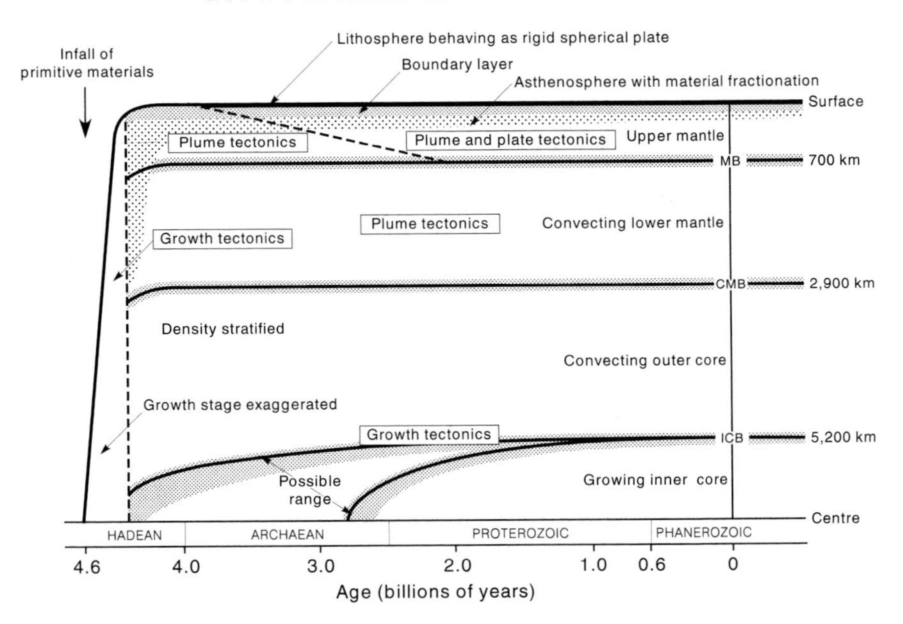

Figure 3.11 Evolving whole Earth tectonics. Abbreviations: MB is the mantle boundary at about 700 km, CMB is the core–mantle boundary; and ICB is the inner-core–outer-core boundary

Source: After Kumazawa and Maruyama (1994)

Early crustal evolution probably ran along the following lines (Condie 1980, 1989a: 337–54):

1 The earliest crust, which was oceanic, might have formed on a magma ocean about 4,500 million years ago, just after, or during, the late stages of planetary accretion. At this time, heat-producing elements were far more abundant, and the geothermal gradient was much steeper, than nowadays. Mantle convection was vigorous. If there were not a longer network of ocean ridges, then ocean crust should have been created much more quickly than at present. The crust was composed largely of komatiite, a rock rich in olivine and similar in composition to undepleted mantle. The thin komatiitic lithosphere was highly folded and faulted, the lower parts probably melted and recycled into the mantle at sink sites (Figure 3.13a). Pressures at the lithosphere base were probably too small to generate heavy-density mineral assemblages. Some authorities claim that, because of this, the lithosphere was not dense enough to subduct into the mantle and plate tectonics did not occur (Baer 1981). Others believe that, at any one time, there were a hundred or more separate plates, all rapidly being created and destroyed by the mantle in a hectic

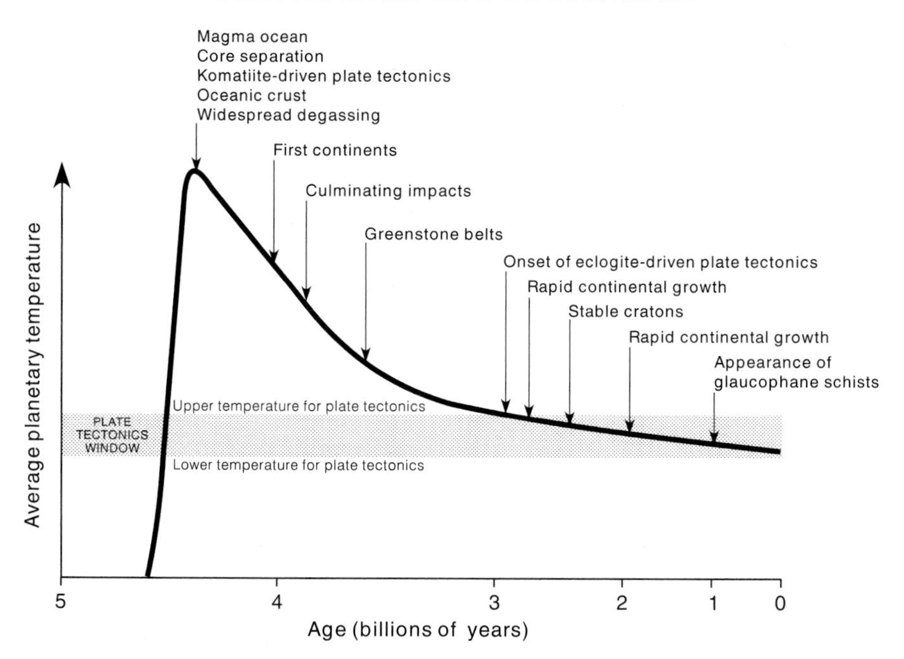

Figure 3.12 Average planetary temperature during Earth history. Key events in crustal evolution are indicated. The 'plate tectonics window' is defined by the upper and lower limiting temperatures at which plate tectonics is possible
Source: After Condie (1989a)

recycling system (Taylor and McLennan 1996). The recycling took place at sinks that had different geometries from modern subduction zones. Possibly, two plates subducted together, a feature seen in modern lava lakes (Duffield 1972). The komatiitic crust was a fraction denser than the underlying mantle and tended to sink into it. Planetesimal impacts aided near-surface recycling.

2 As the geothermal gradient decreased, so the lithosphere gradually thickened. Between about 3.8 and 2.5 billion years ago, basaltic magma was extruded onto the komatiitic crust, possibly through tensional cracks, to create greenstone belts rich in ultramafic volcanic rocks. The oldest of these is the Isua greenstone belt, Greenland, which, at about 3,800 million years old, is the oldest known example of a supracrustal rock assemblage (supracrustal rocks are rocks formed at the Earth's surface). It is an admixture of volcanic rocks, quartzites, iron formations, carbonates, and pelitic rocks (claystones and mudstones). Before about 3,000 million years ago, parts of the lithosphere reached a thickness of 50 km, perhaps in regions of compression above cooler, downwelling areas in the mantle where they formed basaltic submarine plateaux. This

thickening of parts of the lithosphere and continued cooling led by the close of the Archaean aeon to the rapid formation of small continents resistant to subduction (Condie 1986). This occurred catastrophically, some time between 3.0 and 2.7 billion years ago. When the roots of the basaltic plateaux cooled to the temperature that eclogite is stable, they

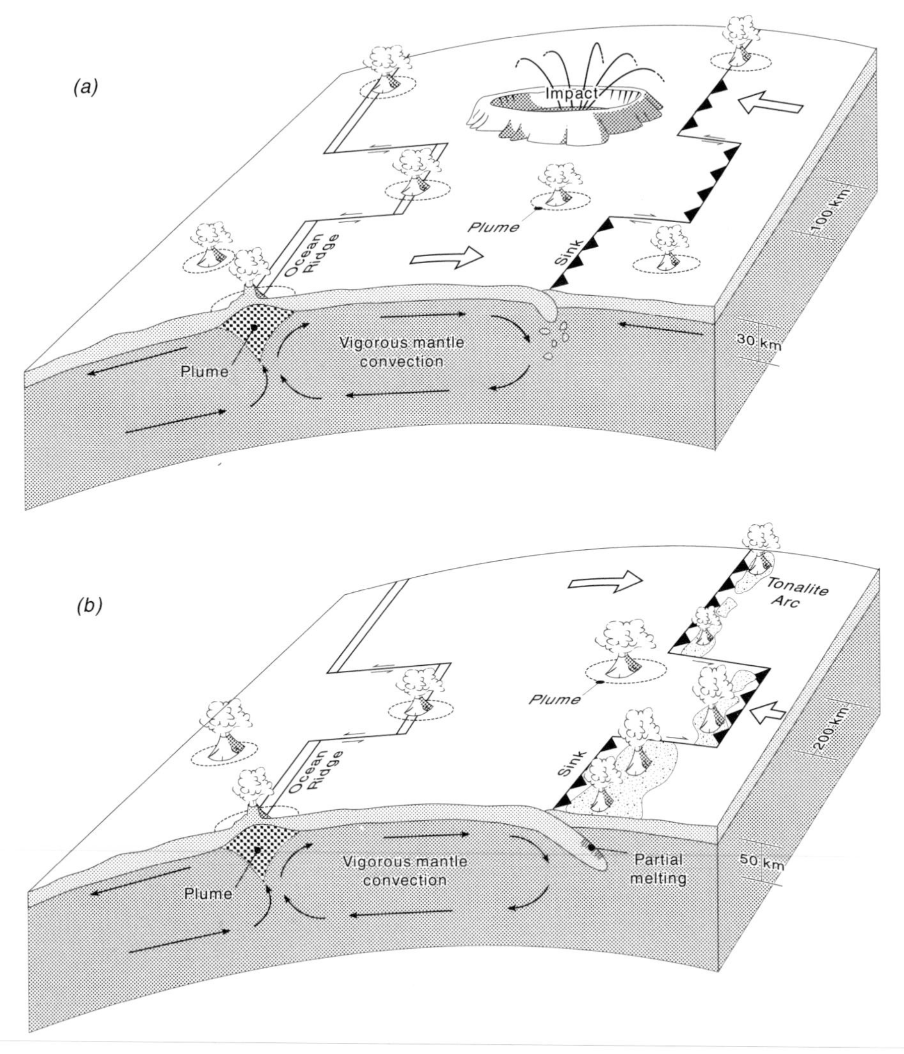

Figure 3.13 Schematic diagrams of the early crust. (a) Primitive komatiitic crust about 3.9 billion years ago. (b) Early continental crust (tonalite arcs) at convergent plate margins about 3 billion years ago
Source: After Condie (1986)

inverted to eclogite and sank into the mantle. Wet partial melting, which happened as the plateaux sank, generated large volumes of tonalite that rose to the surface to form continental crust as tonalite arcs (Figure 3.13b).

3 Between 2.6 and 2.4 billion years ago, collisions of tonalite arcs produced continental regions that became stable cratons (or platforms) and partial melting of the lower crust yielded granites that were intruded into the tonalitic crust. To be sure, by the commencement of the Proterozoic aeon, 2.5 billion years ago, there were a few largish continents.
4 The oceanic lithosphere continued to become denser, owing to an increasing thickness and to decreasing temperature gradients. Sometime in the early to middle Proterozoic aeon, the oceanic lithosphere was dense enough for deep subduction to occur. This produced glaucophane-schist (blue-schist) belts. Subduction became widespread late in the Proterozoic aeon and the Earth passed into the plate tectonic regime that has lasted to the present day. It is possible that the oceanic crust was thinned by subduction from about 12 km to 6 km around 1.0 billion years ago (Moores 1993). This would have increased the freeboard of continents, decreased the fraction of continents flooded by the oceans, and disrupted a stable climate. The current geotectonic regime will cease at some point in the future when the ever-cooling lithosphere is too thick to recycle.

The theory of whole Earth tectonics offers a convincing explanation of crustal evolution (Figure 3.11). During the Archaean aeon, plume tectonics prevailed throughout the mantle, but its role in the upper mantle diminished as the asthenosphere thinned. By the opening of the Proterozoic aeon, plate tectonics was in operation.

The expanding earth

A growing or shrinking Earth radius would automatically lead to several directional changes in Earth history. The most obvious consequences would be changes in the length of day, in the number of solar days in a year, in the number of solar days in a synodic month, and in the number of synodic months per year. Proposals of an expanding or contracting globe are controversial but not without foundation. Earth contraction, once regarded as a good explanation of tectonic episodes, is no longer given credence, but Earth expansion still has several supporters.

Proponents of Earth expansion hold up the improved fit of the Triassic continents on a globe with a reduced radius as one of the strongest pieces of empirical evidence in favour of expansion (Carey 1958, 1976; Owen 1976, 1981; Shields 1979). When Pangaea is reassembled on a globe of modern dimensions, the fit between continents is good at the centre of the reassembly but becomes increasingly bad away from it; but when the reassembly is carried out on a globe with a smaller radius, the fit is much more precise. A lack of extensive oceanic crust during the Proterozoic may be explained by a smaller globe covered almost entirely by a supercontinent (Glickson 1980). A less secure piece of evidence adduced in favour of Earth expansion is the apparent secular emergence of continents (decrease of sea level) during the Phanerozoic aeon (Egyed 1956a, 1956b). The argument is that a progressive decline in the proportion of continents submerged beneath oceans, both individually and collectively, demands that the Earth's radius should have increased, on average, at a rate of 0.5 mm/yr. However, a secular decrease in sea level could have arisen from the 20 per cent reduction in heat production over the last 500 million years (Armstrong 1969). Sea-level rises are ultimately caused by heat production, and a 20 per cent tail-off in production would lead to a sea-level fall by about 80 m, which accounts for most of the progressive Phanerozoic sea-level decline. The area covered by sea in North America suggests a roughly constant relationship between the elevation of continents and sea level since the Cambrian period. But the data are not sufficient to document secular emergence or submergence of the continent (Wise 1973). A reduced radius would call for a sea level 1–1.6 km above the present level in Triassic and Jurassic times, whereas sea level was actually lower then (Hallam 1984; Weijermars 1986).

Many geologists feel that the evidence for expansion is ambiguous and based on *ad hoc* assumptions. But it won't go away. A recent study concludes that the Earth's radius has increased by 17 per cent owing to upper mantle formation, which has been caused by gravitational differentiation of matter within the barysphere and phase transitions (Kozlenko and Shen 1993).

A cyclical world

The geological world is rhythmical. Cycles of geological activity are registered across a great gamut of time-scales, ranging from 'the rapid oscillation of surface waters, recorded in ripple-mark, to those long-deferred stirrings of the deep imprisoned titans which have divided earth history into periods and eras' (Barrell 1917: 746).

The lithosphere displays tectonic megacycles (e.g. Gastil 1960; Dearnley 1965; Sutton 1967; Benkö 1985). Identifying the length of the cycles is problematic owing to the resolution of dating methods and the difficulties of mixing dates derived from different isotope ratios. The best resolution (5–10 million years for Precambrian rocks) is achieved by uranium–lead zircon dating. The frequency distribution of these dates, grouped by geographical areas, from Precambrian rocks displays at least three noteworthy features (Figure 3.14). First, orogenic episodes occurred in each geographical region, but only on two occasions (2.8–2.6 billion and

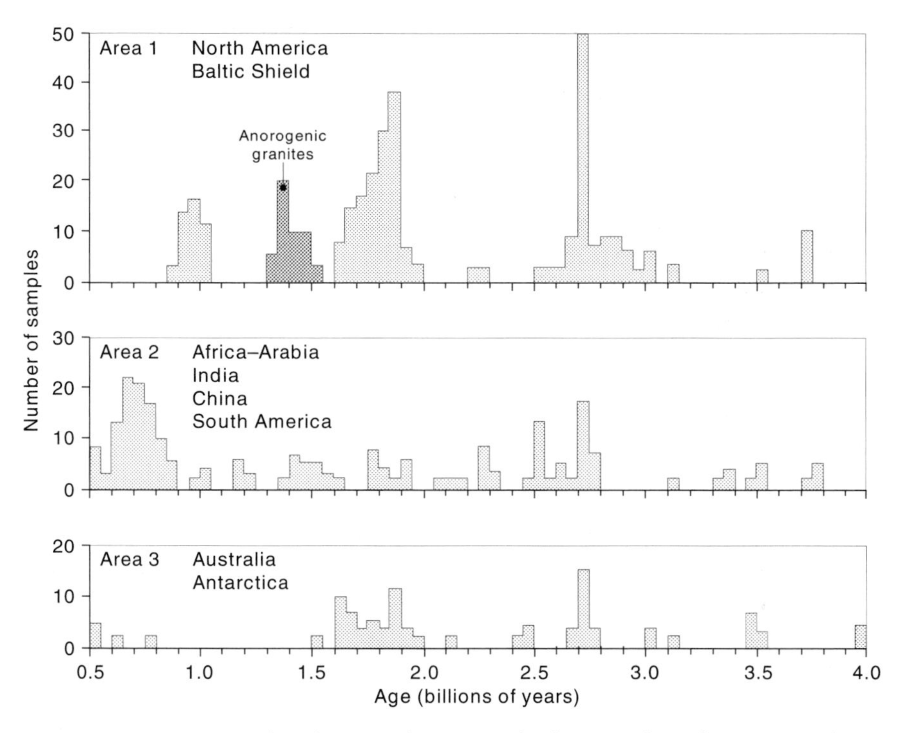

Figure 3.14 Frequency distribution of uranium–lead zircon dates from Precambrian igneous rocks
Source: After Condie (1989a)

1.9–1.7 billion years ago) were these episodes global. These two world-wide orogenies were associated with phases of rapid crust formation. Second, since 3.0 billion years ago, orogenies within geographical regions have lasted about 50 million years. In many cases, broad peaks of orogeny (such as that in North America and the Baltic Shield between 2.0 and 1.6 billion years ago) represent several closely spaced orogenies, often in different parts of the geographical region. Third, orogenic 'gaps' in one region are, in many cases, filled by orogenies in another region. Tectonic episodes in the Phanerozoic aeon appear to coincide with continental collisions. Such collisions need not occur concurrently in geographical distant parts of the planet (though they could do).

The Wilson cycle, named after the Canadian geologist J. Tuzo Wilson, envisages an ocean life-cycle that starts with continental rifting and the opening of a new ocean and ends, approximately 800 million years later, with orogeny and then ocean closure (Wilson 1968; J. F. Dewey and Spall 1975). It thus links tectonic episodes with the plate tectonic model and provides a possible explanation of tectonic episodes. A cycle of similar length occurs in strata-bound ore deposits associated with the opening and closing of the Proterozoic and Phanerozoic oceans (Titley 1993). The Wilson cycle may be explained by the theory of plume tectonics (Figure 3.15). A supercontinent is broken up by a superplume. The supercontinental fragments drift into the super-ocean. Subduction zones develop at random sites. Stagnant, cold slabs of lithospheric material accumulate at about 670 km depth. These megaliths then collapse episodically into the lower mantle. A huge and regular mantle downwelling may form – a cold superplume – that 'attracts' continents and leads to the formation of a massive cratonic sedimentary basin. To form a supercontinent, subduction zones evolve at the edges of the commingled continents. A chain of cold plumes girdles the supercontinent. The downwelling cold slabs (megaliths) squeeze out a superplume by thermally perturbing the D″ layer. This superplume then starts to destroy the supercontinent that created it and the cycle starts anew. The entire process may have run faster during the Proterozoic aeon, possibly taking as little as 400 million years (Maruyama 1994).

Orogenic episodes over the last 250 million years, as listed in the *Geological Time Table* (Haq and Van Eysinga 1987), give a dominant period of 30.6 million years (Rampino and Caldeira 1993). Other major geological changes over the last 250 million years have an approximate 30-million-year pulse. Dominant periods are 28.3 million years for sedimentary sequence boundaries, 23.1, 15.4, and 26.2 for flood-basalt eruptions, and 18.4 and 26.7 for sea-floor spreading rates (Rampino and Caldeira 1993). The highest peak for all data, including mass extinctions, anoxic events in the ocean, and the formation of evaporite deposits, corresponds to a period of 26.6 million years, and a phase with the most recent maximum of the cycle occurring 8.7 million years ago.

The root causes of lithospheric cycles are somewhat elusive. Geological and cosmic explanations are both plausible. Periodic activity at the core–mantle boundary could well produce pulses of rifting, volcanism, and orogenesis in the lithosphere. It would lead also to bouts of rapid change in the geomagnetic field, sea level, atmospheric composition (especially atmospheric carbon dioxide levels), and the biosphere (e.g. Vogt 1979; Sheridan 1987; Rampino and Caldeira 1993). A possible driving mechanism is the quasi-periodic generation of mantle plumes at the core–mantle interface (Courtillot and Besse 1987; Loper *et al.* 1988). On the other hand, the lithosphere itself might produce mid-ocean ridge 'jumps' through sudden changes in subducting oceanic slabs (Stewart and Rampino 1992), or through the changing configuration of the tectonic plates in the super-continent cycle (Worsley *et al.* 1984; Nance *et al.* 1988).

The supercontinent cycle accounts for the very long-term pulse in several related geological processes. According to this theory, the pattern of heat conduction and loss through the crust causes the continents to coalesce as a

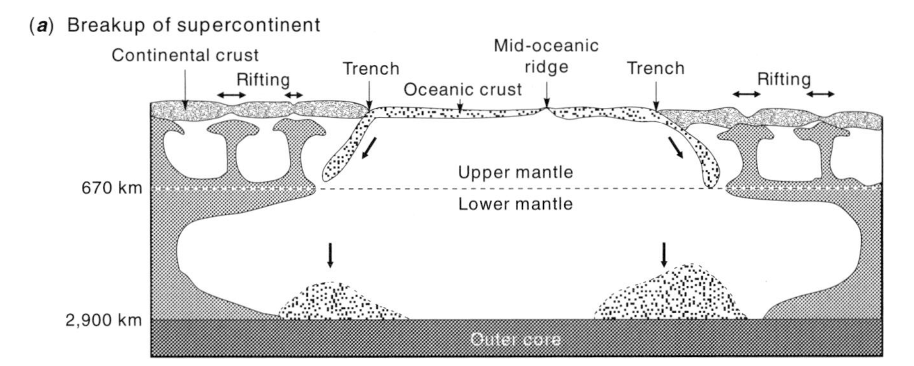

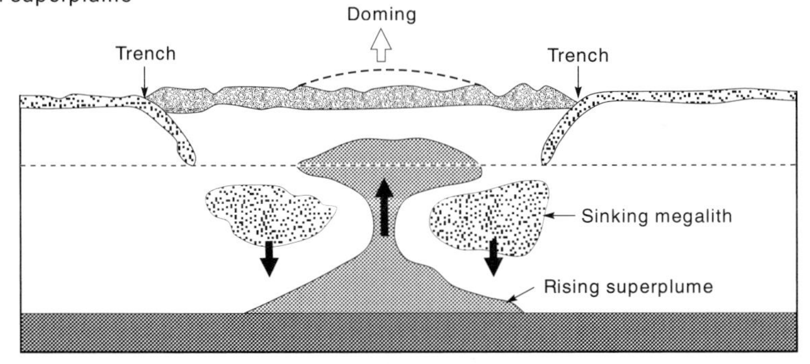

Figure 3.15 Schematic illustration of the Wilson cycle explained by plume tectonics
Source: After Maruyama (1994)

supercontinent, then break into smaller continents, then coalesce again, and so on. The entire cycle was originally estimated to take about 440 million years, but some authorities (Taylor and McLennan 1996) have extended this to 600 million years. When a supercontinent is stationary, heat from the mantle collects underneath it. As the heat accumulates, the supercontinent domes upwards. Eventually, the single land mass breaks apart, and fragments of the supercontinent disperse. The heat that has built up under the supercontinent is then released through the new ocean basins forming between the dispersing continental blocks. When eventually enough heat has escaped, the continental fragments will be driven back together. Thus, the model depicts the surface of the Earth as a sort of coffee percolator: the heat supply is essentially continuous, but because of poor conduction through the continents, the heat is released in relatively sudden bursts (Nance *et al.* 1988: 44).

The supercontinental tectonic megacycle explains why crustal evolution displays growth spurts. It is easy to envisage how it might modulate the tempo of crustal growth. There are two situations when continental crust creation is particularly fast. The first of these is when a supercontinent breaks apart. At this time, oceanic crust is at its oldest and is most likely to form new continental crust after its subduction. The second time is as the separate continents come together again and volcanic arcs are swept up and welded onto continental platforms. The intervening periods see much slower rates of crustal growth. The best known example of a supercontinent cycle is the coming together and breaking up of Pangaea (Figure 3.16). A late Proterozoic supercontinent – Rodinia – formed about 1 billion years ago, and broke up about 700–800 million years ago (e.g. Dalziel 1991). The oldest known supercontinent cycle occurs in a belt of supracrustal Archaean granite and greenstones, Pilbara Block, Western Australia (Krapez 1993).

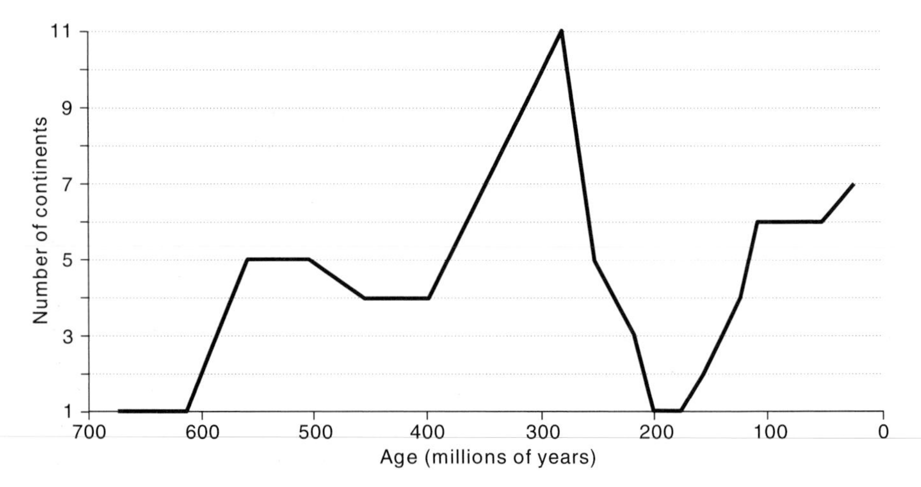

Figure 3.16 The number of continents at various times during the Phanerozoic aeon
Source: After Maruyama (1994)

Several galactic and Solar System cycles might account for cycles of activity in the lithosphere. According to a recent theory (Shaw 1994), impact events ring the lithospheric changes. The argument depends on impact craters non-randomly distributed around the planet and through time. Clusters in space are recognized as the Phanerozoic cratering nodes (p. 49). Clusters in time are strongly suggested by the geological divisions of the Phanerozoic aeon (Shaw 1994: 54–5). The Phanerozoic aeon comprises three eras, each about 190 million years long (possibly corresponding to the largest impact events), eleven periods (with the Tertiary counted as a period), each about 52 million years long (possibly associated with large impacts), some forty-three epochs, each about 13 million years long (possibly associated with intermediate size impacts), and some 130 stages, each about 4.4 million years long (possibly associated with smaller impact events). These 'average' periods within the geological column are quite close to rational-number multiples of the 26-million-year pulse in many geological changes: ~22, ~7, 2, 1/2, and ~1/6, respectively. This finding is consistent with the idea that almost all geological changes reflect a common system of non-linear resonances associated with processes in the Solar System. If this should be true, then the geological column constitutes a non-linear, fractal, or multifractal time series of self-similar geodynamic events. It records a hierarchy of local and global catastrophes. It marks chaotically intermittent geological changes that are resonant with the record of impact-cratering processes, which in turn results from the non-linear dynamics of orbital evolution in the Solar System. In short, catastrophes big and small are all part of the natural geological order. They are part of a highly co-ordinated non-linear system, and are not the outcomes of rare and random events.

A changed world?

Have secular changes in Earth history led to evolution of the geological environment? Has the geological world changed? Is the present truly the key to the past?

Geological processes seen in action today did operate in the past according to the same physical and chemical laws. However, they did so under different circumstances and, on occasions, with greater energy. Some of the parameters in the laws governing geological processes have altered due to irreversible changes in the atmosphere, oceans, biosphere, and lithosphere. So the present-day geological environment is not quite the same as its earlier manifestations. This view is now firmly established in geological thinking. For instance, a primary thrust of modern research into Precambrian strata tries to identify how the early Earth differed from the current geological order.

Different views on the past geological order have emerged. There is, for example, disagreement about the applicability of plate tectonics to crustal change in the Proterozoic aeon. Before the early 1980s, some geologists opined that plate tectonics did not occur in Precambrian times. It was maintained that, while the lithosphere was cooler and thicker during the Proterozoic than during the Archaean aeon, it was hot, partially melted, and therefore ductile; that is, it might have flowed (Wynne-Edwards 1976). When it lay over an upwelling region of the mantle, it was extended by ductile flow to produce a kind of orogeny that characterized Proterozoic times but no longer exists. Only in the Phanerozoic aeon did the lithosphere cool to a point where it became brittle, and regions over mantle upwelling sites cracked, leading to the formation of mid-ocean ridges. In Africa, some cratons have Proterozoic orogenic belts between them (called ensialic belts), but the cratons are connected deep below the surface and do not appear ever to have separate continental masses. The conclusion was that no separate plates existed and the orogenic belts could not be explained by the plate tectonic model (Shackleton 1973; Kröner 1977). This appeared to be the case for Proterozoic geology in Australia, too (Duff and Langworthy 1974). Instead, ensialic orogeny might have resulted from a force similar to that causing lithosphere to break up, but not so strong and only capable of causing tension and rifting, subsidence, and orogenic deformation (McWilliams and Kröner 1981). However, work carried out since about 1980 strongly indicates that plate tectonics did operate in Precambrian times, and it started to do so late in the Archaean aeon, if not before (Condie 1992, 1994; Windley 1993).

There is no disagreement about the unique geological environment of the Priscoan aeon. Little evidence remains of this earliest phase of Earth history, but much can be inferred from current processes, and in particular from the bombardment studies. The history of bombardment has been reconstructed from the lunar maria. It would seem that the intensity and magnitude of impacts have decreased through time. The terrestrial planets were heavily bombarded during their early history, from their formation to about 3.8 billion years ago; the rate and magnitude of bombardment events then steadily declined to the roughly constant, if episodic, value that has obtained throughout the Phanerozoic aeon (Hartmann 1977). During the first 2 million years of the Earth's history, giant bodies, three times the mass of the planet Mars, may have collided with the Earth (Wetherill 1985), one such impact possibly creating the Moon (Newsom and Taylor 1989). The enormous impacts that shook the very young Earth no longer occur, but their effects on the Earth's primitive crust can be inferred by extrapolating the effects of the more modest and more recent impacts recorded in impact craters. For instance, during the phase of heavy bombardment, some 30–200 impact basins a thousand kilometres or more across and several kilometres from rim to floor were formed (Frey 1980; Grieve and Parmentier 1984). Such vast impact events might have produced the early ocean basins and established the basic distribution of oceanic and continental crust (Frey

1980). It is also possible that impacting comets provided the water for the early oceans (Chyba 1987, 1990a), and much organic material (Cronin 1989; Chyba *et al.* 1990; Chyba and Sagan 1992; Chyba and McDonald 1995).

The conclusion is that the lithosphere is ever changing and evolves. For this reason, the geological present is not the key to all aspects of the geological past. It is against the backdrop of ever-changing and evolving geological and cosmic environments that change in the atmosphere, hydrosphere, toposphere, pedosphere, and biosphere will now be explored.

SUMMARY

The solid Earth is layered. The main layers are the core, mantle, and crust. The lithosphere (crust plus part of the upper mantle) consists of minerals and rocks (igneous, metamorphic, and sedimentary). Continental lithosphere is thick and buoyant. Oceanic lithosphere is thin and dense. The swapping of material between the lithosphere and ecosphere produces the rock or geological cycle. This grand circulation of Earth materials has three component cycles – the geochemical cycle, the water cycle, and the tectonic cycle. The lithosphere forms a set of plates and platelets. It changes through plate tectonic (geotectonic) processes. These processes produce characteristic geological features (including earthquakes, volcanoes, and rock types) at constructive, destructive, and conservative plate margins, and in the large areas within plates. Processes in the mantle (plume tectonics) and the core (growth tectonics) largely drive plate tectonics. Large and small rising mantle-plumes explain many features of surface tectonics. Descending slabs of cold lithospheric material influence processes in the core and mantle. They possibly start the growth of hot, ascending superplumes. Impacts by asteroids or comets the size of Mount Everest might perturb core and mantle processes, setting in train processes that will influence plate tectonics. Geological processes act both slowly and quickly – the rock record bears signs of gradual and catastrophic change. Many geological changes are quasi-periodic or episodic. A 26-million-year pulse is found in continental flood-basalt eruptions, continental drift rates, sea-floor spreading rates, sequence boundaries, and orogenic events. Many geological changes, such as crustal evolution, go in a definite direction. Long-term cycles, including the Wilson cycle, and steady states are superimposed on these secular trends. The geological environment has evolved. The present, therefore, is not a master key to the past, but it does help to unlock the mysteries of Earth history.

FURTHER READING

A very basic and well-illustrated account of plate tectonics is found in *Introducing Physical Geography* (Strahler and Strahler 1994). A more detailed description account is located in *Global Geomorphology: An Introduction to the Study of Landforms* (Summerfield 1991a). Geological accounts with historical slants include *Plate Tectonics and Crustal Evolution* (Condie 1989a) and *The Evolving Continents* (Windley 1995). *Volcanoes: A Planetary Perspective* (Francis 1993) and *The Nature of the Stratigraphical Record* (Ager 1993a) are well worth reading. For a sound and stimulating introduction to many aspects of geology, try *Understanding the Earth* (G. Brown *et al.* 1992). For a real challenge with considerable rewards, try *Craters, Cosmos, Chronicles: A New Theory of Earth* (Shaw 1994).

4

ATMOSPHERE

DUSTY GAS: AIR

Gases

The atmosphere is the aeriform fluid that enshrouds the Earth. It is a dusty gas. Much of its mass is contained in its lowermost layer.

The composition of the atmosphere mostly reflects the presence of life on the planet's surface. The chief gases (by volume) are nitrogen (78.08 per cent) and oxygen (20.95 per cent), neither of which is abundant on any other terrestrial-like planet (Mercury, Venus, and Mars). Water vapour (1 per cent, but variable) is a smallish, but highly significant, component. Next comes argon (0.93 per cent), an inert gas. The remaining gases are minor in quantity, though some play leading roles in biogeochemical cycles and in the regulation of atmospheric temperature. They include carbon dioxide (360 ppm and rising), neon (18 ppm), helium (5.2 ppm), methane (1.5 ppm), krypton (1.0 ppm), hydrogen (0.5 ppm), ozone (0.4 ppm), nitrous oxide (0.33 ppm), xenon (0.087 ppm), carbon monoxide (0.12 ppm), ammonia (0.01 ppm), nitrogen dioxide (0.001 ppm), and sulphur dioxide (0.0002 ppm). The figures for the minor constituents are for the lower atmosphere and vary somewhat with altitude and sometimes latitude. Some water vapour is split by ultraviolet light into hydrogen atoms and hydroxyl (OH) radicals; the latter, mainly because it is a catalyst in ozone destruction, has a material influence on atmospheric chemistry.

Particulates are most concentrated in the lower atmosphere. Those found in the upper air materially affect radiation balances. They are found at all places, though some spatial concentrations are observed. Water, in the form of water droplets and ice crystals, is the main particulate. The mean global cloud cover is about 50 per cent. Concentrated sulphuric acid droplets (originating mainly from volcanoes), sulphates, silicates, and carbonates (dust), sea salt, and organic materials are also important. The significance of particulates for energy balances rests in their ability to interact with sunlight. The degree of interaction may be expressed as optical depth, τ. An optical depth much less than 1.0 indicates very little scattering or absorption of sunlight; an optical depth much greater than 1.0 indicates a deal of scattering and absorption of sunlight. Water vapour has an optical depth of 5; dust, sea salt, and organic materials an optical depth of between 0.05 and 3; and sulphuric acid droplets an optical depth of 0.003–0.3.

Gaseous spheres

The atmosphere is divisible into several spheres, each having a characteristic temperature, pressure, and composition: the troposphere, stratosphere, mesosphere, thermosphere, and exosphere (Figure 4.1). Besides these spheres are the chemosphere, ionosphere, and magnetosphere. The chemosphere is the region between about 32 and 92 km where many important chemical reactions take place. The ionosphere is a shell of high electron concentration. It results from very short wavelength sunlight stripping electrons from atoms and molecules, mainly molecular oxygen and nitrogen, to create an ionized layer. The magnetosphere is the constantly changing

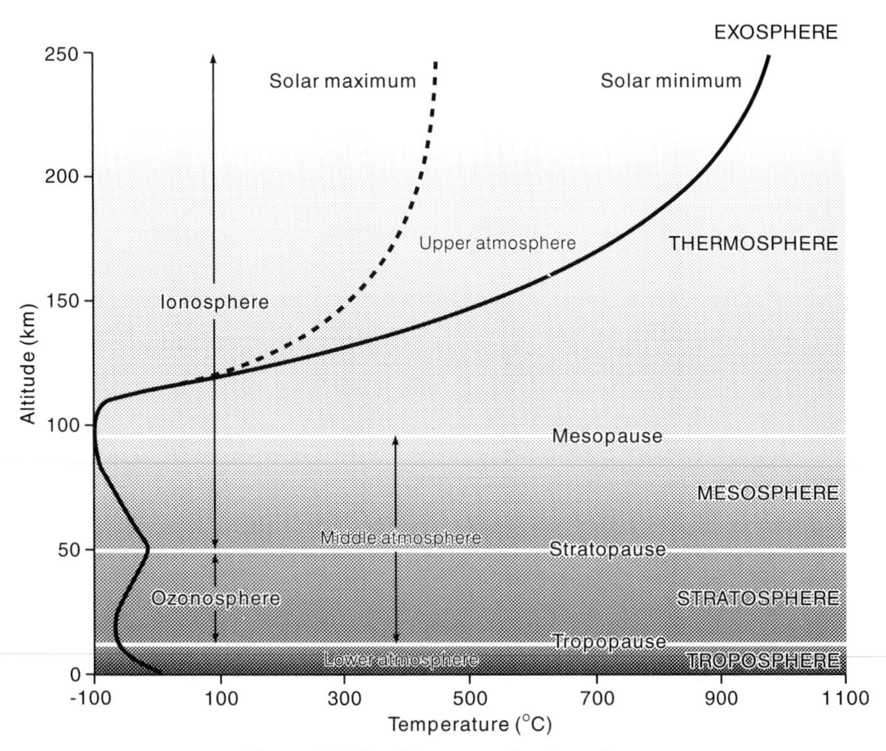

Figure 4.1 Earth's atmospheric regions

magnetic field generated by the Earth's dynamo that influences the behaviour of electrically charged particles. It extends about ten Earth radii (64,000 km) into space on the sunward side.

Air becomes cooler with increasing altitude through the troposphere, except when temperature inversions are present. It warms again in the stratosphere, owing to sunlight absorption by ozone. Air in the mesosphere, where there is less ozone and more carbon dioxide (which at this altitude and density acts as a coolant), chills again. Beyond the mesosphere, atoms and molecules, mainly of oxygen and nitric oxide, are far too sparse to emit thermal radiation and so cool the atmosphere. Consequently, the air warms again. Air density declines with altitude and reflects a balance between gravity, which pulls air down, and gas pressure, which tends to move the gas upwards from high-pressure regions to low-pressure regions. As a rule of thumb, air density falls by a factor of three for every 10 km of altitude.

The weather experienced by life on Earth is confined to the relatively dense troposphere. In the rare upper air, several complex chemical and thermal reactions are powered by the incoming flux of waves and particles from the cosmosphere. These reactions seem to have a far greater influence on the climate at the ground than was once thought possible, though the connections between the two are still a trifle puzzling. All solar emissions may lead to ionization, chemical changes, and heating in the atmosphere. Electromagnetic radiation appears to wield the most substantial and direct effect on the atmosphere.

Circulating air

Air moves perpetually. Atmospheric motion is fundamental to understanding the Earth's climates (Figure 4.2). On a planetary scale, the temperature gradient between the equator and the poles is the foremost driver of air movements. This thermal gradient powers a vast overturning of air, first suggested by Edmund Halley in 1686 and elaborated by George Hadley in 1735, called the Hadley circulation.

On Venus, there is one grand Hadley cell in each hemisphere: a huge convective current of air rises at the Venusian equator, moves polewards, sinks over the poles, then returns at ground level to its equatorial origin. On the Earth, the Hadley circulation breaks down into three component cells in each hemisphere (Ferrel 1856; Bergeron 1928; Rossby 1941). The equatorial Hadley cell is driven largely by the heat released as water evaporated from the tropical oceans condenses, mainly in the inter-tropical convergence zone (ITCZ), which is also called the equatorial low-pressure trough. At ground level, the air returning towards the ITCZ produces the trade winds. The middle or Ferrel cell flows in the reverse direction, that is, equatorwards aloft and polewards at the surface. The third cell, known as the polar cell, is rather weak.

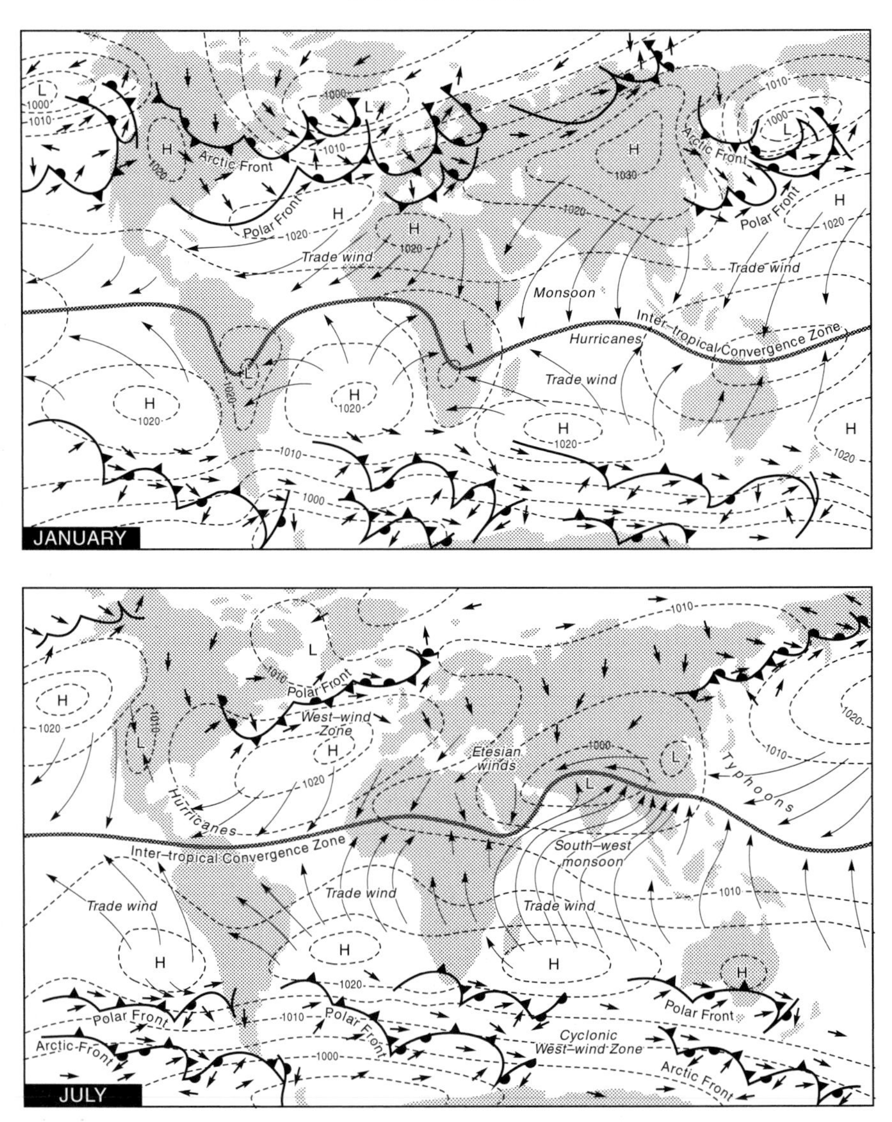

Figure 4.2 Atmospheric pressure and circulation pattern near sea level in January and July
Source: After Weischet (1991)

Because the Earth rotates, winds in the tropics tend to blow north-east to south-west (Northern Hemisphere) or south-east to north-west (Southern Hemisphere). In middle and high latitudes, where Corioli's force is strongest, they have a strong east-to-west component. In addition, easterly winds predominate in the upper troposphere in the tropics, and westerly winds elsewhere. The upper air westerlies are the main flow of the atmosphere. They form a circumpolar vortex, the fastest flowing ribbons of which are called jet streams. This vortex progresses from shallow zonal waves to great meandering meridional loops. The full sequence of the 'index cycle', so named because the zonal index (the average hemispherical pressure gradient between 35° and 55° N) runs from low to high and back again over a four- to six-week period (Namias 1950). The tropical easterlies may also form jet streams, especially in summer.

The tripartite Hadley circulation has a smaller 'eddy' circulation superimposed upon it. This eddy circulation is created by baroclinic instabilities and the development of standing waves. Baroclinic instabilities arise when steep temperature and pressure gradients cross one another. The result is that baroclinic waves develop, creating the disturbances known as the travelling cyclones and anticyclones characteristic of mid-latitudes. These disturbances play a major role in transferring heat to the upper troposphere and towards the poles. In tropical latitudes, large and intense wave cyclones are absent because of the weak Corioli's force. Instead, disturbances take the form of easterly waves, which are slow-moving troughs of low pressure within the trade wind belt (Dunn 1940; Riehl 1954); weak equatorial lows, which form at the heart of equatorial troughs and serve as foci for individual convective storms along the ITCZ; and powerful tropical cyclones. Tropical cyclones, which are called hurricanes in the Atlantic Ocean, typhoons in the Indian and Pacific Oceans, and willy-willies in Australia, are another large-scale, though seasonal, feature of the general circulation. They begin as weak low-pressure cells over very warm ocean water (27°C and more) between 8° and 15° N and S that grow into deep circular lows. Monsoon circulations also develop in parts of the tropics. In southern Asia, these involve a changeover from north-east winds in winter to south-west monsoon winds in summer, probably associated with the seasonal shift of the equatorial low-pressure trough.

Stationary eddies (standing waves) are generated by such topographic features as massive mountain ranges, and by temperature differences between oceans and continents. The Rockies, for example, tend to anchor the westerly jet stream over North America. A land mass will have hotter summers and colder winters than an ocean occupying the same latitude. This is the effect of continentality or, viewed from the opposite perspective, of oceanicity. It arises because land has twice to thrice the heat capacity of sea, and because the conduction of heat from the land surface is much lower than the downward transfer of heat from the ocean surface by turbulent mixing. For these reasons, the annual and diurnal ranges of surface and air temperatures are much larger in continental climates than in oceanic climates. The influence of oceans is extended over continental areas lying next to oceans, especially non-mountainous areas on western seaboards where maritime air masses move inland. The degree of continentality may be described by various indices. The Northern Hemisphere displays more marked effects of continentality than does the Southern Hemisphere. This is because the Northern Hemisphere contains about twice as much land as the Southern Hemisphere.

Smaller circulations of air include individual eddies revealed by the swirling of autumn leaves, thunderstorms, and land and sea breezes. These are superimposed upon the larger-scale hurricanes, fronts, cyclones and anticyclones, and hemispheric and global circulations.

BLOWING HOT AND COLD: CLIMATIC CHANGE

What is climate?

Climate is the average atmospheric conditions at a particular place. It is often defined by mean temperatures, pressures, winds, and so forth, over a thirty-year period. Planetary fields of temperature and wind create three basic climatic zones: the torrid zone of low latitudes, the temperate zones of middle latitudes, and the frigid zones of high latitudes (Figure 4.3; Plates 4.1–4.3). This fundamental zonary arrangement of climates is hugely distorted by the distribution of land and sea, and by the presence of large-scale topographic features, chiefly mountain ranges and plateaux. At present, the distortions to the three basic climatic zones produce some seven climatic regions, though the exact number varies from one authority to the next. The regions are:

1 Humid tropical climate (equatorial rain zone)
2 Savannah climate (tropical margin zone with summer rains)
3 Desert climate (subtropical dry zone)
4 Mediterranean climate (subtropical zone of winter rain and summer drought)
5 Temperate climate (temperate zone with precipitation all year round). This extensive zone is divisible into maritime climate (warm temperate subzone), nemoral climate (typical temperate subzone – a short period of frost), continental climate (arid temperate subzone – cold winter), and boreal climate (cold temperate subzone)
6 Tundra climate (subpolar zone)
7 Polar climate (polar zone).

Past climates are less easy to define than modern climates. Their description is restricted by the time resolution of stratigraphic sequences. An annual calendar of palaeoclimatic events is seldom available. Another difficulty is the lack of modern analogues for all past climates. The disposition of land and sea, the position of high land, and other physical conditions were different in the past and, in some cases, produced climates

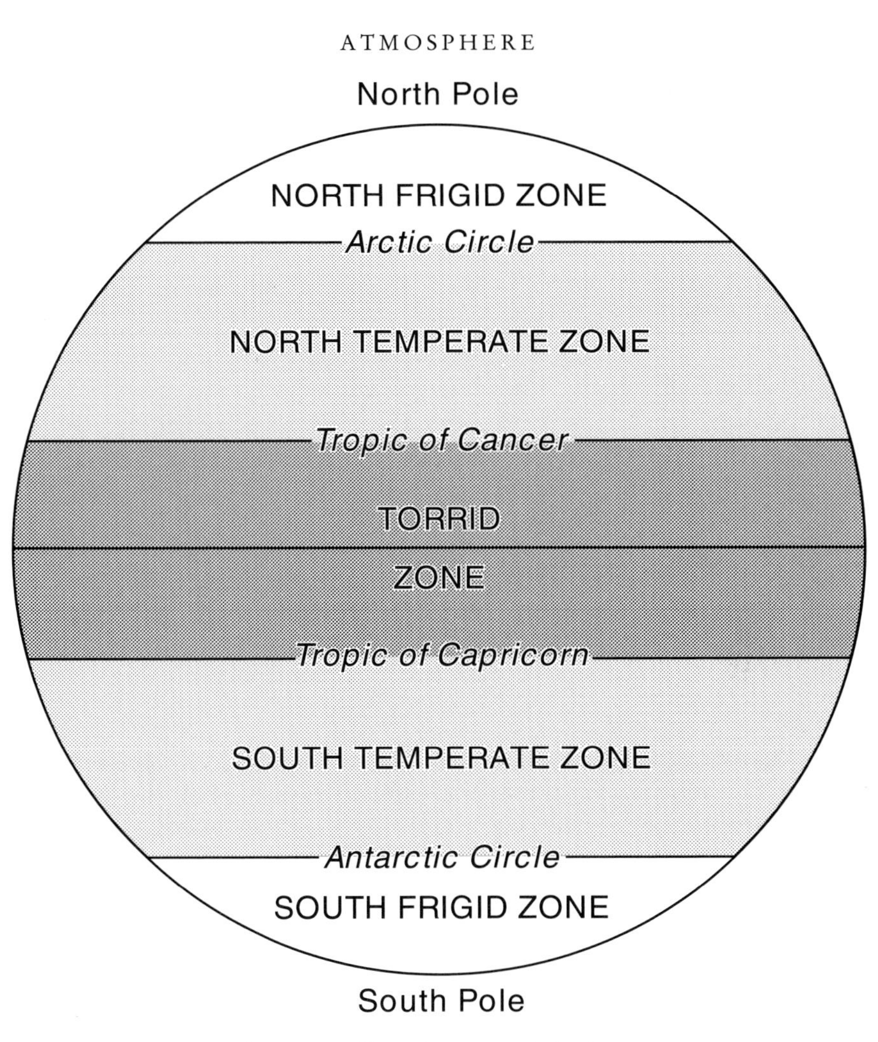

Figure 4.3 Climatic zones of the Earth

that do not exist today. A prime example is the 'boreal grassland' climate that occupied a broad belt immediately south of the Northern Hemisphere ice sheets (p. 272).

Today, climate is taken to embody more than just the atmospheric system. It is assumed, with much justification, that the atmosphere, oceans, and ice sheets act as a single body. This atmosphere–ocean–cryosphere unit is the world climate system (see Huggett 1991: 4–6). Some workers add the active layer of the land as an element of this system (Monin 1986: 2), and some extend its range to include the lithosphere and biosphere as well (e.g. Yeh and Fu 1985) to produce the 'atmosphere–hydrosphere–cryosphere–

Plate 4.1 Arid climate: a dry landscape (plain eroded in schist), central Namibia
Photograph by C. R. Twidale

lithosphere–biosphere' climate system (Kondrat'ev 1993) – there seems no end to the compass of the world climate system! But the matter is straightforward: all systems of the ecosphere, which includes the atmosphere, are interdependent; it is impossible fully to comprehend atmospheric change without taking into account attendant changes in the lithosphere, toposphere, pedosphere, and biosphere.

What is climatic change?

The three-cell general atmospheric circulation has probably persisted throughout the Phanerozoic aeon. Variations have been played on the basic theme. Temperatures have at times changed dramatically (Figure 4.4). The position and intensity of such key components as the mid-latitude cyclones and the subtropical high-pressure cells have altered, and some present-day components, such as the Afro-Asian monsoons, were subdued or even inactive (Figure 4.5). Precambrian climates may have been nothing like modern-day climates. Physical conditions were so different – the Earth rotated much faster, for instance – that even the general circulation may have followed an unfamiliar pattern. Details of these changes are elucidated later. The point to make here is that climatic change does not entail a simple shift of boundaries between present climatic regions – it is more likely to involve complex reorganizations of the atmosphere–cryosphere–ocean system.

Plate 4.2a Frigid climate: a glacial landscape, Corneliussensbreen, Okstindan, Norway
Photograph by Wilfred H. Theakstone

Plate 4.2b Frigid climate: a glacial landscape, Commonwealth Range, near Beardmore Glacier, Antarctica
Photograph by Robin L. Oliver

Plate 4.3 Humid climate: a temperate landscape, Doubtful Sound, South Island, New Zealand
Photograph by Brian S. Kear

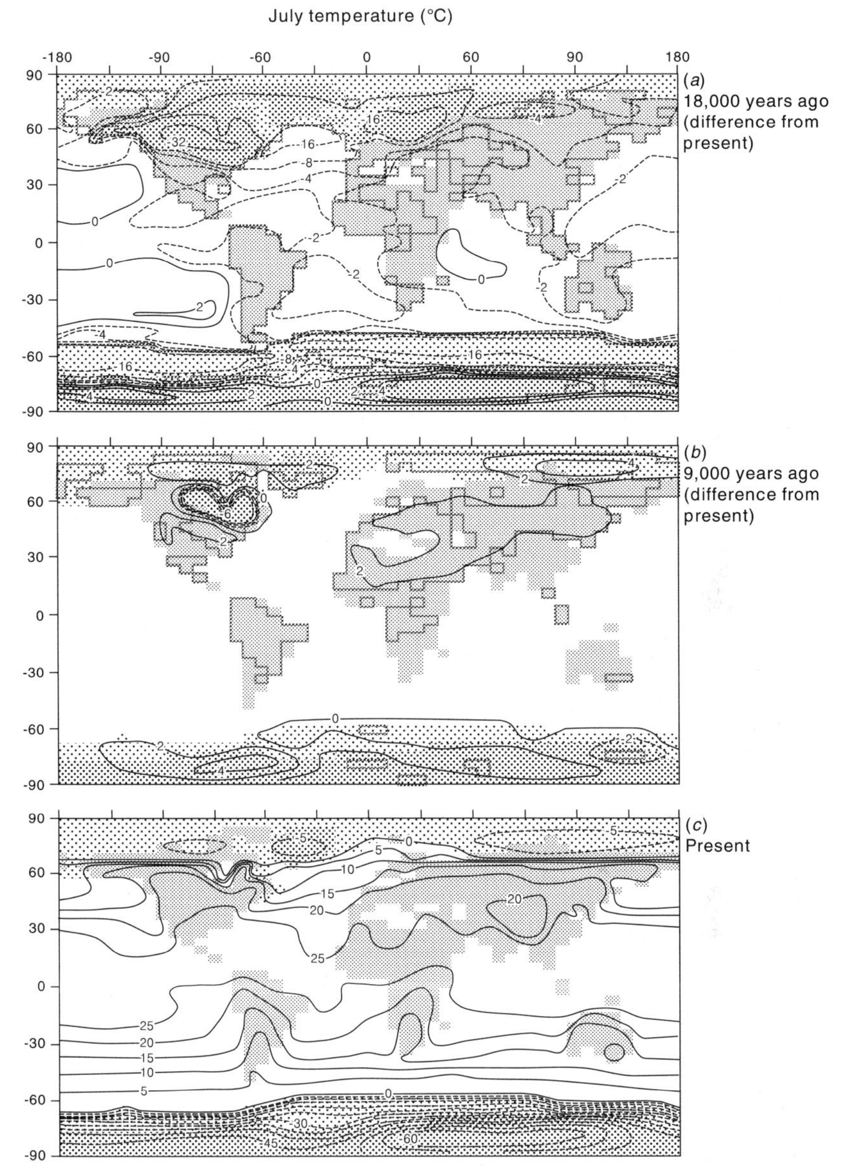

Figure 4.4 July temperatures (°C), 18,000 years ago, 9,000 years ago, and at present. The figures are predicted by the COHMAP climate-model experiments. The 'present' is a control simulation. The July temperatures for 18,000 and 9,000 years ago are expressed as differences from present July temperatures. Regions where the differences are statistically significant at or above a 95 per cent level of confidence are outlined by the dark grey lines. Light hatching shows sea ice, and dense hatching shows land ice

Source: After Kutzbach *et al.* (1993)

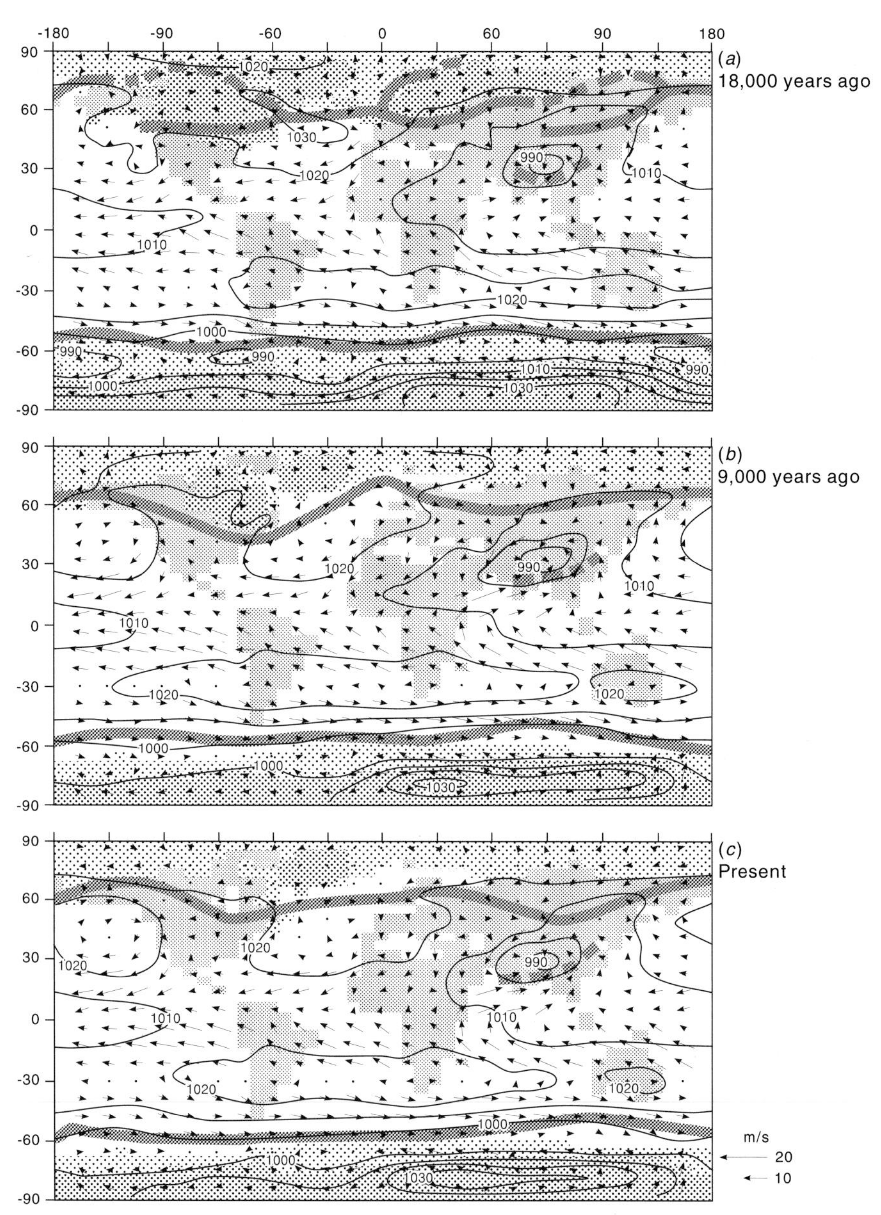

Figure 4.5 July surface wind direction and strength (m/s), sea-level pressure (mb), and surface storm tracks 18,000 years ago, 9,000 years ago, and at present. The figures are predicted by the COHMAP climate-model experiments. The 'present' is a control simulation. Light hatching shows sea ice, and dense hatching shows land ice
Source: After Kutzbach *et al.* (1993)

Climatic change is a change in average atmospheric conditions at a particular place or in a particular region. It is often difficult to distinguish from climatic variability, especially in strongly seasonal climates. Climatic variability is the differences between mean atmospheric states of the same kind; for instance, between mean July temperatures or mean winter precipitation in different years. In the Mediterranean region, seasonal conditions fluctuate wildly from year to year. Mean summer temperatures vary by some 5–7°C, and mean winter temperatures by some 8–13°C, from one year to the next, while seasonal rainfall departs 80 per cent from the long-term mean.

Paradoxically, change in modern climates is not easily perceived – it is commonly hidden within the welter of direct measurements taken over short time periods; whereas change of past climates is unmistakable, owing to limited observations of proxy climatic indicators taken from rocks, sediments, and soils. Modern humans may lack a sense of climatic change because the weather over the last 10,000 years has been more consistent and equable than at any time in the past hundred millennia. None the less, classical Greek and Roman philosophers and natural historians were aware that the climate in one year was not always like the climate in the previous year. Geological changes of climate were recognized during the seventeenth century: tropical fossils unearthed in temperate climes suggested that the present climate was unlike the ancient climate. In the nineteenth century, the recognition of glacial and interglacial regimes indicated that the climate had changed drastically in the relatively recent past. A multitude of theories attempted to explain these long-term changes of climate. These theories, many of them highly inventive, were difficult to appraise because hard palaeoclimatic evidence was meagre, the interpretation of climatic indicators was problematic, and a sound chronology of events did not exist. And most of the theories were rather simplistic because many of the complex interactions and forcings of the climate system were only guessed at or remained unimagined, and because knowledge of the Solar System and Milky Way Galaxy was limited.

Over the last thirty years, the problems faced by earlier generations of theorists in trying to evaluate the worth of their conjectures have been remedied, to some extent at least. Today, geochronometric dating techniques provide an absolute time-scale of events. Ice cores, deep-sea sediments, cave sediments, bog and lake cores, tree rings, and palaeosols yield up evidence of the climatic past. A better understanding of the relation between climate and Earth surface phenomena means that climatic indicators can more confidently be used to reconstruct ancient climates. Computer models of the general circulation allow climates of the past to be simulated, and allow the effect of Earth's changing orbital motions on solar input to be calculated quickly and accurately. Used in combination, all these new palaeoclimatological methods have unveiled a far more colourful spectrum of climatic changes than earlier generations of climatologists were aware of. It is now confirmed that, for a variety of reasons, weather and climate change over time-scales ranging from less than an hour, as in short-lived but severe meteorological phenomena, to over tens of millions of years, as in protracted phases of global warming and cooling (Table 4.1). In addition, they follow secular trends lasting aeons.

THE WORLD WEATHER MACHINE: CAUSES OF CLIMATIC CHANGE

The basic mechanisms that cause the atmosphere to change are fairly well understood in broad terms, though the relative importance of various factors over different time-spans is still uncertain. The chief processes involved in atmospheric change are summarized in Figure 1.3. In the short term, over years, decades, and centuries, the atmosphere may change owing to external forcings (cosmic and geological) or to internal atmospheric dynamics. Cosmic forcing arises from changes in gravitational forces, by variations in electromagnetic and particulate radiation receipt from the Sun and from space, and by the impact of asteroids and comets. Gravitational stresses may emanate from three sources – the Solar System, our Galaxy, and other galaxies – but only interactions within the Solar System will force the atmosphere in the short term when the delivery of energy is modulated by the overall motions and alignments of the planets in the Solar System and by Earth–Moon motions. Geological forcing is, in the short term, caused by the injection of volcanic dust and gases into the stratosphere. Internal dynamics of the atmospheric system involve short-term, cyclical components. Thus climatic change can occur without the aid of external forcing. Sorting out signals from these potential sources of climatic change is no mean feat. None the less, climate modelling experiments show that each has a characteristic signature in time and in space by which it may be recognized in the climate record (Rind and Overpeck 1993).

In the medium and long term, over thousands to millions of years, atmospheric change is caused by cosmic and geological forcings, and by internal atmospheric dynamics. Cosmic forcing over medium and long time-scales involves orbital variations, secular changes in solar output, and, indirectly, large-body impacts that can influence geological processes. Geological forcing over medium and long time-scales involves plate tectonic processes, which in turn are driven by processes going on in the core and mantle. In the long term, plate tectonics leads to changes in palaeogeography – the redistribution of continents and oceans, the formation of mountain ranges, changes in the volume of the oceans, changes of sea level, and so on – all of which may induce atmospheric change. Redistribution of mass within the Earth can lead to secular trends in true polar wander that will have repercussions in the atmosphere, too. Processes in the core and

Table 4.1 Major climatic variations, excluding secular trends

Variation	*Period (yr, unless otherwise stated)*	*Nature of period*	*Nature of variation*
Short-term variations			
Diurnal	1 day	Spike[a]	Daily cycle of climate
Weekly	3–7 days	Peak[b]	Synoptic disturbances, chiefly in middle latitudes
Annual	1	Spike	Annual cycle of climate
Quasi-biennial	~26 months	Peak	Wind shift (east phase to west phase) in tropical stratosphere. Called the quasi-biennial oscillation (QBO)
Quinquennial	2–7 years	Peak	Rapid switch of pressure distribution across the southern Pacific Ocean. Linked with changes in temperatures in same region, and together termed El Niño–Southern Oscillation (ENSO). Average period is 4–5 years; 2–7 years is the range
Undecennial	~11	Peak	Quasi-periodic variation corresponding to the sunspot cycle
Nonadecennial	~18.6	Peak	Quasi-periodic variation corresponding to the lunar nodal cycle
Octogintennial	~80	Peak	Quasi-periodic variation corresponding to the Gleissberg cycle
Bicentennial	~200	Peak	Quasi-periodic variation, possibly corresponding to the solar orbital cycle
Bimillennial	~2,000	Peak	Quasi-periodic variation of uncertain correspondence
Medium-term variations			
Precessional	~19,000	Spike	A periodic component of the Earth's precession cycle
Precessional	~23,000	Spike	A periodic component of the Earth's precession cycle
Tilt	~41,000	Spike	Main periodic component of the Earth's axial tilt cycle
Short eccentricity	~100,000	Spike	A periodic component of the orbital eccentricity cycle
Long eccentricity	~400,000	Spike	A periodic component of the orbital eccentricity cycle
Long-term variations			
Thirty megayear	~30,000,000	Peak	Quasi-periodic fluctuations corresponding to a tectonic cycle
Warm–cool mode	~150,000,000	Peak	Quasi-periodic fluctuations from warm to cool climatic modes, possibly connected with half galactic year
Hothouse–icehouse	~300,000,000	Peak	Quasi-periodic fluctuations from hothouse to icehouse conditions

Notes: [a] Strictly periodic variation dictated by astronomical cycles
[b] Quasi-periodic variation with preferred time-scales of recurrence

mantle, which drive many lithospheric processes, are themselves influenced by cosmic forcing: large-body impacts may force changes at the core–mantle boundary that, eventually, are felt in the atmosphere. The entire planet is subject to gravitational forcing that may effect long-term changes of axial tilt (astronomical pole shift), again with concomitant effects on the atmosphere. Gravitational forces also gradually change the position of the Solar System relative to the centre of the Milky Way Galaxy. This may lead to very long-term changes of climate. In addition, the atmosphere may change over thousands to millions of years because of processes going on inside it, such as changes in the salinity of the oceans, changes of sea-surface temperatures, the growth and decay of ice sheets and sea ice, the eustatic change of sea level, biological evolution, and changes in terrestrial biomass.

Geological causes

Volcanoes and lava flows

Volcanoes, by injecting dust and gases into the air, force atmospheric changes. The situation is complicated. The expectation is that violent volcanic explosions will cause a temporary reduction of global temperatures. This was the case, for example, after the eruption of Mount Agung, Bali, in March 1963. By late 1964 and early 1965, the temperature of the tropical middle and upper troposphere was depressed by 1°C (Newell 1970, 1981). Contrarily, several great eruptions, such as that of Coseguina, Nicaragua, in 1835 (Self *et al.* 1989) and that of Mount St Helens, United States, in 1980 (Kerr 1981; Deepak 1983), had no detectable effect on atmospheric temperatures. The explosions on Mount St Helens lofted roughly half the amount of material into the stratosphere as the Mount Agung eruption, but despite this, they caused virtually no long-term climatic change, only short-term effects (Robock and Mass 1982). This dilemma now appears to have been solved. The key factors in understanding the relation between eruptions and climate are the composition of the ejecta, and particularly the amount of sulphur volatiles released, and the location, time of year, and prevailing climatic conditions at the time of the eruption which determine the spread and lifetime of clouds of volcanic aerosols (Rampino *et al.* 1988; Palais and Sigurdsson 1989; Sigurdsson 1990). The estimated temperature decrease in the Northern Hemisphere following the eruptions of Mount Agung, Fuego, Mount St Helens, Katmai, Krakatau, Laki, Santa Maria, and Tambora was positively correlated with the estimated yield of sulphur (Figure 4.6). A large portion of the global climate change over the last century may be due to volcanic gas emissions (Robock 1991).

There is a suggestion that volcanic activity is, in the short term, locked into cosmic cycles. Maximum entropy methods of spectrum analysis, when applied to Lamb's Northern Hemisphere dust-veil indices from 1500 to 1968, revealed strong peaks with periods of 18.6 and 10.8 years (Currie 1994).

Without doubt, none of the historical eruptions was anything like as awesome as some of the eruptions in the geological past. Toba, in Sumatra, exploded some 73,500 years ago (Figure 4.7). This gigantic explosion is estimated to have injected some 1,000–10,000 Mt of sulphuric acid aerosols, and equal amounts of fine ash, some 27–37 km into the atmosphere (Rampino *et al.* 1988; Rampino and Self 1992). The Tambora eruption of 1815 was tame in comparison – it released a mere 100 Mt of sulphuric acid aerosols. Whereas the Tambora event would have caused a dimming of the Sun, the higher estimate for the Toba event would have led to the cessation of photosynthesis and a 3–5°C temperature drop in the Northern Hemisphere.

Explosive supereruptions the size of Toba, and even bigger, can be expected to have produced conditions analogous to those which would follow a major nuclear exchange, though volcanic aerosols have a longer residence time than smoke from fires ignited by nuclear explosions would have. The largest eruptions, termed supereruptions, may have global consequences, producing 'volcanic winters' similar to the recently proposed 'nuclear winters' (Rampino *et al.* 1985, 1988). The sheer magnitude of some past volcanic eruptions makes it tempting to speculate that supereruptions

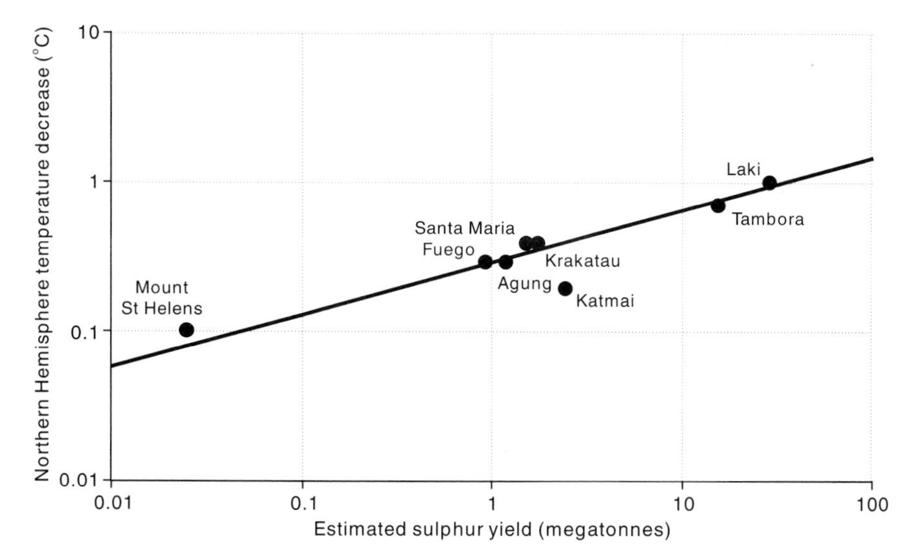

Figure 4.6 Estimated Northern Hemisphere temperature decrease plotted against estimated sulphur yield of recent volcanic eruptions
Source: After Palais and Sigurdsson (1989)

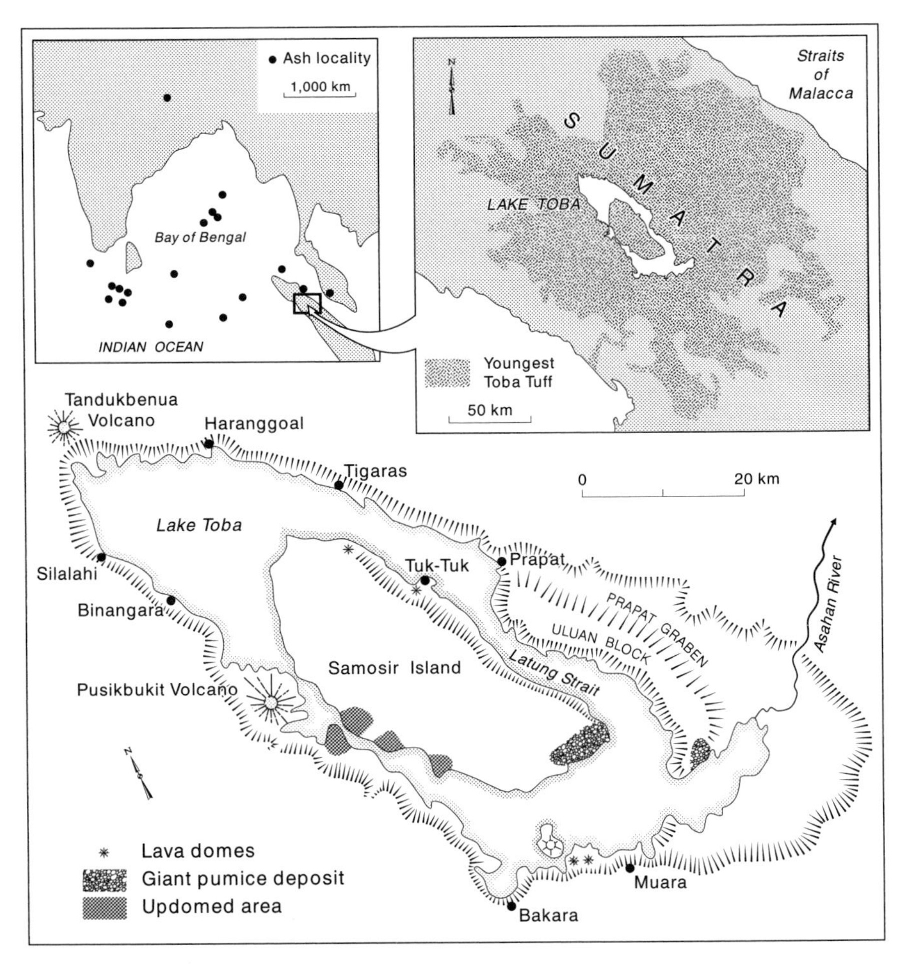

Figure 4.7 Tectonic and location map of the Toba Caldera Complex. The top left inset shows localities where ash from the Youngest Toba Tuff eruption is known to occur; the top right inset shows the present distribution of the Youngest Toba Tuff around Lake Toba on Sumatra

Source: After Rose and Chesner (1990)

(or possibly a large number of smaller eruptions) might have played a pivotal role in the initiation and timing of glacial–interglacial cycles (e.g. Gentilli 1948; Wexler 1952; Bray 1974, 1977). To be sure, Quaternary climatic transitions are times when feedback between volcanism and climate are particularly strong: falling sea levels trigger volcanic eruptions, which in turn lead to further cooling (Rampino and Self 1993). It may not be coincidental that the Toba eruption occurred at a time of rapid ice growth and falling sea level, and might have greatly helped the shift to glacial conditions that was under way (Rampino and Self 1992).

A possible link between geological heat production and climate has recently been brought to light. Large submarine lava flows, with apparent volumes exceeding 10 km^3, imaged on the bottom of parts of the Pacific Ocean, may produce thermal anomalies large enough to perturb the cyclical processes of the oceans, and could be a factor in the genesis of the El Niño phenomenon (Shaw and Moore 1988). Mid-ocean magma production would be capable of generating repeated thermal anomalies as large as 10 per cent of the El Niño sea-surface anomaly at intervals of about five years, which happens to be the mean interval of El Niño events between 1935 and 1984. Estimated rates of eruption, cooling of lava on the sea floor, and transfer of heat to the ocean surface could reasonably create a thermal anomaly comparable to that associated with El Niño.

Relief

Lofty mountain ranges, particularly those with a north–south trend, influence atmospheric circulation patterns and so induce climatic change. This has been convincingly demonstrated in a series of computer experiments that test the climatic effect of major mountain-building over the past 10–40 million years. That orographical changes might have caused global changes of climate during this time seems feasible given the active orogenic environment (Ruddiman *et al.* 1989). The Tibetan Plateau has been raised by up to 4 km over the last 40 million years and at least 2 km in the last 10 million years. Two-thirds of the uplift of the Sierra Nevada has occurred in the past 10 million years. Similar changes have taken place (and still are taking place) in other mountainous areas of the North American west, in the Bolivian Andes, and in the New Zealand Alps. The climatic effects of such orographical changes have been simulated using the Community Climate Model (Kutzbach *et al.* 1989; Ruddiman and Kutzbach 1991). The simulations showed present climatic features to result from mountain ranges (Ruddiman and Kutzbach 1989). One such feature is the Mediterranean climate in the western United States, which has dry and hot summers. This is caused by the conversion of the westerly winds that would flow over the region if the Rockies were not there, into northerly winds blowing southwards from British Columbia to the Mexican border and associated with a deepened low-pressure cell over the Colorado Plateau. On the other side of the Rockies, both seasons are presently wetter than earlier in the Cenozoic era, winter because the jet stream is forced south and winds that were formerly westerly are north-westerly, and summer because monsoon flows are created by the Colorado low-pressure centre. Some simulated changes in climate point to global orographic effects. European winters are now colder than once they were because the Icelandic low has been displaced westwards. And Mediterranean summers are now drier than once they were owing to cyclonic flow around the Tibetan Plateau and a high-pressure cell above the subtropical Atlantic.

The growth of skyscraping mountains and plateaux influences chemical weathering rates as well as airflow. Global cooling during the Cenozoic era may have been instigated by the uplift of the Tibetan Plateau. Increases in chemical weathering associated with this uplift have caused a decrease in atmospheric carbon dioxide concentrations over the last 40 million years (Raymo and Ruddiman 1992). The carbon dioxide combines with rocks during weathering and is carried to the sea.

Geography

The disposition of land and sea has a profound effect upon climate. There are at least three reasons for this. First, land provides a surface on which snow, with a high albedo, may accumulate. The more land there is around the poles, the greater the chances of glaciation occurring. Second, each arrangement of land masses produces a circulation in the atmosphere and oceans different from that found today, and thus alters the poleward transport of heat. Third, owing to the different reflective and thermal properties of land and sea, the placement and relative areas of continents and oceans will modulate climate.

The effects of geography on climate are often subtle and only just beginning to be appreciated (e.g. Barron 1989; Useinova 1989; V. P. Wright 1990). Experiments with a coupled climate–sea-ice model produce subzero summer temperatures when a supercontinent lies in polar regions (Hauglustaine and Gérard 1992). This suggests that large polar continental masses might initiate glaciations. The effects of Late Permian geography were to maintain a temperate climate in Gondwana (Yemane 1993). Atmospheric general circulation and energy-balance models suggest that the climate at the time should have been strongly seasonal. Climatic indicators gainsay this suggestion. The discrepancy may be explained by the finding that southern Africa was dominated by a series of giant lakes, up to 2,000 km long, connected by rivers. This large water body would have had a moderating influence on regional climate, and so would account for the palaeoclimatic evidence.

True polar wander

A slow drift of the geographical poles is suggested by palaeomagnetic data – true polar wandering does seem to occur (e.g. Gordon 1987). The rate at which it operates is a matter of debate, but the possibility of 5° per million years has been raised (Sabadini and Yuen 1989). True polar wander would lead to climatic change. It has been suggested that, during the Jurassic period, the position of the poles might have led to most of the continents lying in tropical latitudes (Donn 1987). This would at once account for the generally uniform distribution of temperature, and the anomalous warmth of the polar regions, from the Triassic period to the Eocene epoch. After the Eocene epoch, the meridional temperature gradient steepened because of increased cooling in Arctic and Antarctic latitudes until, by the late Cenozoic, glacial conditions had developed.

Glacial conditions have been explained by true polar wander. On a viscoelastic planet, like the Earth, true polar wander may be induced by the presence of polar ice sheets. If the poles should wander, then the ice sheets would be carried to latitudes at which they would melt. Thus true polar wander could explain the ending of ice ages. It may be that the length of a typical ice-age period (about 10 million years) is controlled by the time required for the poles to wander far enough for irreversible melting to occur and the astronomical forcing of glacial–interglacial cycles to become inoperative (Sabadini *et al.* 1983).

Cosmic causes

Solar forcing

Solar activity varies over days and years. These short-term solar cycles leave their mark in many parts of the environment (Table 4.2). For this reason, a connection between the sunspot cycle and ecospheric processes seems undeniable, though the causal links remain elusive. The details of how solar emissions affect the atmosphere are very hazy but rapid advances in aeronomy and solar-terrestrial physics, given a big fillip by the Geosphere–Biosphere Programme, are providing information on the matter (see Board on Global Change 1994).

Periods of prolonged solar minima have occurred during the last several centuries (Figure 4.8). Some climatologists argue that they tend to be associated with periods of low temperatures (Eddy 1977a, 1977b, 1977c). The Maunder Minimum, for example, coincides with the Little Ice Age, though the correlation between these two events is questionable (e.g. Landsberg 1980; Eddy 1983; Legrand *et al.* 1992). The matter is unresolved. However, recent observational and theoretical studies indicate a definite connection between periods of low solar activity and atmospheric properties. Observations of radiocarbon in tree rings strongly suggest reduced solar activity throughout the Maunder Minimum (Kocharov *et al.* 1992). On the theoretical front, a study made with a simple energy-balance climate model demonstrated that cool periods like the Maunder Minimum could have been produced by a 0.22–0.55 per cent reduction in solar irradiance (Wigley and Kelly 1990). And it is possible that the kinetic energy

Table 4.2 Climatic cycles

Forcing cycle	*Approximate period (years)*	*Examples in climatic data*
Lunar		
Lunar nodal	18.6	Air temperature records, USA (Currie 1993a) Air pressure in Japan, USSR, southern Europe, southern Africa, and South America (O'Brien and Currie 1993) Tree-ring width in Atlas cedar (*Cedrus atlantica*), Morocco (Dutilleul and Till 1992), in tree-ring chronologies from Argentina and Chile (Currie 1991a), from North America (Currie 1991b, c), from Europe (Currie 1992a), and from Tasmania, New Zealand, and South Africa (Currie 1991d)
Solar		
Sunspot	~11	Air temperature records, USA (Currie 1993a) Permafrost temperatures and snowfall, northern Alaska (Osterkamp *et al.* 1994) Annual minimum temperatures, US Gulf Coast region (Kane and Gobbi 1992) Tree-ring width in Atlas cedar (*Cedrus atlantica*), Morocco (Dutilleul and Till 1992), in tree-ring chronologies from Argentina and Chile (Currie 1991a), from North America (Currie 1991b, c), from Europe (Currie 1992a), and from Tasmania, New Zealand, and South Africa (Currie 1991d)
Hale	~22	Varve thickness in Elk Lake, Minnesota (R. Y. Anderson 1992)
Gleissberg	~80	Northern Hemisphere land and air temperatures over the last 130 years (Friis-Christensen and Lassen 1991)
Solar orbital	~180	Air temperatures since Little Ice Age (Baliunas and Jastrow 1993) Radiocarbon in tree rings over last 9,000 years (Sonett 1991)
Bimillennial	~2,000	Radiocarbon in tree rings over last 9,000 years (Sonett 1991; Damon and Jirikowic 1992)
Orbital		
Tilt	41,000	Oxygen isotope and dust flux in Atlantic DSDP Site 659 during Pliocene (Tiedemann *et al.* 1994)
Precession	19,000 and 23,000	Magnetic susceptibility variations in eastern tropical Atlantic and Arabian Sea deep-sea cores (Bloemendal and DeMenocal 1989), and in loess deposits at Halfway House, Alaska (Begét and Hawkins 1989)
Short eccentricity and orbital plane inclination	100,000	Diatom temperature records at western North Pacific DSDP sites 579 and 580, 0–780,000 years ago (Koizumi 1994)
Long eccentricity	400,000	Diatom temperature records at western North Pacific DSDP sites 579 and 580, 0–780,000 years ago (Koizumi 1994)

of the Sun increased during the Maunder Minimum, leading to a reduction in solar radiation (Nesme-Ribes and Mangeney 1992). Solar cycles lasting millennia might explain all the climatic cycles that have occurred since the start of the Pleistocene epoch (Willett 1949, 1980).

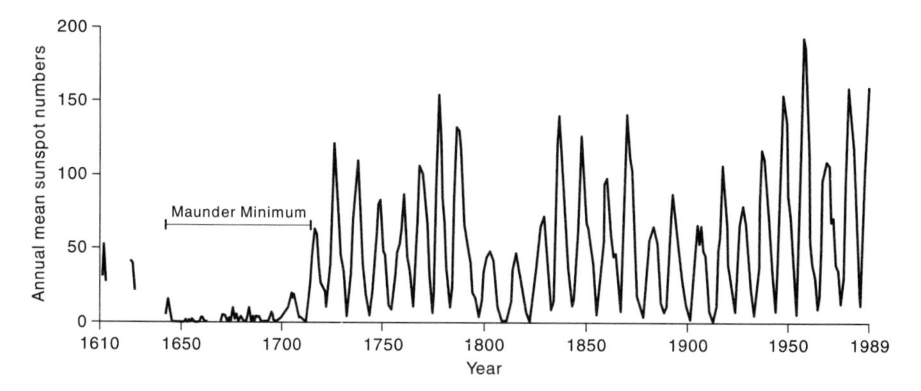

Figure 4.8 Annual mean sunspot numbers, 1610–1989. The period between about AD 1645 and 1715, when the Sun seems to have been quiet and sunspots to have been scarce, is called the Maunder Minimum. However, this dearth of sunspots might reflect poor historical records. The 11-year sunspot cycle and the 80-year Gleissberg cycle may be picked out in the post-1800 data

Possible changes in solar activity over geological time-spans might have influenced climate. The Sun may from time to time change state and energy emissions vary accordingly. A temporary reversion of the Sun to an earlier, cooler state may explain glaciations of the last 500 million years, and the 'flickerings' of a disturbed Sun may have induced glacial–interglacial cycles (Öpik 1958a, 1958b).

Gravitational forcing

The strong, 18.6-year beat given by the lunar nodal precession cycle appears to exert an influence on the atmosphere. It is present in many atmospheric variables (Table 4.2).

The Earth's orbital cycles wield a considerable influence over climate (Table 4.2). They do so by changing seasonal and latitudinal patterns of solar radiation receipt, and possibly by altering endogene processes. Climatic change during the Pleistocene and Holocene epochs appears to have been driven by orbital variations in the Croll–Milankovitch frequency band. Orbital forcing has led to climatic change in middle and high latitudes, where ice sheets have waxed and waned, and to climatic change in low latitudes, where water budgets and heat budgets have marched in step with high-latitude climatic cycles. The 100,000-year cycle of eccentricity is recorded in Quaternary loess deposits, sea-level changes, and oxygen isotope ratios of marine cores. The precessional cycle (with 23,000 and 19,000 year components) and the 41,000-year tilt cycle are superimposed on the 100,000-year cycle. They, too, generate climatic changes that are registered in marine and terrestrial sediments. Oxygen isotope ratios in ocean cores normally contain signatures of all the Earth's orbital cycles, though the tilt cycle, as it affects seasonality, is more pronounced in sediments deposited at high latitudes.

Variations in orbital parameters do not explain all aspects of Quaternary climatic change. They do not explain why the Quaternany ice age began. No more do they explain why glacial stages occurred at the same time in the Northern and Southern Hemispheres, nor why the observed strong beat of the 100,000-year short cycle of eccentricity is so strong. The recent 'discovery' of a 100,000-year cycle of the Earth's orbital inclination may solve the last of these enigmas (Muller and MacDonald 1995).

The Solar System's passage about the centre of the Galaxy may account for the approximately 150-million-year pulse of glaciations (e.g. Steiner 1967, 1973, 1978, 1979; Steiner and Grillmair 1973; G. E. Williams 1975a). Exceedingly long intergalactic cycles may influence terrestrial processes. The tidal action of the Large and Small Magellanic Clouds may modulate the properties of the interstellar medium, or the energy output of the Sun (or both), and have far-reaching climatic consequences (G. E. Williams 1975a). It is just possible that the pole of the ecliptic and the plane of the Solar System may track the Large Magellanic Cloud in its orbit around the Galaxy, thereby causing secular changes in the Earth's axial tilt and consequent very long-term changes in the Earth's palaeoclimates and tectonism (Table 4.3).

Bombardment

Clouds of fine dust raised by large impact events would stay in suspension for months or years and spread globally. Impacting asteroids or comets with energies over about 600,000 Mt should produce a dust cloud with a global equivalent optical depth of about 3 (Covey *et al.* 1990). During the first two

Table 4.3 Axial tilt and Earth's climate

Obliquity (°)	*Ratio of annual radiation of either pole to the insolation at the equator*[a]	*Latitude of Arctic and Antarctic Circles*	*Latitude of Tropics of Cancer and Capricorn*	*Seasonality*	*Dominant latitudes of glaciation*[b]
0	0	90	0	Nil	Very high (if any)
23.50	0.4278	66.50	23.50	Moderate	High
54	1.0	36	54	Strong	All (?)
90	1.5708	0	90	Very strong	Low to middle

Notes: [a] Croll (1875); Milankovitch (1930)
[b] G. E. Williams (1975b) except 0° obliquity

weeks after the impact, global average land temperature would fall by about 8°C. The effects would diminish as the dust cloud spread, so diluting its power. Impacts with energies above about 5,000,000 Mt would block sunlight sufficiently to curtail photosynthesis, while those with energies exceeding about 10,000,000 Mt would cause intense darkness and severe and long-lasting cooling, especially in continental interiors. In addition, large injections of dust might produce world-wide drought, with precipitation decreasing 95 per cent for several months (Toon *et al.* 1994). After a spell of cold and dry weather, long-term warming would probably occur. It would result from the rise of atmospheric carbon dioxide levels following a continental or an oceanic impact (O'Keefe and Ahrens 1989), and from the rise of atmospheric water-vapour content following oceanic impact (Roddy *et al.* 1987). The extra water vapour and cloud in the stratosphere would, in the manner of a very efficient greenhouse, generate a large rise in global temperature, which at the surface might exceed 10°C.

Work with atmospheric general circulation models shows that it is difficult to freeze large portions of the Earth's surface for more than a few months, largely owing to the ameliorating effect of heat stored in the oceans. Presumably impacts of asteroids and comets tens of kilometres in diameter would lift enough dust to induce global freezing, but it is unclear whether this has happened since the heavy bombardment in early Earth history (Covey *et al.* 1990). It is possible that such enormous impacts, should they occur, might trigger an ice age (Clube and Napier 1982).

Internal causes

Short-term dynamics

El Niño–Southern Oscillation events are driven by short-term instabilities in the atmosphere–ocean system. Every few years, exceptionally warm waters appear off Ecuador and Peru and extend far westwards in equatorial regions. This is an El Niño event. It is associated with an air-pressure fall over much of the south-eastern Pacific Ocean, and an air-pressure rise over Indonesia and northern Australia. When cold water returns to the western South American seaboard – a La Niña event – the pressure gradient reverses over the Pacific and Indian Oceans. This flip-flopping of pressure is called the Southern Oscillation. The combined oceanic and atmospheric changes are the El Niño–Southern Oscillation, sometimes called ENSO events. ENSO events affect climates world-wide (Bjerknes 1969; Ropelewski and Halpert 1987). They do so by causing wavelike disturbances of air flow in the upper troposphere that extends into mid-latitudes. Mid-latitude mountains tend to amplify these wave disturbances (DeWeaver and Nigam 1995). The effects are not always the same for each event. The intervals between strong El Niño events range from 2 to 7 years and average 4–5 years. This unusual variability may result from the interaction of two pulses in the Pacific equatorial winds, one with a 2-year beat and the other with a 4–5-year beat (Rasmussen *et al.* 1990). A strong El Niño event may occur when these two cycles are in phase, and a weak El Niño event when they are out of phase.

Correlations between climate and the sunspot cycle might be spurious. This possibility was raised by running a simple atmospheric circulation model with nothing longer than annual forcing (James and James 1989). Although the forcing was annual, the non-linear dynamics of atmospheric flow produced variability with periods in the range 10–40 years.

Instabilities in the atmosphere–ocean system may also cause abrupt climatic changes within Pleistocene glacial and interglacial stages. These changes, which include Dansgaard–Oeschger cycles and Heinrich events, are discussed in the next section.

Medium-term dynamics

Greenhouse gases and atmospheric dust may have contributed to glacial–interglacial temperature changes. The biggest contributor is probably carbon dioxide. Reliable measurements of past carbon dioxide levels in the atmosphere come from air bubbles entrapped in Arctic and Antarctic ice cores. Carbon dioxide levels in the Vostok ice core, Antarctica, vary from between 190 and 200 ppmv during a glacial stage, to between 260 and 280 ppmv during an interglacial stage (Barnola *et al.* 1987). Carbon dioxide changes, which might themselves be partly driven by orbital variations, could magnify orbitally induced insolation changes and bring a switch of climate. This would help to explain the synchroneity of major glaciations in the Northern and Southern Hemispheres.

Climate models have clarified the role of carbon dioxide in climatic change. A three-component, dynamic model governing global ice mass, atmospheric carbon dioxide content, and mean ocean temperature, possesses solutions that replicate the major features of climatic variations implied by the oxygen isotope record for the last 2 million years (Saltzman and Sutera 1984, 1987; Saltzman *et al.* 1984; Saltzman 1987). The model represents only internal dynamics: no orbital forcing is prescribed. Climatic variations replicated include a major 'transition' between a low-ice, low-variance mode before roughly 900,000 years ago to a high-ice, high-variance mode with almost a 100,000-year period from 900,000 years ago to the present. The chief features exhibited by the ice record derived from oxygen isotope data, and the atmospheric carbon dioxide record derived from the Vostok ice core – including a rapid deglaciation during which a spike of high carbon dioxide and a rapid surge in North Atlantic deep water production occurs – can be deduced as a free oscillatory solution of the model (Saltzman and Maasch 1988). The remaining variance is likely to result from external (orbital) forcing (Saltzman and Maasch 1990). If the physical aspects of the model should be correct, then the ice ages prevailing over the last 2 million years result from a sensitive dynamical balance between water and ice (Saltzman and Sutera 1987). A critical aspect of the balance is that the oceans are deep, with a large thermal and chemical capacity, and exercise some control on the carbon dioxide content of the atmosphere. Positive feedbacks involving carbon dioxide introduce a fundamental instability into the system that drives the ice mass variations. The 100,000-year climatic cycles during the Pleistocene epoch were possibly forced by the downdraw of atmospheric carbon dioxide, perhaps because of the weathering of rapidly uplifted mountains, to levels low enough for the 'slow climatic system', which includes the mass of glacial ice and the state of the deep ocean, to become unstable (Saltzman and Maasch 1990). In addition, ice-calving catastrophes associated with bedrock variations might have led to oscillations of around 40,000 years that were independent of tilt-cycle forcing (Saltzman and Verbitsky 1993; see also Birchfield and Ghil 1993).

QUICK FREEZE: RATES OF CLIMATIC CHANGE

How fast does climate change? This question has taken on a new significance with global warming threatening. Past climates provide some answers. The consensus is now that changes may be alarmingly swift, possibly within a decade or two. Evidence of rapid climatic change comes mainly from the Pleistocene and Holocene epochs, though geologically fast changes are recorded in much older rocks.

Pleistocene pulses

The terrestrial and marine records both attest to long periods of glacial expansion (climatic cooling) abruptly ended by rapid deglaciations (climatic warming) throughout the Pleistocene epoch. The Younger Dryas termination, which marks the end of the last glacial stage, is a splendid example of such abrupt climatic mode switches. Other abrupt climatic changes are recorded during the Pleistocene. These include the stadial–interstadial climatic shifts, known as Dansgaard–Oeschger cycles, that occurred during the last glacial period, and the fast climatic shifts that occurred during the last interglacial involving speedy, short-lived climatic deteriorations. Orbital forcing is unlikely to drive these swift climatic changes. A more likely causal candidate is internal instabilities in the atmosphere–ocean system.

Orbital forcing since the last glacial maximum

Before considering rapid Pleistogene climatic changes, it may be helpful to look at the relatively gradual changes driven largely by slow variations in the Earth's orbit and orientation.

The configuration of the Earth's orbit and orientation 18,000 years ago, at the time of Last Glacial Maximum, was similar to its present configuration. From 18,000 to 9,000 years ago, the perihelion date changed from January to July, and the axial tilt increased. From 9,000 years ago to the present, axial tilt has decreased, and the date of perihelion has advanced through the second half of the year, so that it now occurs during northern winter again (Kutzbach and Webb 1993). These changes have altered the seasonal cycle of solar radiation receipt (Figure 4.9). From 12,000 to 6,000 years ago, summer insolation was considerably greater than today in the northern summer. At its peak around 9,000 years ago, northern summer insolation was about 30 W/m^2 (8 per cent) higher than today, and winter insolation was lower by about the same amount (Figure 4.9). At the same time in the Southern Hemisphere, the seasonality of solar radiation was less marked than today. The increased northern seasonality 9,000 years ago was exacerbated by the contrasting heat capacities of land and water. North America, North Africa, and Eurasia have relatively smaller heat capacities

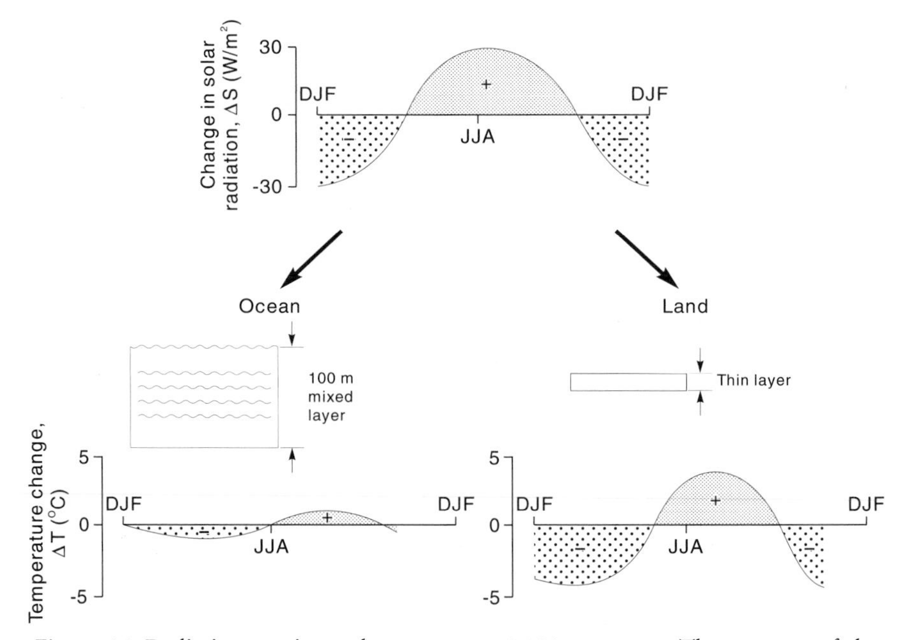

Figure 4.9 Radiation receipt and temperature 9,000 years ago. The top part of the diagram shows the seasonal change in Northern Hemisphere solar radiation receipt, ΔS, compared with the present. The lower part of the diagram depicts schematic temperature responses in ocean and on land to the same change in solar radiation. DJF is December, January, February; JJA is June, July, August

Source: After Kutzbach and Webb (1991)

than the North Pacific and North Atlantic Oceans. These differences led to a larger land-surface temperature response to seasonal radiation receipt. Northern continents were probably about 5°C warmer than at present in summer and the oceans only about 1°C warmer. In addition, the seasonal ocean warming would have lagged about a month behind the land warming. Northern winters were correspondingly colder than at present.

The effect of the northern land–ocean temperature contrast in summer was to increase the air temperatures over land, to increase advection over land, to lower pressure over land, and to boost the convective circulation of air from land to ocean at high levels and from ocean back to land at low levels. The more vigorous airflow increased the summer monsoon. Conditions along coasts and for some distance inland were moister. Far inland, in those parts that the increased moisture flux could not reach, conditions were drier owing to raised evaporative losses in the higher temperatures. Similarly, monsoon circulations blowing from land to ocean at low levels in the northern winter were also intensified. However, the net annual moisture (precipitation minus evaporation) was greater. This is because the increase in summer moisture outweighed any decrease in winter moisture.

These orbitally induced changes to monsoon circulations were modified by the size of continents (Figure 4.10). The maximum temperature response occurred in continents large enough to have interiors lying beyond the moderating influence of airstreams from the cooler ocean. For this reason, monsoon circulations 9,000 years ago were more responsive to changed insolation receipt in North Africa and Eurasia than in North America. Another complicating factor in North America was the effect of residual ice patches, which lingered till about 9,000 years ago (Figure 4.11).

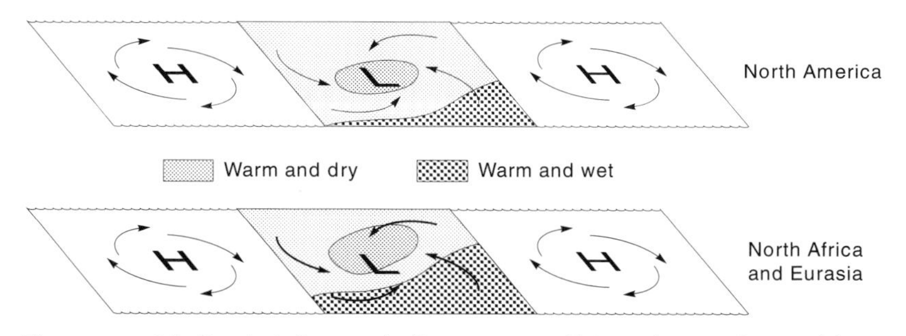

Figure 4.10 Idealized airflow and climate on small (North America) and large (North Africa and Eurasia) continents 9,000 years ago. The conditions differ from present conditions owing to the intensified northern summer (June, July, and August) monsoons produced by orbitally induced changes of solar radiation. The monsoon circulation is more intense, and the temperature increase in the interior greater, on the larger continent. H is high pressure; L is low pressure
Source: After Kutzbach and Webb (1991)

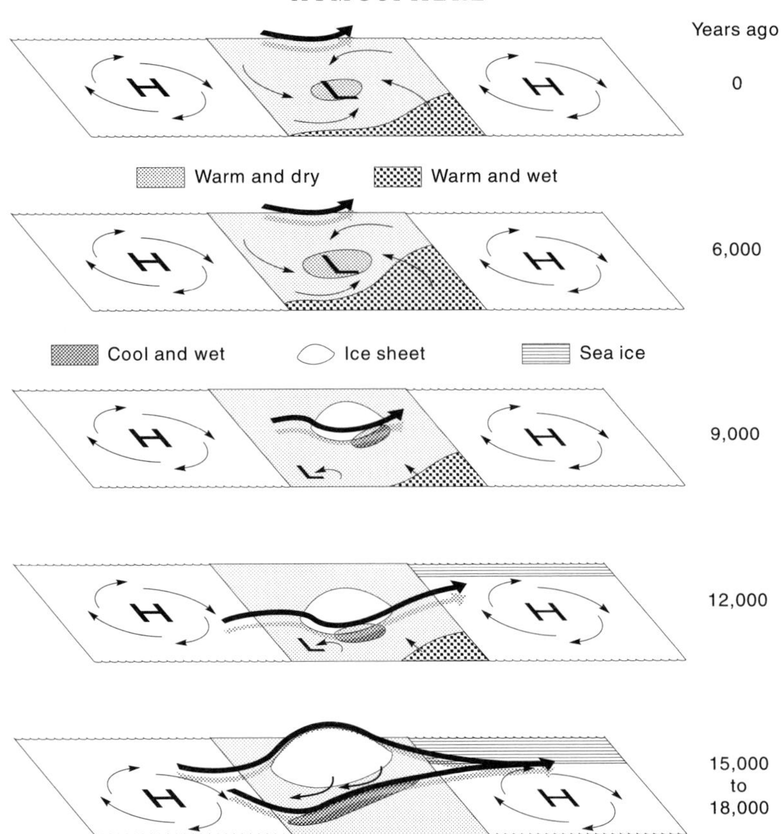

Figure 4.11 Idealized changes in the summertime circulation over North America from 18,000 years ago to the present. Between 18,000 and 15,000 years ago, the circulation of air was strongly influenced by the large North American ice sheet and by extensive sea ice in the North Atlantic. The jet stream split into two limbs over North America and extended far downstream over the Atlantic Ocean. The southern limb was shifted far south. It was associated with frequent cyclonic storms. Along the southern margin of the ice sheet, conditions were generally dry under a dominating glacial anticyclone. By 12,000 years ago, the ice sheet was smaller, sea ice less extensive, and the jet stream weaker and positioned along the southern ice margin, to which it brought storms and precipitation. Increased summertime radiation, resulting from orbital changes, helped to establish warmer conditions to the west of the ice sheet, a low-pressure centre in the south-west, and increased precipitation in the south and south-east. By 9,000 years ago, the ice sheet was very small and the summertime increase in solar radiation had an effect. By 6,000 years ago, the northern summer monsoon over land and anticyclonic circulation over the oceans (both brought about by orbital changes) were stronger than at 9,000 years ago (when the small, lingering ice sheet moderated orbital effects), and they were also stronger than today. The interior of the continent was warmer than at present, and the region of increased moisture extended further into the continent on the southern and eastern margins. Present summertime climatic features are not as marked as they were 6,000 years ago

Source: After Kutzbach and Webb (1991)

Younger Dryas

The Younger Dryas was the most recent millennium-long cold period. It lasted from about 12,000 to 11,000 years ago. Cold conditions characterized north-west Europe (where the tundra plant *Dryas* expanded its range) and eastern North America. Some evidence points to global cooling: postglacial cooling in Antarctica appears to have halted during the Younger Dryas, New Zealand's mountain glaciers advanced considerably, planktonic communities in the South China Sea changed, and the atmospheric methane level fell by 30 per cent. The termination of this event appears to have been rapid. Ice core data from Greenland suggest that it happened very fast indeed – oxygen isotope ratios indicate that it ended within about seventy years; dust and deuterium excess (proxies for winds and sea-surface conditions) indicate that it took less than twenty years; and the accumulation rates of snow indicate that it happened in just a few years (Dowdeswell and White 1995).

The last glaciation

During the last glaciation, oxygen isotope ratios display twenty-four alternations between relatively high and low values (Grootes *et al.* 1993). Each high–low interval lasted between several hundred and a few thousand years and involved variations of 4–6 per thousand, implying a temperature change of 7–8°C. The higher values correspond to stadials when full glacial conditions prevailed. They suggest temperatures 10–13°C lower than during the Holocene. The lower values testify to warmer interstadials. These lasted some 500–2,000 years. The switch from stadial to interstadial climates was remarkably quick, perhaps taking places in as little as a few decades (Johnsen *et al.* 1992). The return to stadial conditions was less abrupt, commonly consisting of a gradual cooling followed by a more rapid slide into stadial conditions (Grootes *et al.* 1993). The repeated alternations of stadial and interstadial climates, which display a saw-tooth pattern (fast warming and more gradual cooling), are called Dansgaard–Oeschger cycles.

The sudden shifts to warmer conditions during the last glaciation may be associated with the periods of maximum North Atlantic iceberg production (Bond *et al.* 1993). During the last glaciation, several rapid episodes of iceberg production, debris rafting, and sediment deposition – Heinrich events – are recorded in North Atlantic sediments (Heinrich 1988; Dowdeswell *et al.* 1995). These events appear to have involved the rapid discharge of icebergs, and the melting out and sedimentation of debris held within them, probably from an ice stream lying within the Hudson Strait and draining much of the central Laurentide ice sheet. Detailed studies of the last two events, based on analysis of more than fifty North Atlantic cores, indicate that the most likely duration of a Heinrich event is 250–1,250 years (Dowdeswell *et al.* 1995).

The Eemian interglacial

Ice cores from Greenland – the Greenland Ice-Core Project (GRIP) and Greenland Ice Sheet Project 2 (GISP2) cores – reveal a series of climatic 'mode switches': cold snaps alternate with warm spells, each lasting 70–5,000 years (e.g. Figure 4.12a). These climatic changes are suggested by the records of stable isotopes (oxygen), atmospheric dust (as measured by calcium content), methane content, and electrical measures. The transitions between cold and warm states took place in as little as a decade, and involved a lowering or raising of temperature of up to 10°C.

Marine isotope stage 5e (MIS-5e) in the GRIP Summit core contains a spectacular series of climatic oscillations (Figure 4.12c–d). Details of two such oscillations show the enormous speed with which large climate shifts have taken place. 'Event 1' took place around 115,000 years ago, at the culmination of the Eemian interglacial. Oxygen isotope levels rose to mid-glacial levels, acidity fell rapidly, and atmospheric dust content shot up. The event appears to have lasted about seventy years (given a calculated annual ice-layer thickness 2.5 mm). 'Event' 2 is one of a lengthy series of huge and sustained oscillations that characterized the first 8,000 years of the Eemian interglacial, and the end of the previous deglaciation, sequence. Stable isotope values rose to Younger Dryas levels and were held there for about 750 years. Oxygen isotope data suggest temperature decreases of 14–10°C!

Pre-Quaternary shifts

Changes in geological climates over the last 570 million years have fluctuated considerably with quasi-periods of ~150 million years (warm–cool mode cycle) and ~300 million years (hothouse–icehouse cycle) (Figure 4.13). Mean global temperatures display periods of relative stability characterized by gradual and minor changes, and occasional bouts of more rapid change. The switching of warm and cool climatic modes during the Phanerozoic aeon will be discussed in the next section, but relevant here is the similarity between rates of warm-mode–cool-mode switching and rates of glacial–interglacial switching. As with interglacial terminations, warm-mode terminations are gradual and geographically variable; and as with the onset of interglacial conditions, the onset of warm modes is sudden and somewhat puzzling – it is difficult to see why some cool modes ended (Frakes *et al.* 1992: 195).

Palaeoclimatic indicators provide considerable details of change in geological climates, at least in parts of the stratigraphical column, and reveal times of relatively sudden shifts. For instance, in the Late Permian

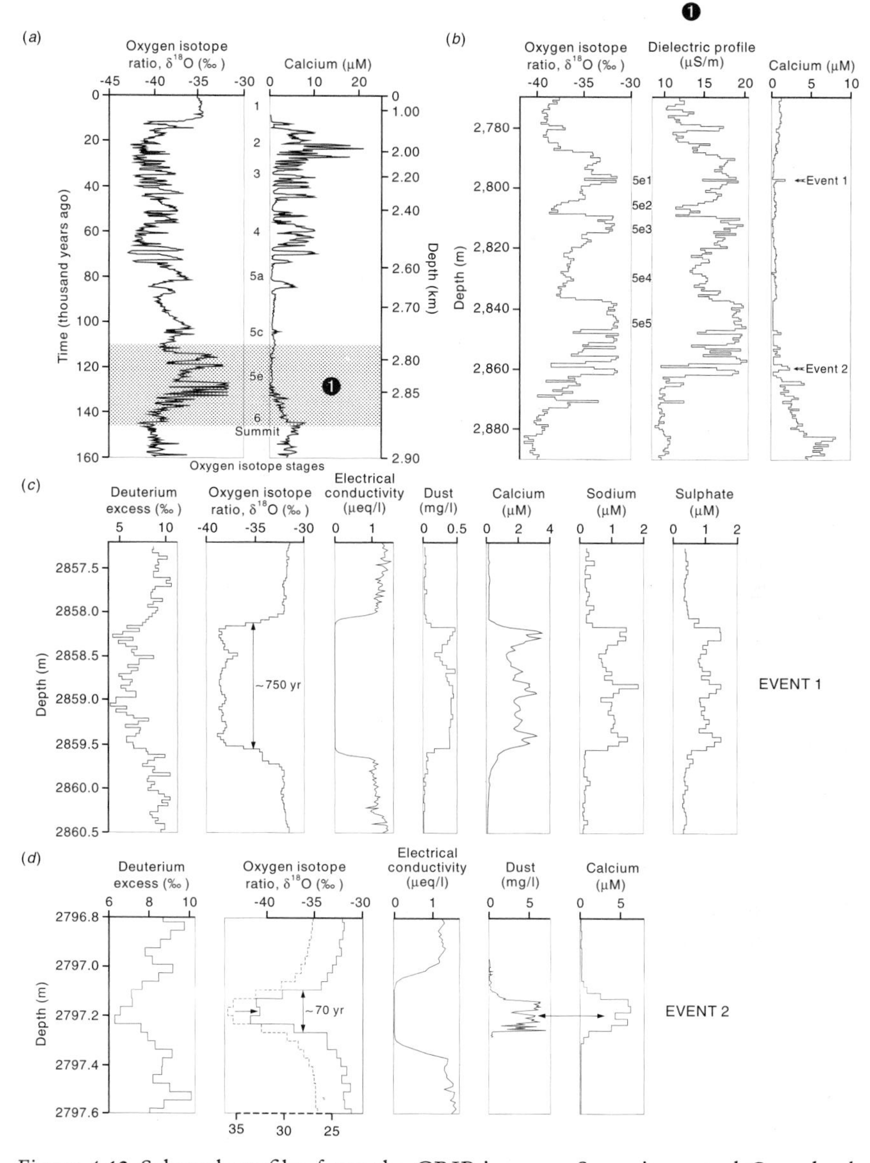

Figure 4.12 Selected profiles from the GRIP ice core, Summit, central Greenland. (a) Oxygen isotope ratio and calcium ion concentration (200-year means). (b) Enlargement of shaded timeslice 1 in (a), with a dielectric profile added. This corresponds to the last interglacial (Eemian). Three main warm periods (5e1, 5e3, and 5e5) are separated by sustained cool periods (5e2 and 5e4). Two extreme cooling events are indicated. (c) High-resolution profiles through 'Event 1'. (d) High-resolution profiles through 'Event 2'. Notice that dust concentrations increase during cold spells, suggesting drier conditions

Source: After Greenland Ice-Core Project (GRIP) Members (1993)

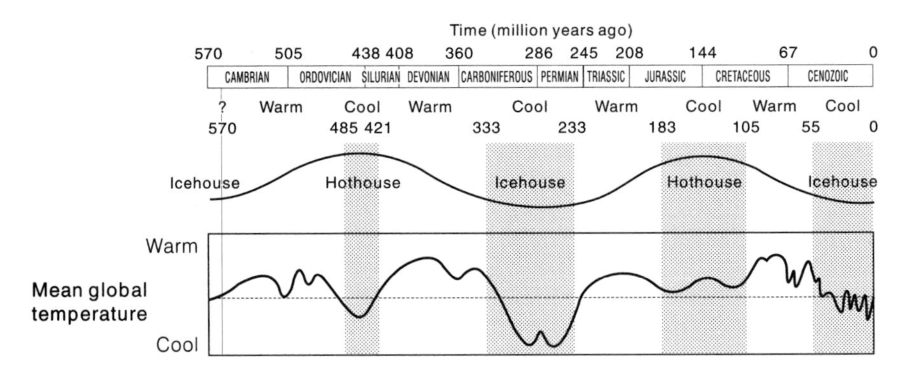

Figure 4.13 Climatic megacycles during the Phanerozoic aeon. The hothouse–icehouse cycle has a period of about 300 million years; the warm-mode–cool-mode cycle has a period of about 150 million years. Notice that the two cycles 'conflict' in places. For instance, cool modes occur during the hothouse phase prevailing during Late Cambrian, Ordovician, Silurian, and Devonian times, and during the hothouse phase prevailing during Jurassic, Cretaceous, and Late Eocene times

Sources: Hothouses and icehouses after Fischer (1981, 1984a); warm and cool modes after Frakes *et al.* (1992); generalized temperature curve after Martin (1995)

period, coal very abruptly becomes far less common in the stratigraphic record, and stays uncommon until the Middle Triassic period (Faure *et al.* 1995; Retallack *et al.* 1996). This large reduction in global coal production is associated with a decrease of atmospheric and oceanic carbon-13. The anomalous changes may have resulted from the deformation and uplift of coal-bearing foreland basins around the margins of Pangaea, which then were all under compressional stress (except for the southern shores of Tethys). Once exposed to the air, the vast peat deposits would have oxidized, so effecting a massive flux of carbon dioxide into the atmosphere and forcing global warming.

HOTHOUSES AND ICEHOUSES: DIRECTIONS OF CLIMATIC CHANGE

The atmosphere probably originated by degassing through volcanoes. It may have accumulated slowly over geological time (Rubey 1951), or rapidly, within 2 billion years after planetary accretion (Fanale 1971). The rapid-growth interpretation is called the 'big burp' hypothesis. Atmospheric concentrations of hydrogen and carbon dioxide display a secular decrease, while those of oxygen and nitrogen display a secular increase. The overall state of the atmosphere has, broadly speaking, passed through four distinct stages that correspond very approximately to the Priscoan, Archaean, Proterozoic, and Phanerozoic aeons. Several processes – some cosmic, some geological, some biological – may account for these directional transitions. Biological evolution, large changes of obliquity, the increasing luminosity of the Sun, and the Earth's spin rate all seem particularly important in explaining very long-term, directional changes in the atmosphere. Since Precambrian times, the composition and temperature of the atmosphere have held fairly steady, though considerable fluctuations about a mean condition have occurred that express themselves as climatic swings from warm to cool modes and from hothouse to icehouse states (Figure 4.13).

Biological effects

Release of oxygen as a waste product of early photosynthetic bacteria is the key to understanding the early Proterozoic transition from an anoxic to an oxic atmosphere. The oxygen titrated the primitive, reduced oceans and was responsible for the deposition of banded iron formations. Then, about 2 billion years ago, it began to accumulate in the atmosphere, doubling its concentration every 250 million years up to about 250 million years ago when saturation point was reached (Worsley and Nance 1989).

Carbon dioxide has been withdrawn from the atmosphere through Earth history. The decrease in atmospheric carbon dioxide concentration is about a hundredfold in the past 3 billion years, with a halving time of about 0.4 billion years (Worsley and Nance 1989). The carbonate–silicate cycle accounts for about 80 per cent of this carbon dioxide downdraw, while photosynthesis and the burial of organic carbon currently account for 20 per cent. The rate of decrease was especially fast between about 350 and 300 million years ago (Figure 4.14).

A brighter Sun

The slow increase in solar luminosity through geological time creates a climatic paradox. Energy-balance models and radiative-convective models predict that if the solar constant were to fall just a few per cent, then, as a result of the feedback between ice cover and albedo, the Earth would be totally covered with ice (Figure 4.15). A world-wide glaciation did not occur until the Huronian glaciation (about 2.35 billion years ago), and other factors, particularly carbon dioxide and methane levels, must have counteracted the effect of low solar luminosity during Precambrian times (Gérard 1989). There seems little doubt that volcanic activity in the Archaean aeon, which ended about 2.5 billion years ago, would have been much greater than today with carbon dioxide emissions possibly as much as a hundred times their present values. The greenhouse effect produced by such enormous emissions of carbon dioxide might explain why glaciation is unknown until later in Precambrian times (Wyrwoll and McConchie 1986). However, as oceans can freeze faster than carbon dioxide can accumulate in

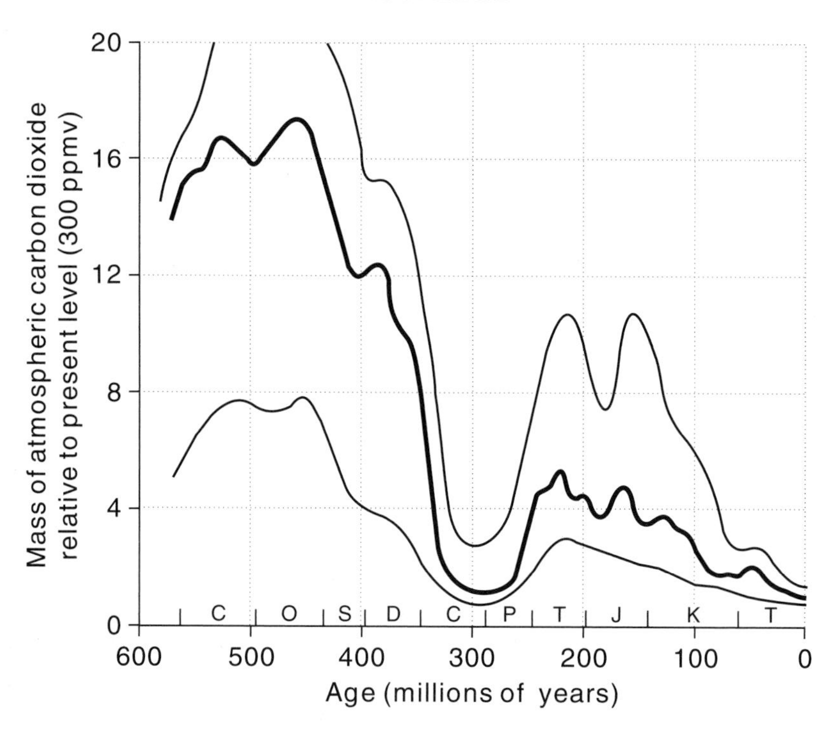

Figure 4.14 Carbon dioxide levels over the last 600 million years as predicted by GEOCARB II, a store-and-flux simulation model of the long-term geochemical carbon cycle. The steady decline in atmospheric carbon dioxide levels is due to the increasing solar constant. Fluctuations about this steady decline reflect geological and topographical factors. The high Mesozoic carbon dioxide levels are due largely to low relief prevalent at the time, whilst the low Cenozoic levels result mainly from mountain uplift and decreasing metamorphic and volcanic degassing of carbon dioxide
Source: After Berner (1994)

the atmosphere, another cause must be sought for a lack of early global refrigeration (Caldeira and Kasting 1992). A dense atmosphere of carbon dioxide, nitrogen, and water vapour may have remained hot after the accretion of the planet and prevented the growth of dense carbon dioxide clouds that would have reflected much sunlight and, paradoxically, initiated an irreversible glaciation. Alternatively, additional greenhouse gases, such as ammonia and methane, might have kept the atmosphere warm enough to prevent condensation of carbon dioxide. Experiments with a seasonal energy-balance climate–sea-ice model indicated that, if solar luminosity were to drop more than about 5 per cent below its present level, ice–albedo feedback would cause irreversible glaciation, though this value was decreased by nearly 10 per cent when continents were not near the poles

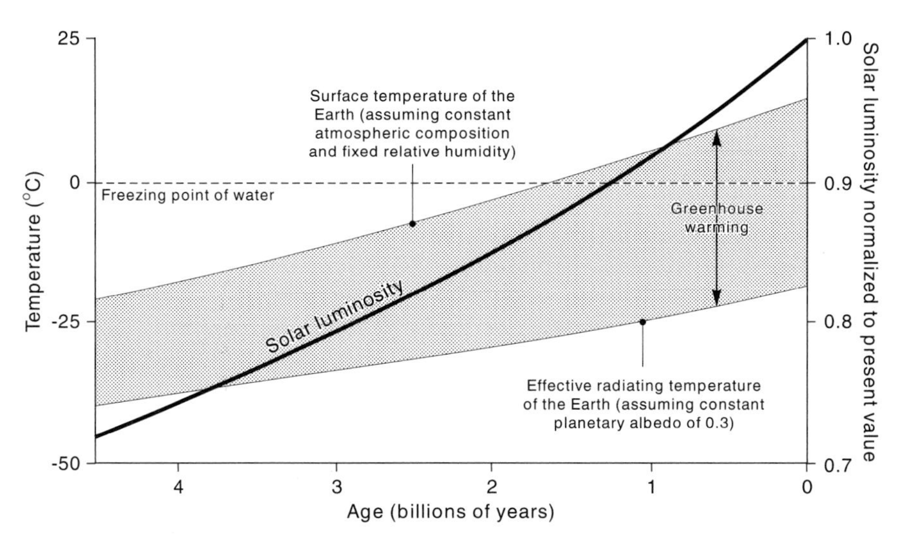

Figure 4.15 The estimated increase in solar luminosity through geological time, and its effect upon the Earth's surface temperature. The bottom of the shaded region is the Earth's surface temperature with no greenhouse gases in the air. The top of the shaded region is the Earth's surface temperature with greenhouse gases present. The difference between the two is the degree of greenhouse warming. Current concern over the increasing burden of greenhouse gases in the atmosphere is understandable, but the diagram shows how lucky it is for the biosphere that some greenhouse warming does occur!
Source: After Condie (1989a, 1989b)

(Gérard *et al.* 1992). They also showed that a 30 per cent fall in solar output requires a 20,000-fold increase in atmospheric carbon dioxide to prevent global refrigeration. This figure is plainly unrealistically high, even for the Archaean aeon. However, with no continents, a 2,000-fold increase is required to stabilize the climate. It seems possible, therefore, that reduced continental area, palaeogeographical changes, and higher carbon dioxide levels combined to prevent the Earth from becoming an iceball planet.

Rotation rate

Changes in the Earth's rotation rate appear to have caused climatic changes (Whyte 1977). Periods of spin acceleration are associated with climatic amelioration, while periods of spin deceleration are associated with climatic deterioration. Spin minima were preceded by ice ages with strong seasonal and zonal climatic contrasts, while spin maxima were preceded by climatic optima with warm and equable climates. However, the link between climate and spin velocity is not direct. Spin velocity may affect climate through its effect on sea level. There is much evidence suggesting that tectono-eustasy is largely controlled by the volume of oceanic ridges and thus the spreading rate of crustal plates, and it is a distinct possibility that tidal and rotational forces would influence plate tectonic mechanisms. Increased plate generation leads to a greater volume of oceanic ridges, an upping of sea level, and an amelioration of climate; it also appears to coincide with an acceleration of mantle-spin velocity.

The effects of increased rotation rate on climate have been investigated using general circulation models. One model set all radiative terms at current values and boosted the rate of rotation fivefold, a situation representative of Precambrian times (B. G. Hunt 1979). The fivefold increase in rotation rate greatly reduced the extent and intensity of the Hadley cell, almost halved the intensity of mean zonal winds, shifted the westerly jet stream in the troposphere to latitude 10°, and all but eradicated the stratospheric westerly jet. The upshot was a reduced poleward transport of sensible and latent heat, owing to the lack of dominant large-scale eddies, and thus warmer low latitudes and colder high latitudes than at present. The climatic importance of the Earth's slowing rotation rate was confirmed in a later simulation (Kuhn *et al.* 1989; see also Jenkins 1996). The time evolution of the Earth's surface temperature was predicted from known histories of solar luminosity, atmospheric carbon dioxide content, rotation rate, and continent formation. Mean surface temperatures were 5°C higher than today through much of geological time, and the variation of mean surface temperatures never exceeded 15°C. Perhaps surprisingly, the Earth's rotation rate had a considerable effect upon surface temperature distribution as late as 500 million years ago. For example, 3.5 billion years ago, the much faster rotation rate made little difference to equatorial temperatures relative to their present value, but it depressed polar temperatures by 15°C. Indeed, the model indicated that rotation rate has wielded a major influence on surface temperature for much of Earth history, and that continental growth and albedo changes have played a secondary role.

Large changes of obliquity

Any departure of the axial tilt from its present posture will cause a change in the world climatic zones, and a consequent change in the seasonal swing of climate (Table 4.3). The climatic conditions associated with these different obliquities have long been a matter of conjecture. Palaeomagnetic data suggest that the Neoproterozoic glaciation, which occurred 1,000–540 million years ago, saw ice sheets in low latitudes (Harland and Bidgood 1959; Schmidt *et al.* 1991; G. E. Williams *et al.* 1995). This curious finding is explicable by a global glaciation (Harland 1964), dubbed the 'snowball Earth hypothesis' (Kirschvink 1992), or else by a large obliquity (G. E. Williams 1993).

Experiments using atmospheric general circulation models indicated the likely consequences of increased and decreased tilt (B. G. Hunt 1982). For an obliquity of 0°, where the rotatory axis stands bolt upright, the simulated climate was more vigorous than present climate, drier and colder by about 10°C at the surface at high latitudes, slightly warmer by about 1°C in the tropics, with an overall slight warming of the troposphere and an overall cooling of the stratosphere at all latitudes (cf. Barron 1984). Latitudinal precipitation rates were almost identical to present rates, but evaporation rates reflected changes in mean temperatures, being a little higher in the tropics and lower poleward of latitudes 45° N and S. The response to a 0° obliquity was thus complex, the results implying that the Earth's climatic zones would not change much from their present arrangement, but that the climate at very high latitudes would be much harsher in view of the greatly reduced temperatures, so causing the habitable zone to contract.

Experiments were also made with an obliquity of 65°. An axial tilt of this magnitude, and even a little more, might have obtained during almost all Precambrian time (G. E. Williams 1993; see Figure 2.6). The results of the high obliquity experiments are difficult to interpret because of the extreme seasonality that would occur in practice. If the simulations should be adequate representations of annual conditions, then it appears that low latitudes would be much colder and drier throughout the year, but not glaciated, even if there were high (glacial) albedos at all latitudes. Polar regions would experience long, cold winters but, because of the coldness of the winter atmosphere and its inability to hold water vapour, snowfall would be much reduced and melt during the long, hot summers. On the other hand, for current geography, sea ice should form, but, with its depth being limited by the conductivity of the ice itself, it would not grow thick enough to survive during the summer. If the ice were to persist all year, then the pattern of climate would be drastically altered, with maximum summer temperatures in middle latitudes. As with a reduced obliquity, the habitable zone of the Earth would contract.

Many researchers are opposed to the idea of large obliquity changes. None the less, several geologists find that it provides the most parsimonious explanation of many long-term environmental changes (e.g. Allard 1948; G. E. Williams 1975b; Wolfe 1978, 1980; Xu 1979, 1980; J. G. Douglas and Williams 1982).

Very long climatic cycles

Hothouse–icehouse and warm-mode–cool-mode cycles characterize Phanerozoic climatic fluctuations. These dual pulses are reflected in the periodicity of glaciations. The 300-million cycle appears to be recorded in the thickness of storm-bed deposits known as tempestites (Brandt and Elias 1989). Thick tempestites, suggestive of more intense storms, are associated with icehouse phases, while thin tempestites, suggestive of less intense storms, are associated with hothouse phases.

The two climatic megacycles may be driven by geological processes or by galactic processes. The Solar System's galactic orbit might produce alternations between cold and hot periods (e.g. Steiner 1979; Steiner and Grillmair 1973; G. E. Williams 1975a). The interplay of carbon dioxide addition and withdrawal rates from the atmosphere–ocean system, which is driven by the supercontinent cycle (Veevers 1990), may produce similar climatic shifts (Fischer 1981, 1984a). Carbon dioxide is added to the system by volcanism, and withdrawn from the system by weathering (as gaseous carbon dioxide is converted to carbonates). These two processes are governed by very different factors, but their action will always strive towards a steady-state level of atmospheric carbon dioxide. Volcanism increases during bouts of accelerated mantle convection when plate fragmentation and plate movement occur (Figure 4.16). At these times, the supply of carbon dioxide into the atmosphere–ocean system increases. Associated with plate activation is an increase in the volume of mid-ocean ridges leading to marine transgressions. Less land area then being available, the atmosphere cannot lose carbon dioxide by weathering so fast as previously. The net effect of increased mantle convection is thus to pump up carbon dioxide levels in the atmosphere–ocean system, until a new balance is reached wherein the greater intensity of weathering offsets the smaller area being weathered to counterbalance the volcanic additions. The carbon dioxide level of the atmosphere may rise to three or four times its present level by this process, so creating a supergreenhouse effect and a much warmer climate. The mid-Cretaceous superplume may have pumped up atmospheric carbon dioxide levels to 3.7–14.7 times their modern pre-industrial value of 285 ppm (Caldeira and Rampino 1991). During times of sluggish mantle convection, the number of plates becomes smaller, the volume of mid-ocean ridges diminishes, and the continents become aggregated. Volcanism becomes subdued and consequently carbon dioxide emissions decline. Sea level drops, so exposing more land to the atmosphere and increasing the withdrawal of carbon dioxide from the air. When the carbon dioxide content of the atmosphere has fallen low enough, the hothouse state is broken, and the climate system assumes an icehouse state with ice sheets and glaciers.

Shorter megayear climatic cycles have run their course several times during the Phanerozoic aeon. A 30-million-year cycle was detected in pelagic environments, and can be expected to occur in other environments as well (Fischer 1981). In the inorganic world, it is expressed in temperature fluctuations, variations in carbon isotope ratios, and carbonate compensation depths (Fischer 1984a). A link with cyclical carbon dioxide fluctuations seems likely. Tertiary climatic cycles lasting 10 million years (Wolfe and Poore 1982) and 20 million years (McGowran 1990) have been detected.

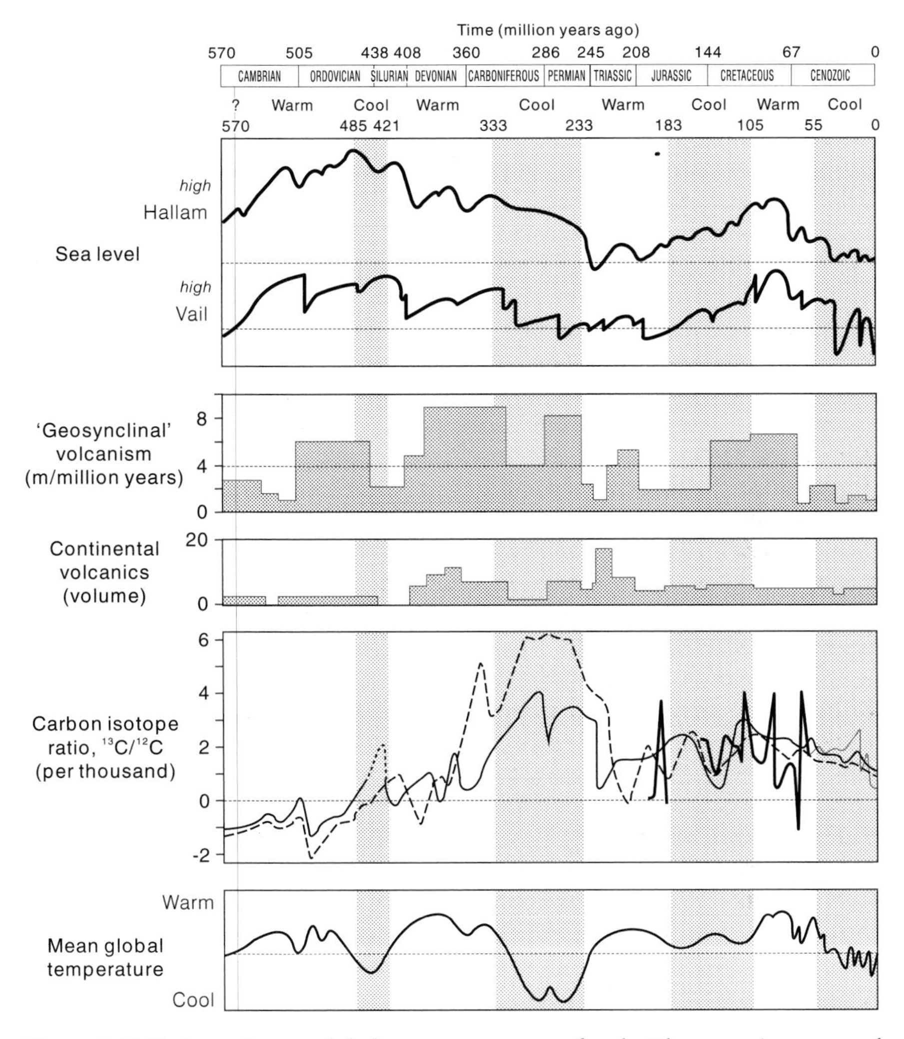

Figure 4.16 Estimated mean global temperature curve for the Phanerozoic compared with key conditions in the lithosphere and hydrosphere – sea level, 'geosynclinal' volcanism, the volume of continental volcanic rocks, and carbon isotope ratios ($^{13}C/^{12}C$) in carbonates (several different estimates are shown). Carbon isotope ratios rose during warm modes. This could have resulted from increased biological productivity or from increased burial of carbon-12 under anoxic conditions

Source: After Frakes *et al.* (1992)

SUMMARY

The atmosphere is a dusty gas. Its composition is unique among terrestrial planets due to the action of organisms. Air constantly moves in grand convective overturnings and great swirling eddies, collectively called the general circulation of the atmosphere. Climate is defined by 'climatic norms' (30-year averages). Climatic stability is abnormal – climatic change is the norm. The spectrum of climatic change bears spikes and peaks corresponding to the dominant periods and quasi-periods of atmospheric fluctuations. These are short-term (daily, weekly, annual, and 2, 5, 11, 18.6, 80, 200, and 2,000 years), medium-term (19,000, 23,000, 41,000, 100,000, and 400,000 years), and long-term (30 million, 150 million, and 300 million years). The atmosphere responds to changing cosmic forces (solar cycles, lunar cycles, planetary cycles, bombardment) and geological forces (volcanoes and lava flows, mountain ranges and plateaux, the placement of continents, and the position of the poles). Internal mechanisms in the weather machine cause climatic change. Climatic change occurs slowly and quickly. Fast changes include deglaciation, stadial–interstadial shifts (Dansgaard–Oeschger cycles), climatic 'events' during the Eemian interglacial, and the switch from cool to warm modes in geological climates. The atmosphere has changed composition during Earth history. Hydrogen and carbon dioxide concentrations have decreased. Oxygen and nitrogen concentrations have increased. Atmospheric composition and temperature have held fairly steady throughout the Phanerozoic aeon. Fluctuations about a mean condition express themselves as climatic swings from warm to cool modes and from hothouse to icehouse states.

FURTHER READING

Several books offer basic accounts of atmosphere, weather, and climate. *Introducing Physical Geography* (Strahler and Strahler 1994) is well worth reading if a grounding in meteorology and climatology is required. *Atmosphere, Weather and Climate* (R. G. Barry and Chorley 1992) provides a more thorough account. Readers interested in climatic change in historical times could do no better than read *Climate, History and the Modern World* (Lamb 1995). Climatic change in modern times is well covered by *Climate since AD 1500* (Bradley and Jones 1992), and *The Little Ice Age* (Grove 1988) is essential reading. Climate during the last ice age is explained clearly in *Ice Ages: Solving the Mystery* (Imbrie and Imbrie 1986). A more recent and relatively simple account appears in *Climate, Earth Processes and Earth History* (Huggett 1991). *Global Climates since the Last Glacial Maximum* (Wright *et al.* 1993) should be consulted. Geological changes of climate are discussed at length in *Climate Modes of the Phanerozoic* (Frakes *et al.* 1992).

5

HYDROSPHERE

DIRTY LIQUID: WATER

Planet ocean

The hydrosphere is all the Earth's waters. It includes liquid water, water vapour, ice and snow. Water in the oceans, in rivers, in lakes and ponds, in ice sheets, glaciers, and snow fields, in the saturated and unsaturated zones below ground, and in the air above ground is all part of the hydrosphere. Some people set the ambits of the hydrosphere to exclude the waters of the atmosphere.

The hydrosphere presently holds about 1,384,120,000 km^3 of water in various states (Table 5.1). By far the greatest portion of this volume is stored in the oceans. A mere 2.6 per cent (36,020,000 km^3) of the hydrosphere is fresh water. Of this, 77.23 per cent is frozen in ice caps, icebergs, and glaciers. Groundwater down to 4 km accounts for 22.21 per cent, leaving a tiny fraction stored in the soil, lakes, rivers, the biosphere, and the atmosphere.

Water is the chief component of the hydrosphere. Concentrations of constituents dissolved in water vary in different parts of the hydrosphere (Table 5.2). Sodium and chlorine are the main constituents of sea water. The acidity of waters in the hydrosphere is highly variable. The most acid natural waters on Earth occur in volcanic crater lakes. These may have a pH less than 0 (e.g. Brantley *et al.* 1993). Water free of carbon dioxide in contact with ultramafic rocks may have a pH of 12. Soils in desert basins and rich in sodium carbonate and sodium borate may be equally alkaline. In most environments, 9 is the upper limit of pH.

Circulating water

Water, even in its solid state, seldom stays still for long. Meteoric water circulates through the hydrosphere, atmosphere, biosphere, pedosphere, and upper parts of the crust. This, the water cycle, involves evaporation,

Table 5.1 Water volumes in the hydrosphere

Water	*Volume (km^3)*	*Volume (%)*
All water (fresh and salt)		
Oceans	1,348,000,000	97.39
Ice caps, icebergs, glaciers	227,820,000	2.01
Groundwater and soil moisture	8,062,000	0.58
Lakes and rivers	225,000	0.02
Atmosphere	13,000	0.001
Sum	1,384,120,000	100
Fresh water (2.6 per cent of all water)		
Ice caps, icebergs, glaciers	27,818,246	77.23
Groundwater, 0–0.8 km	3,551,572	9.86
Groundwater, 0.8–4 km	4,448,470	12.35
Soil moisture	61,234	0.17
Freshwater lakes	126,070	0.35
Rivers	1,081	0.003
Hydrated earth minerals	360	0.001
Atmosphere	14,408	0.04
Life	1,081	0.003
Sum	36,020,000	100

Source: After Baumgartner and Reichel (1975)

Table 5.2 Major dissolved constituents of the hydrosphere

	Concentration		
Constituent	*Sea water (g/kg at 35‰ salinity)*[a]	*Rivers (with human inputs deducted) (mg/l)*[b]	*Precipitation*[c] *(mg/l)*
Sodium (Na^+)	10.765	5.15	0.55
Magnesium (Mg^{2+})	1.294	3.35	0.2
Calcium (Ca^{2+})	0.412	13.4	0.65
Potassium (K^+)	0.399	1.3	0.15
Strontium (Sr^{2+})	0.0079	–	–
Chloride (Cl^-)	19.353	5.75	1.0
Sulphate (SO_4^{2+})	2.712	8.25	1.35
Bicarbonate (HCO_3^-)	0.142	52	1.25[d]
Bromide (Br^-)	0.0674	–	–
Fluoride (F^-)	0.0013	–	–
Boric acid (H_3BO_3)	0.0256	–	–
Silica (SiO_2)	–	10.4	–

Notes: [a] Derived from Millero (1974)
[b] Meybeck (1979)
[c] Meybeck (1979)
[d] Computed from ionic balance

condensation, precipitation, and runoff. It is not a closed system. Deep-seated, juvenile water associated with magma production may issue into the meteoric zone; while meteoric water held in hydrous minerals and pore spaces in sediments (connate water) may be removed from the meteoric cycle at subduction sites.

The Earth's surface water balance is written as

$$\text{Precipitation } (P) = \text{Evaporation } (E) + \text{Runoff } (D) + \text{Storage } (S)$$

In the long term, storage can be taken as constant and ignored in drawing up a global inventory of annual water fluxes. It is reasonable to assume that world water reserves have not changed substantially in recent times. This being the case, it follows that the quantity of water evaporated from the entire planetary surface must be replaced by an equal amount of precipitation; and it must be true that the annual global precipitation equals the annual global evaporation, so there is no net runoff. Estimates of annual global precipitation and evaporation fix them both at 973 mm, which converts to 496,100 km^3 of water. But precipitation, evaporation, and runoff are unequal on land and in oceans (Table 5.3). Annual estimates show that 111,100 km^3 of water fall over land, of which 71,400 km^3 evaporate and 39,700 km^3 runoff; in the oceans, 385,000 km^3 of water fall, 424,700 km^3 evaporate, and –39,700 km^3 'runoff' (this is simply the numerical difference between precipitation and evaporation). This fundamental inequality in the water balance results in a transfer of water vapour ('runoff') from the oceans, through the atmosphere, to the continents, and a counter-transfer from the continents to the oceans in runoff. These two flows are thus the driving forces behind the water circulation among the latitudinal zones and between continents and oceans. The only regions that do not possess this inequality are continental regions of internal drainage where runoff is zero and precipitation equals evaporation. The prime importance of the oceans in the global water cycle is suggested by the fact that the oceanic water exchanges account for 80 per cent of the total. Notice that, expressed as water volume, runoff from the land exactly balances the loss from the oceans.

The potential of the water cycle to influence environmental change is made evident by considering the turnover of atmospheric moisture.

Table 5.3 Water balances for land and sea

	Continents (148,900,000 km^2)			*Oceans (361,100,000 km^2)*		
	Precipitation	*Evaporation*	*Runoff*	*Precipitation*	*Evaporation*	*Runoff*
Water volume (km^3)	111,100	71,400	39,700	385,000	424,700	–39,700
Water depth (mm)	746	480	226	1,066	1,176	–110

Source: After Baumgartner and Reichel (1975)

Although atmospheric water is but a minute part of the hydrosphere, its significance to the other spheres is disproportionately large. The volume of water stored in the atmosphere is 13,000 km^3. Since the area of the Earth's surface is 510,000,000 km^2, it is a matter of simple arithmetic to work out that, if all the water-vapour in the atmosphere were to condense, it would form a layer 2.54 cm (exactly one inch) deep. Globally, the mean annual precipitation is 97.3 cm. So, there must be $97.3 \div 2.54 \approx 38$ precipitation cycles per year, and the average life of a water molecule in the atmosphere is therefore $365 \div 38 \approx 10$ days. Furthermore, the global store of surface fresh water, if not replenished, would be lost by evaporation in as little as five years and drained by rivers in ten.

The world pattern of precipitation displays five chief features. First, more rain falls over the oceans (1,066 mm) than over the land (746 mm). This is because the oceans supply the atmosphere with more water than do the continents, and much of the extra supply precipitates over, and falls back into, the oceans. Second, there are three latitudinal belts of prodigious precipitation: the tropical zone and two zones of west winds, one in each hemisphere (Figure 5.1). These belts are separated by two subtropical low-precipitation zones. Poleward of the west-wind belt, precipitation decreases to the polar zone, which has low precipitation. Third, and to some extent beclouding the zonal pattern described above, precipitation shows an overall

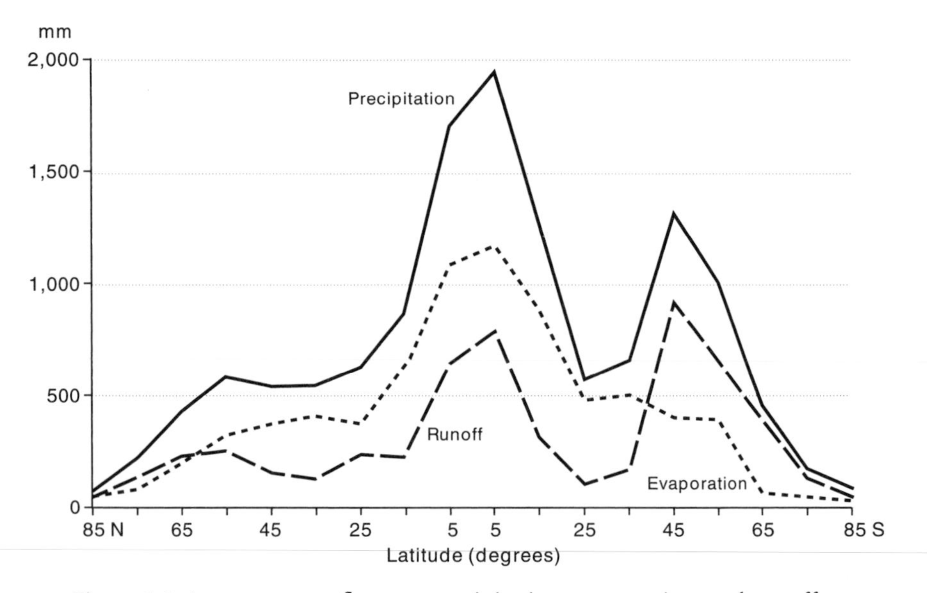

Figure 5.1 Average water fluxes – precipitation, evaporation, and runoff – for 10°-latitude zones on the land surface
Source: After data in Baumgartner and Reichel (1975)

Table 5.4 Precipitation, evaporation, and runoff in holospherical zones

Holospherical zone	*Precipitation (mm)*	*Evaporation (mm)*	*Runoff (mm)*
60°–90°	380	176	202
30°–60°	908	812	96
0°–30°	1,180	1,307	–127

Source: Data from Baumgartner and Reichel (1975)

decline from the equator to the poles owing to the poleward decrease in the temperature-dependent water-vapour content of the air – hot tropical air can hold more water vapour than can cold polar air. This is evident in mean annual precipitation values for holospherical, 30° latitude zones (Table 5.4). Fourth, the humid tropical zone extends all around the Earth, though with varying amounts of precipitation. The other zones are interrupted, the moist west-wind zones by arid continental regions, and the subtropical arid zones by regions with rainy monsoon seasons. Fifth, superimposed on the above patterns are areas of very high precipitation and very low precipitation on windward and leeward sides respectively of mountain chains.

The world pattern of evaporation displays four chief features. First, evaporation over oceans is over twice as great as over land (1,176 mm versus 480 mm). This is because water is always to hand over oceans and evaporation may run at its potential rate. Over land, a shortage of water may cause evaporation to fall below its potential level. Second, evaporation decreases greatly from the equator to the poles (Figure 5.1 and Table 5.4). This poleward decline in evaporation stems largely from the poleward decrease of net radiation and air and water temperature. Factors that play a secondary role are, over land, the poleward decrease in precipitation, and, over the oceans, the decrease in the vapour-pressure difference between sea and air. Both these factors are a result of the lower temperature levels towards the poles. The poleward decrease of evaporation is not uniform. Increased cloudiness affects net radiation and reduces evaporation in the ITCZ and west-wind belts. Third, the highest evaporation values are associated with high-precipitation regions over land, and with high-radiation regions over oceans. Fourth, over land, the relationship between evaporation and net radiation is obscured by precipitation. The pattern of precipitation over land is partly mirrored in the evaporation figures.

There are two main features of global continental runoff. First, in contrast to precipitation and evaporation, runoff over land and oceans combined increases from equator to poles with decreasing temperature (Table 5.4). Second, continental runoff displays a zonal pattern (Figure 5.1). There is a tropical belt with high runoff, shifted to the south a little, and a subtropical belt of lower runoff. There is a mid-latitude belt of higher runoff (west-wind belt) shifted polewards from the high-precipitation zone owing to a poleward decrease in evaporation caused by radiation receipt diminishing in that direction. And there is a polar zone within which runoff decreases towards the poles but not at such a high rate as the precipitation.

The water balance of the continents varies considerably (Table 5.5). South America is in a class of its own with by far the highest precipitation, evaporation, and runoff depths. This uniqueness is due to the high precipitation in the tropical lowlands, in the cordilleras, and in the southern Andes. Australia (excluding islands like New Guinea) is the driest continent except for Antarctica. The water balances of the other continents are alike.

Much insight into the water cycle should be gained from the Global Energy and Water Cycle Experiment (GEWEX) and Programme of the World Climate Research Programme (e.g. McBean 1991; Chahine 1992a, 1992b; Schaake 1994).

Ocean currents

The oceans are an enormously important component of the climate system. They can store vast quantities of heat and are reluctant to give it up. They are one of the three main routes for transporting surplus tropical energy towards the poles (the other two routes are latent heat transfer and sensible heat transfer in the atmosphere). They also hold sixty times more carbon dioxide than does the atmosphere, and because the gas is readily diffused between surface water and the overlying air, they regulate the atmospheric concentration of carbon dioxide (though organisms regulate the surface concentration).

Ocean water, like air in the atmosphere, circulates. The dominant paths of flow are called ocean currents. Nearly all large currents at the top of the

Table 5.5 The water balance of the continents

Continent	*Area (millions km^2)*	*Water depths (mm)*			*Water depths (per cent global values)*		
		Precipitation	*Evaporation*	*Runoff*	*Precipitation*	*Evaporation*	*Runoff*
Antarctica	14.1	169	28	141	23	6	53
Australia	7.6	447	420	27	60	86	11
Europe	10.0	657	375	282	88	78	106
North America	24.1	645	403	242	86	84	91
Africa	29.8	696	582	114	93	121	43
Asia	44.1	696	420	276	93	88	104
South America	17.9	1,564	946	618	209	197	232

Source: After Baumgartner and Reichel (1975)

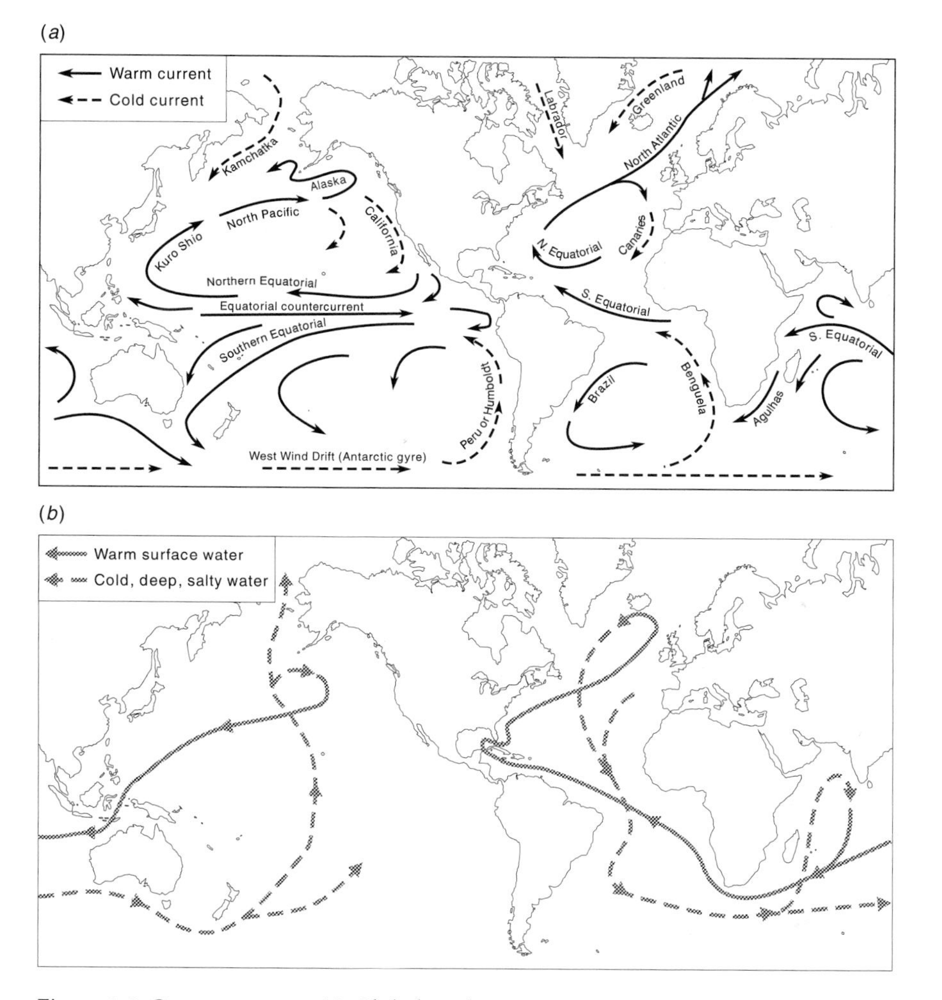

Figure 5.2 Ocean currents. (a) Global surface ocean currents. (b) Major deep-water and surface currents between oceans

ocean are driven by prevailing surface winds. The main feature of the surface-ocean circulation is a set of gyres (Figure 5.2a). These are slow-moving vortices that occur in the oceans of the Northern and Southern Hemispheres. They rotate clockwise in the Northern Hemisphere and anticlockwise in the Southern Hemisphere, tracking the passage of air around the subtropical high-pressure cells. A large gyre, or circumpolar current, also moves around Antarctica. Near the equator, ocean currents travel westwards. On approaching land, they are diverted to the north and south to form warm currents that run parallel to the coastline. In the Pacific

Ocean, some of the water returns eastwards, approximately along the equator, as an equatorial counter-current. Once in middle latitudes, the warm currents are forced eastwards by the westerly winds. They travel across the oceans until they reach the western edges of continents, which they keep relatively warm in winter. They then either join the Arctic or the Antarctic circulation, or else they return towards the equator as cool currents. They are often associated with cold-water upwellings along continental margins. This is the case of the Humboldt Current that runs down the coastline of Peru and Chile.

The ocean deep-water circulation is somewhat different from the surface currents (Figure 5.2b). In the ocean deeps, a salty current threads its way around the world's ocean basins (e.g. Broecker and Denton 1990; Broecker 1995). It starts in the North Atlantic Ocean where northward-flowing warm and normally salty water is chilled by cold Arctic air and by evaporation. This already fairly dense water then sinks to the ocean bottom and flows southwards, through the North Atlantic and South Atlantic Oceans. It is warmer and less dense than the frigid surface waters off Antarctica, so it rises to the surface, where it is chilled again and plunges back to the ocean depths. It then turns eastwards between South Africa and Antarctica. One branch moves into the Indian Ocean and another continues eastwards, passing by Australia and New Zealand, and northwards into the Pacific Ocean. In the Indian and Pacific Oceans, the northward flow of cold bottom waters is compensated by a southward flow of surface waters in the gyres. Counterflow of warm surface water in the Atlantic Ocean is rapidly caught up in the strong southwards current of the deep, cold water. Disruption of this 'global conveyor' system can cause severe and sudden climatic change.

WATER CYCLES: HYDROSPHERIC CHANGE

The volumes of water stored in the atmosphere, in the oceans, in lakes and rivers, and beneath the ground all fluctuate. These volumetric variations include changing lake levels, sea levels, water-table heights, ice cover, and, to a lesser extent, average cloudiness. Water fluxes – evaporation, advection, condensation, precipitation, runoff, lake currents, and ocean currents – also fluctuate. For instance, the changing balance between evaporation and precipitation leads to wet and dry phases. All these fluctuations of stored and flowing water occur over time-spans ranging from days to aeons.

Rainfall and runoff

Rainfall and runoff vary in the short term. Precipitation commonly displays daily and seasonal cycles, though the nature of these differs according to climatic regime. It exhibits quasi-periodic changes, the most pronounced of which follow 11-year and 18.6-year beats. Cycles of discharge lasting hundreds of years are known. Quasi-cyclic variations in mean annual discharge into San Francisco Bay, as recorded in strontium isotopic compositions of carbonate mollusc shells from estuarine sediments, occur naturally with a period of about 200 years or less (Ingram and DePaolo 1993).

Longer-term alternations between the wet and dry conditions lead to sustained phases of aridity and humidity. This happened in the Saharan region of Africa during the Holocene epoch (Lézine 1988a, 1988b, 1989; Lézine and Casanova 1989, 1991; Lézine and Vergnaud-Grazzini 1993). Vegetation belts wandered in the wake of shifting climatic zones. About 9,000 years ago, conditions were wetter and the Sahelian wooded grassland, which was confined to latitude 10° N during the height of the last ice age, rapidly migrated northwards. Guinea (humid) vegetation elements reached latitude 16° N and Sahelo-Sudanian (arid and semiarid) vegetation elements reached 21° N, which is the margin of the modern Sahara Desert. Conditions became drier around 5,000 years ago and the vegetation quickly assumed its modern distribution.

The development of glaciation in the Northern Hemisphere around 2.5 million years ago led to aridification in several places – Asia (Dersch and Stein 1994) and Spain (Harvey 1990), for example. Indeed, sustained cooling through the Tertiary period produced widespread aridification in Australia (Locker and Martini 1989), southern Africa (T. C. Partridge 1993), and elsewhere.

During the Pleistocene epoch, the strength of the monsoons changed. They were weaker during the Last Glacial Maximum, some 18,000 years ago, and stronger during the mid-Holocene epoch, about 9,000 to 10,000 years ago (Figure 4.4). The changing monsoon strength altered the water balance in affected regions (e.g. T. Webb *et al.* 1993). At the height of the last ice age, 18,000 years ago, July precipitation in India and south-west Asia was up to 8 mm/day lower than at present (Figure 5.3a). The annual effective precipitation (precipitation minus evaporation) was also significantly less than at present in this monsoon region (Figure 5.4a). By 9,000 years ago, the monsoon circulation was stronger than at present. Consequently, July precipitation and annual effective precipitation were significantly greater than they are now (Figures 5.3b and 5.4b).

Throughout the Phanerozoic aeon, the occurrence and distribution of evaporites, carbonates, and coal provide much information on dry and wet conditions. For instance, evaporite deposition during the Triassic period, and especially in Late Triassic times, indicates widespread aridity (Figure 5.5). The pattern of occurrence, which crosses the equatorial zone in Africa, the Arabian peninsula, and south-eastern North America, implies that a tropical rain belt did not exist (Frakes *et al.* 1992: 59). Coal accumulation during the Triassic was relatively low, although during the Late Triassic

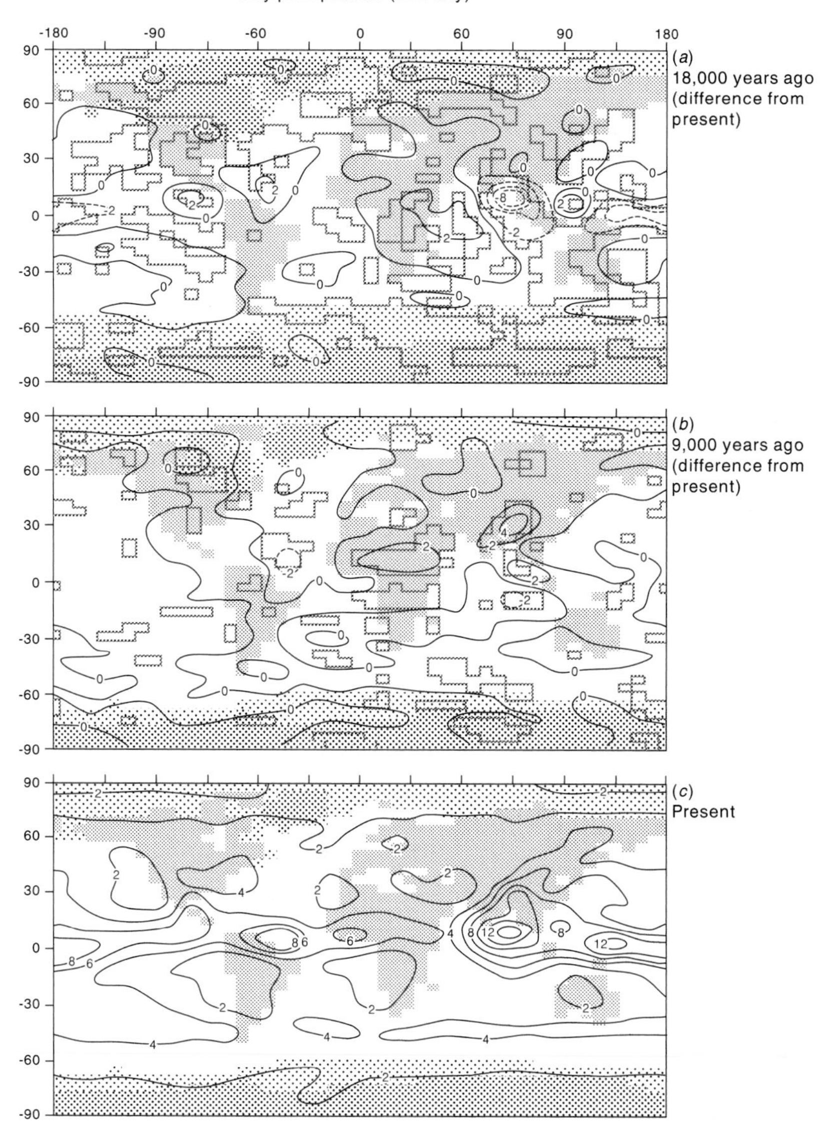

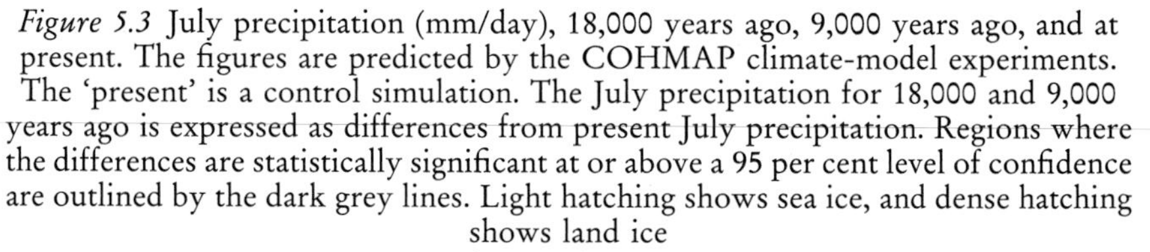
Figure 5.3 July precipitation (mm/day), 18,000 years ago, 9,000 years ago, and at present. The figures are predicted by the COHMAP climate-model experiments. The 'present' is a control simulation. The July precipitation for 18,000 and 9,000 years ago is expressed as differences from present July precipitation. Regions where the differences are statistically significant at or above a 95 per cent level of confidence are outlined by the dark grey lines. Light hatching shows sea ice, and dense hatching shows land ice

Source: After Kutzbach *et al.* (1993)

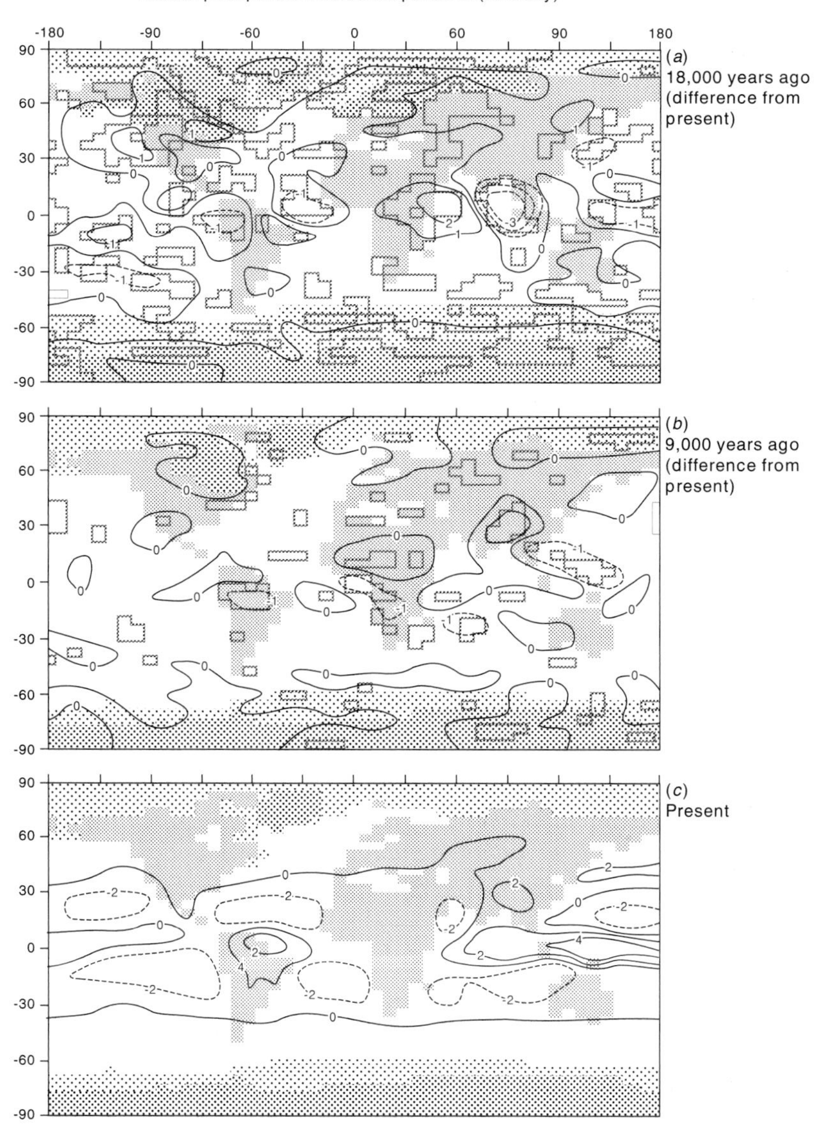

Figure 5.4 Annual effective precipitation (mm/day), 18,000 years ago, 9,000 years ago, and at present. The figures are predicted by the COHMAP climate-model experiments. The 'present' is a control simulation. The annual effective precipitation for 18,000 and 9,000 years ago is expressed as differences from present annual effective precipitation. Regions where the differences are statistically significant at or above a 95 per cent level of confidence are outlined by the dark grey lines. Light hatching shows sea ice, and dense hatching shows land ice

Source: After Kutzbach *et al.* (1993)

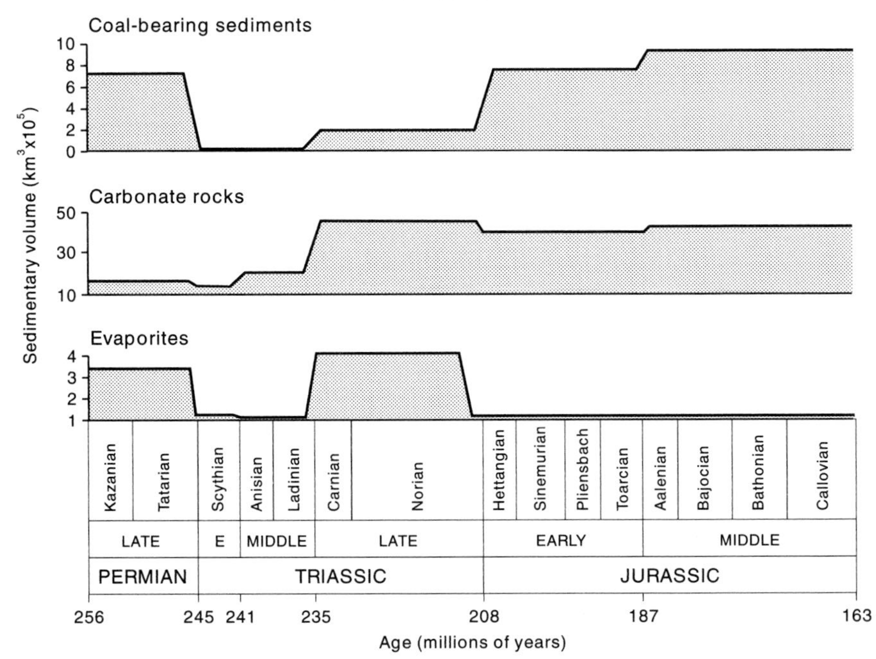

Figure 5.5 Indicators of dry and wet geological climates from Late Permian through to Middle Jurassic times. Note the clear signs of aridity during the Late Triassic period – considerable evaporite deposition and a reduction in coal-bearing rock production

Source: After Frakes *et al.* (1992)

phase of major evaporite deposition, coal abundance increased, perhaps because monsoons developed around the margins of Tethys (P. L. Robinson 1973; Parrish *et al.* 1982; Kutzbach and Gallimore 1989; Parrish 1993).

The piecing together of evidence from palaeohumidity indicators and from computations of past precipitation and evaporation patterns suggests a succession of arid and humid periods existed during the Phanerozoic aeon (Tardy *et al.* 1989). The Cretaceous, Devonian–Silurian, and Cambrian were humid periods, while the Tertiary and Permo-Triassic were dry periods (Figure 5.6).

Lake levels

Lake levels are determined by the balance between water inputs and outputs, plus changes in lake-basin volume resulting from deposition or erosion. Inputs are precipitation and incoming flows of river water and groundwater. Outputs are evaporation, outgoing flows of river water and groundwater, and extraction for domestic, agricultural, and industrial use.

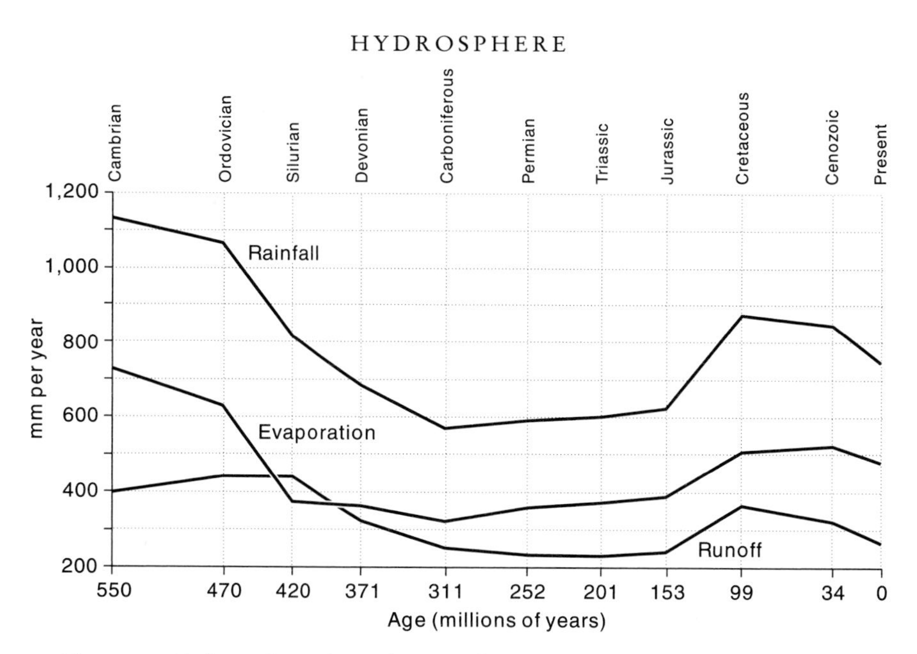

Figure 5.6 Estimated continental water fluxes through the Phanerozoic aeon. The times are mid-ages of geological periods
Source: From data in Tardy *et al.* (1989)

Changes in precipitation and evaporation rates are known to effect a speedy response in lake levels, though the rate of response depends on many factors, including local geomorphology and climate.

The effect of a sustained water-balance change on lake levels may be gauged by calculating the time it takes for the lake level to adjust to the new conditions. This, the equilibrium-response time, varies from 1.5 to 350 years for closed lakes, which have no outlets, and from 2.9 days to 2 years for open lakes (Mason *et al.* 1994). Great Salt Lake, Utah, in the United States, rose 6.25 m from 1963 to 1987 (Stephens 1990). About 60 per cent of this increase occurred after 1982 due to a greater than average precipitation and a less than average evaporation. This century, record high lake levels for all the Great Lakes, except Ontario, occurred in 1985 and 1986, causing considerable economic losses around the lake margins. The peak water-level in the Lake Michigan–Huron was 50 cm higher than the 1986 record (Quinn and Sellinger 1990). In future, with global warming, they may fall by 0.5–2.5 m, depending on the scenario used (Sanderson 1989).

Medium-term directional changes in the hydrosphere track climatic fluctuations. For instance, Medicine Lake, South Dakota, responded to climatic change from 10,800 to 4,500 years ago (Kennedy 1994). In summary, while spruce occupied the Medicine Lake catchment, 10,800–10,000 years ago, dark-grey massive basal sediments were deposited in a freshwater lake. They had a low total-sulphur content, a low carbonate content, organic carbon content increased upwards, and they had a high magnetic susceptibility. The catchment vegetation changed from spruce to oak and elm and finally to prairie 10,000–9,200 years ago. The lake became shallower and salinity increased from less than 2 to more than 10 per thousand. Light-grey and dark-grey calcareous and organic-carbon-rich sediments were deposited. These had a low total-sulphur content and low magnetic susceptibility. During the early prairie phase, some 9,200–5,500 years ago, alternating sections of aragonite-rich layers and grey massive sediments with high total-sulphur content and multiple gypsum layers were laid down in a meromictic environment (the lake water remained unmixed) under fluctuating lake levels and salinity. Continued aridity from 5,500 to 4,500 years ago maintained low lake levels and high salinity.

Some lake-level fluctuations result from non-climatic changes. A palaeohydrological study of hypersaline Lake Asal, Djibouti, revealed that about 6,000 years ago a sudden change of drainage occurred causing a drastic drop in lake level of about 300 m (Gasse and Fontes 1989). This change was linked to an exceptional episode of rift-valley opening and subsequent volcanism and to a decrease in continental groundwater supply as argillaceous sediments plugged the bottom of upstream Lake Abhé.

Sea levels

Sea-level change is caused by volumetric and mass distribution changes in the oceans (Table 5.6). Ocean volume changes are eustatic or steric. Eustatic change results from water additions to or extractions from the oceans (glacio-eustatic change), and from changes in ocean-basin volume (tectono-eustatic change). Steric change results from temperature or density changes in sea water. Much of the predicted sea-level rise during the next century will result from the thermal expansion of sea water. Ocean thermal expansion is about 20 cm/°C/1,000 m (Mörner 1994).

Glacio-eustatic change is tightly bound to climatic change. Globally, inputs from precipitation and runoff normally balance losses from evaporation. (Gains from juvenile water probably balance losses in buried connate water.) However, when the climate system switches to an icehouse state, a substantial portion of the world's water supply is locked up in ice sheets and glaciers. Sea level drops during glacial stages, and rises during interglacial stages. Additions of water to and subtractions of it from the oceans, other than that converted to ice, may cause small changes in ocean volume. This minor process might be termed hydro-eustasy.

Tectono-eustatic change is driven by geological processes. Even when the water cycle is in a steady state, so that additions from precipitation are balanced by losses through evaporation, sea level may change owing to volumetric changes in the ocean basins. An increasing volume of ocean basin

Table 5.6 Causes of eustatic change

Seat of change	*Type of change*	*Approximate magnitude of change (m)*	*Causative processes*
Ocean basin volume	Tectono-eustatic	50–250	Orogeny
			Mid-ocean ridge growth
			Plate tectonics
			Sea-floor subsidence
			Other Earth movements
Ocean water volume	Glacio-eustatic	100–200	Climatic change
	Hydro-eustatic	Minor	Changes in liquid hydrospheric stores (water in sediments, lakes, clouds)
			Additions of juvenile water
			Loss of connate water
Ocean mass distribution and surface 'topography'	Geoidal eustatic	Up to 18	Tides
		A few metres	Obliquity of the ecliptic
		1 m per millisecond of rotation	Rotation rate
		Up to 5	Differential rotation
		2 (during Holocene)	Deformation of geoid relief
	Climo-eustatic	Up to 5 for major ocean currents	Short-term meteorological, hydrological, and oceanographic changes

Source: Partly after Mörner (1987, 1994)

would lead to a fall in sea level and a decreasing volume to a rise in sea level. Decreasing volumes of ocean basin are caused by sedimentation, the growth of mid-ocean ridges, and Earth expansion (if it should have occurred); increasing volumes are caused by a reduced rate or cessation of mid-ocean ridge production.

Geoidal eustasy results from processes that alter the Earth's equipotential surface, or geoid. The ocean geoid is also called the geodetic sea level. The relief of the geoid is considerable: there is a 180-m sea-level difference between the rise at New Guinea and the depression centred on the Maldives, which places lie a mere 50–60 degrees of longitude from one another. There is also a geoid beneath the continents. The configuration of the geoid depends on the interaction of the Earth's gravitational and rotational potentials. Changes in geoid relief are often rapid and lead to swift changes of sea level (e.g. Mörner 1993a, 1993b).

On a short time-scale, local changes in weather, hydrology, and oceanography produce relatively tame fluctuations of sea level. These fluctuations might be called climo-eustasy. They may involve up to 5 m of sea-level change for major ocean currents, but less than half that for meteorological and hydrological changes.

PUMPING WATER: CAUSES OF HYDROSPHERIC CHANGE

Hydrological change is an integral part of climatic change. Phases of aridification and humification, the growth and decay of ice sheets, changing river regimes and groundwater storage, droughts and floods, and eustatic sea-level fluctuations are all related chiefly to climatic factors. These climatic factors may alter due to internal mechanisms within the atmosphere–ocean system, and due to cosmic and geological forcing. Plate tectonic processes are particularly effective in promoting climatic change. They cause tectono-eustatic and glacio-eustatic sea-level changes, They also lead to the repositioning of continents. This has a direct impact on all aspects of climate, especially the water balance, and may be a dominant factor in starting and stopping ice ages. Humans impound and divert rivers, extract water from aquifers, and alter the land cover by such activities as forest clearance, wetland drainage, and urbanization. Beavers dam streams.

Changes in water flux

Flood and drought cycles

Floods are caused by weather systems delivering more precipitation to a drainage basin than it can absorb or store (e.g. Hirschboeck 1991). Intense convective thunderstorms, tropical storms and hurricanes, extratropical cyclones and fronts, and rapid snowmelt are common causes of floods. A considerable body of evidence suggests that the intensities of these weather systems are responsive to short-term solar and, especially, to lunar, cycles. It is not surprising, therefore, that flood and drought records in many parts of the world contain periodic lunar and solar elements (Table 5.7).

Monsoons

Monsoons are sensitive to geography. A model of Triassic climate unveiled the presence of megamonsoons on the Pangaean supercontinent (Kutzbach and Gallimore 1989). In all simulation experiments, the conjoined land masses of Laurasia and Gondwana (Figure 5.7a) exhibited extreme continentality with hot summers, cold winters (Figure 5.7b), and large-scale summer and winter monsoon circulations. Some components of the resulting water budgets are shown in Figure 5.7c–e. The pattern of annual precipitation (Figure 5.7c) shows that rainfall exceeded 2 mm/day (720 mm/yr) along the Tethyan coasts, which were affected by the summer monsoon, and along the eastern coast of middle and high latitudes, which received year-round rain. Rainfall totals averaged 2 mm/day (720 mm/yr) along the west coast in middle latitudes because westerly winds blew there during the winter, and along the west coast at the equator owing primarily to equinoctial wind convergence. The annual rainfall was less than 720 mm over almost all Pangaea. Effective rainfall (precipitation less evaporation) revealed a water deficit in low latitudes (in a band lying between latitudes 30° N and 30° S), and a water surplus in middle and high latitudes. Annual soil-moisture levels were consistent with the precipitation and evaporation rates, being below 5 cm in the band circumscribed by latitudes 30° N and 30° S, and above 10 cm in areas of year-round rainfall (eastern Pangaea bordering the Tethys Sea). Only where the annual soil-moisture levels exceeded 10 cm would significant amounts of runoff have been generated.

Table 5.7 Examples of 18.6-year lunar signals in rainfall and runoff data

Data	*Region*	*Reference*
Rainfall	South Africa	Currie (1993b)
	United States	Currie and O'Brien (1990, 1992)
Flood–drought cycles	Western North America	Currie (1981)
	India	Campbell *et al.* (1983), Currie (1984)
	North-eastern China	Hameed *et al.* (1983), Currie and Fairbridge (1985)
	Mid-latitude South America	Currie (1983)
	Africa drained by the Nile River	Hameed (1984), Currie (1987)

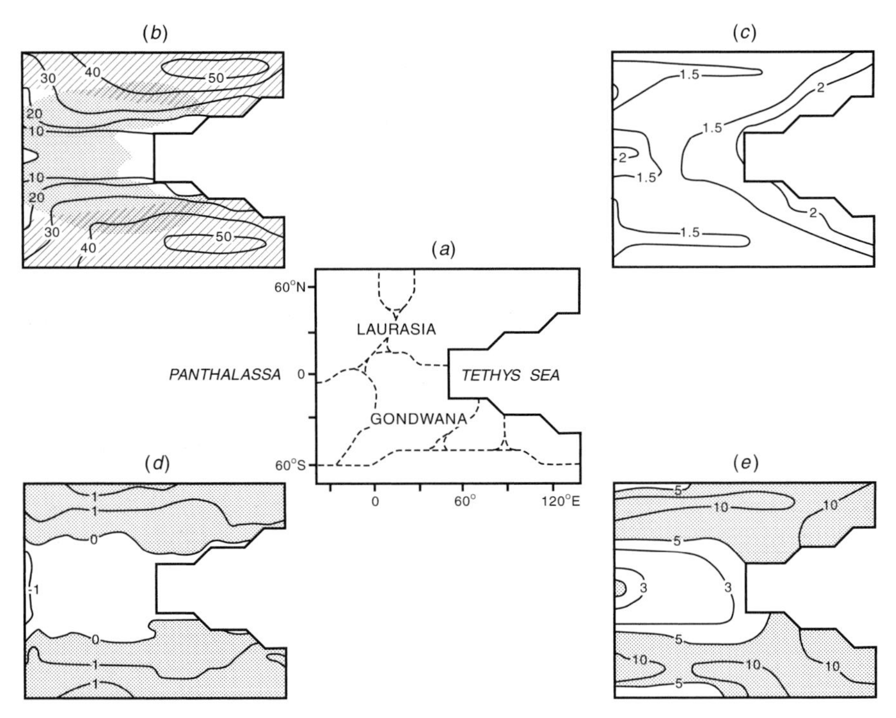

Figure 5.7 Simulated climate on an idealized Pangaea. The boundary conditions include Triassic geography, no relief, modern heating, and modern snow cover. (a) Triassic geography. (b) Surface temperature range (°C) calculated as summer temperature minus winter temperature. Seasonal temperature extremes are indicated: the fine stippled area demarks summer temperatures greater than 30°C; hatched areas show winter temperature less than 0°C. (c) Annual precipitation (mm/day). (d) Annual effective precipitation (precipitation minus evaporation) (mm/day). Fine stippled areas are regions with an excess of precipitation over evaporation (runoff-generating regions). (e) Annual soil moisture (cm). Fine stippled areas have soil moisture in excess of 5 cm

Source: After Kutzbach and Gallimore (1989)

The intensity of monsoons is sensitive to changes in orbital forcing (e.g. Kutzbach and Otto-Bliesner 1982; Kutzbach and Guetter 1986). A major study of monsoon variability over the last 150,000 years used palaeoclimatic records of the Afro-Indian monsoon and a range of general circulation models (Prell and Kutzbach 1987). Orbitally produced changes in solar radiation and the boundary conditions prevailing during glacial stages emerged as the main factors explaining changes in major monsoon climates. Four distinct monsoon maxima, which could be widely correlated across Africa and Asia during interglacial stages, were present in marine sediment cores. These maxima coincided with Northern Hemisphere summer-radiation peaks associated with the precessional cycle. Experiments with general circulation models showed that during interglacial stages, regional monsoon indicators (land–ocean pressure gradients, precipitation, and winds) were closely related to the departure of solar radiation from modern values, the greater the departure the greater the strength of the monsoon. The simulated changes in the water cycle were surprisingly large. They were consistent with geological evidence of a major change in the tropical monsoons. The regional monsoon response to glacial-stage boundary conditions was variable. Around 18,000 years ago, the monsoon in southern Asia was greatly weakened but precipitation increased in the western Indian Ocean and in equatorial North Africa. But, taken as zonal average values, tropical monsoon precipitation was fairly insensitive to glacial boundary conditions, whereas it was highly sensitive to orbitally induced variations in solar radiation receipt. This accords with the findings of a study of monsoon variability in the Arabian Sea and Indian Ocean over the last 350,000 years, which, though confirming the relative weakness of monsoons 18,000 years ago, suggested that snow and ice are not a primary control of monsoon strength and timing (Clemens *et al.* 1991). Rather, latent heat appears to be a key factor in modulating monsoon strength.

Changes in water stores

Soil moisture and groundwater change

Soil moisture exercises a strong control over evaporation. It thus influences the water cycle, as well as the temperature and circulation patterns over land. Experiments using an atmospheric general circulation model with prescribed ocean surface temperatures and cloud showed that the present global water cycle is sensitive to the water-holding capacity of the plant-root zone (Milly and Dunne 1994). Increased global soil-storage capacity increases evaporation and decreases runoff. In addition, atmospheric feedbacks associated with the resulting higher precipitation and lower potential evaporation drive further changes in evaporation and runoff. Most of the changes in evaporation and runoff occur in the tropics and in the northern middle-latitude rain belts. Global evaporation from land increases by about 7 cm for each storage-capacity doubling in the range 1–60 cm, but sensitivity is almost zero beyond 60 cm.

The significance of soil moisture in past water cycles was demonstrated in a general circulation model of climate 9,000 years ago (Kutzbach and Guetter 1986). When soil moisture was included as an interactive variable, conditions in the tropics were similar to a ‘fixed’ soil-moisture model, but summer warming and aridity occurred over northern middle latitudes. Later experiments used an interactive moisture version of a low-resolution general circulation model to examine the climatic role of soil moisture 9,000 years ago (Gallimore and Kutzbach 1989). Of particular interest was the degree to which soil-moisture feedback in the climate system modulates the climatic response to orbital changes in radiation. The results of the experiments showed that soil moisture and runoff increased over northern tropical lands, and over eastern Asia in mid-latitudes, both regions of increased precipitation. Soil moisture was decreased over much of the northern middle latitudes, the biggest decreases occurring in areas that are wet in the control run for present conditions – north-western and south-eastern North America, and western and eastern Eurasia. Runoff in these regions was reduced, too.

Lake-level change

Lake levels during the late Pleistocene and Holocene epochs appear to have been influenced by orbital variations. This is suggested by the running of an atmospheric general circulation model to simulate the January and July climates at 3,000-year intervals, starting 18,000 years ago (Kutzbach and Street-Perrott 1985). The orbital changes in the seasonal distribution of solar radiation receipt and the changing surface boundary conditions assumed to accompany deglaciation are shown in Figure 5.8. The simulated changes of radiation receipt, surface temperature, and water balance in the latitude belt from 8.9° to 26.6° N over the last 18,000 years are shown in Figure 5.8b–d. From 15,000 years ago onwards, the model predicted a strengthening of the monsoon circulation and increased precipitation and effective precipitation in the Northern Hemisphere tropics, which features reached their acme in the period

Figure 5.8 (opposite) (a) Schematic diagram of changing external forcing (Northern Hemisphere solar radiation in June, July, and August, S_{JJA}, and December, January, and February, S_{DJF}, as a percentage different from present solar radiation) and internal boundary conditions (land ice as a percentage of the ice volume 18,000 years ago; global mean sea-surface temperature as a departure from the present value; excess glacial aerosol, including sea ice on an arbitrary scale; and atmospheric carbon dioxide concentration) over the last 18,000 years. Up-pointing arrows correspond to seven sets of simulation experiments using the climate model. (b) Simulated net radiation, expressed as departures from the modern case, in the latitude belt 8.9°–26.6° N. (c) Simulated surface temperature, expressed as departures from the present case, in the latitude belt 8.9°–26.6° N. (d) Simulated values of annual precipitation and effective precipitation in the latitude belt 8.9°–26.6° N. The tops of the columns represent precipitation, the lightly shaded areas evaporation, and the darkly shaded areas effective precipitation (which is available for runoff). The stars indicate that values of P and $P–E$ are significant at $p < 0.05$ in a two-sided t-test. (e) Lake–area ratios, z, assuming a constant ratio of land area to internal drainage, e, and constant lake evaporation, E_L. (f) Lake–area ratios, z, assuming variable values of e and E_L. (g) Temporal variations in the percentage of lakes with low, intermediate, or high levels. Maximum sample size = 73

Source: After Kutzbach and Street-Perrott (1985)

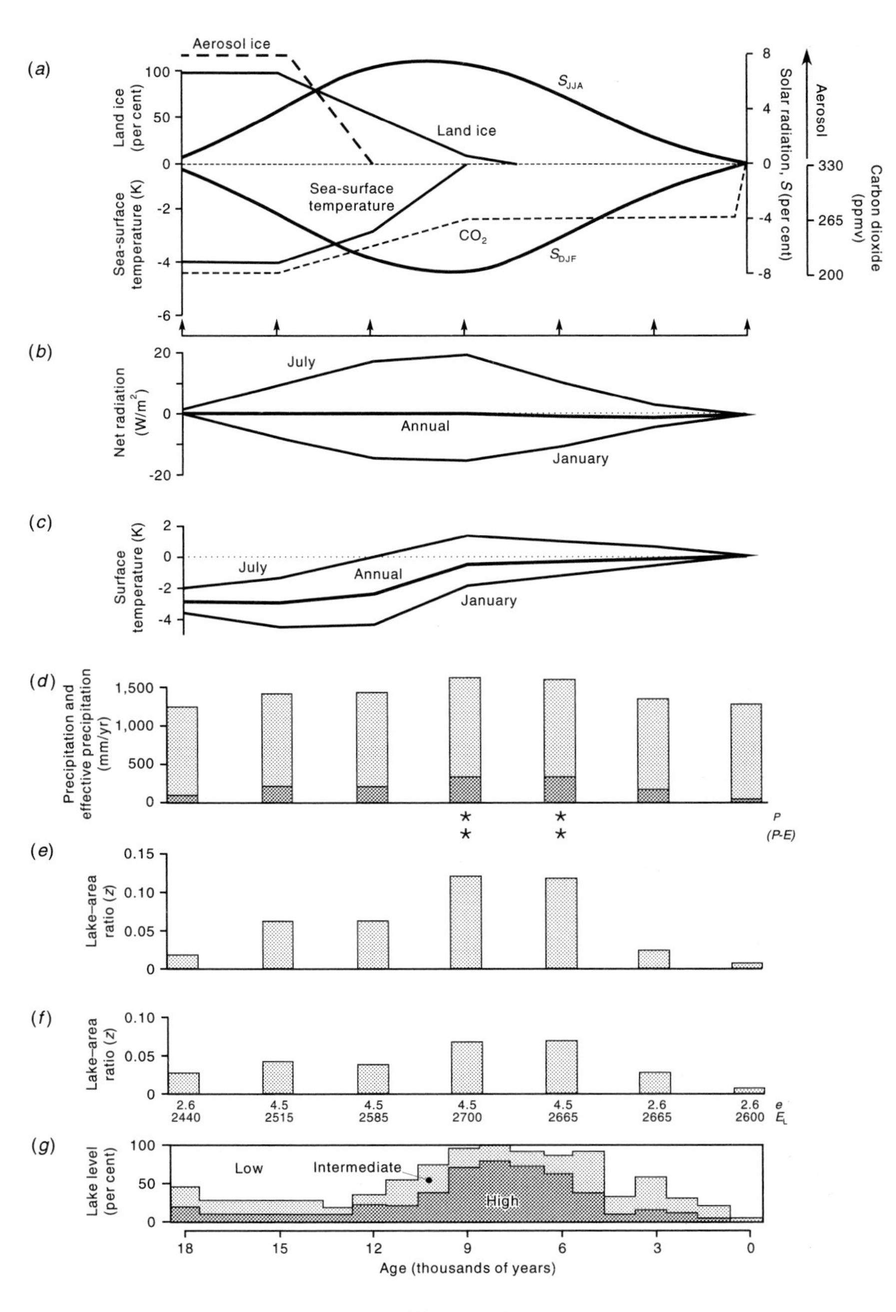

(a)
Aerosol ice
Land ice (per cent)
100
50
0
S_{JJA}
Land ice
Sea-surface temperature
Sea-surface temperature (K)
-2
-4
-6
CO_2
S_{DJF}
Solar radiation, S (per cent)
8
4
0
-4
-8
Aerosol
Carbon dioxide (ppmv)
330
265
200
(b)
Net radiation (W/m²)
20
0
-20
July
Annual
January
(c)
Surface temperature (K)
2
0
-2
-4
July
Annual
January
(d)
Precipitation and effective precipitation (mm/yr)
1,500
1,000
500
0
P
(P-E)
(e)
Lake-area ratio (z)
0.15
0.10
0.05
0
(f)
Lake-area ratio (z)
0.10
0.05
0
2.6 2440
4.5 2515
4.5 2585
4.5 2700
4.5 2665
2.6 2665
2.6 2600
e
E_L
(g)
Lake level (per cent)
100
50
0
Low
Intermediate
High
18
15
12
9
6
3
0
Age (thousands of years)

Figure 5.8

9,000–6,000 years ago (Figure 5.8d). The water budgets predicted by the model were used to estimate the area of closed lakes located between latitudes 8.9° and 26.6° N. Predicted lake levels closely matched actual lake levels as reconstructed from geological evidence (Figure 5.8e, f, and g; Plate 5.1).

Eustatic change

Short-term eustatic changes are locked into solar and lunar cycles. This applies to regional eustatic sea-level changes, such as in northern Europe (Currie 1976), and to global eustatic sea-level changes (Currie 1992b). The differences involved range from about 20 to 40 mm. Sea-level variations in nine stations from eastern Canada yielded spectral peaks at 18.6, 10.8, and 8.47 (lunar perigee) years (El Sabh and Murty 1993). The 18.6-year lunar nodal cycle, and a 14-month pole tide, are present in tide-gauge sea-level data from around the globe (Trupin and Wahr 1990).

Glacio-eustatic processes produce appreciable fluctuations of sea level during ice ages. These fluctuations are locked into orbital pulses. High-

Plate 5.1 Lake Chew Bahir (formerly Lake Stefanie), southern Ethiopia. The middle-ground feature is a fossil spit. The lake level was much higher around 9,000–6,000 years ago, which accords with the predicted climate for that time
Photograph by Andrew S. Goudie

stands and low-stands of sea level during Quaternary ice ages are recorded in marine terraces and drowned landscapes (e.g. Butzer 1975; Aharon 1984; Bloom and Yonekura 1985, 1990; Gallup *et al.* 1994; Ludwig *et al.* 1996). They accord with peaks and troughs in the Milankovitch radiation curves. For example, thorium dates of ancient coral reefs in Eniwetok atoll, the Florida Keys, and the Bahamas showed that four maxima for Milankovitch's radiation curve for latitude 65° N correspond to three high-stands of sea level – 120,000 years ago, 80,000 years ago, and today (Broecker 1965). Similarly, thorium dates for three coral-reef terraces on Barbados matched interglacial episodes predicted by a revised version of the Croll–Milankovitch theory (Broecker *et al.* 1968). The terraces were dated at 125,000, 105,000, and 82,000 years old. The 105,000-year-old terrace was puzzling. The Milankovitch radiation curve for latitude 65° N had no maximum at that date. However, examination of the radiation curve for lower latitudes, in particular 45° N, did contain peaks near all three of the Barbados terrace dates. The lack of a radiation peak 105,000 years ago at latitude 65° N results from the pulse of the tilt cycle being felt more strongly at higher latitudes. In lower latitudes, the precessional pulse is more marked.

Glacio-eustatic cyclicity is recorded in many pre-Quaternary sediments. It is found, for example, in middle Pennsylvanian (Desmoinesian) shelf carbonates of the south-western Paradox Basin, which lies beneath south-east Utah and south-west Colorado, United States (Goldhammer *et al.* 1994).

Tectono-eustatic processes, especially changes in the volume of the mid-ocean ridge system, lead to marine transgressions and regressions (e.g. Hallam 1963; Russel 1968). An increase in the volume of the mid-ocean ridge system would cause the oceans to overflow their basins and spill onto continental lowlands, producing a transgression. Conversely, a decrease in the volume of the mid-ocean ridge system would cause the oceans to retire to their basins, leaving the previously flooded portions of continents dry, once the regression was complete. The key factor might be the rate of sea-floor spreading (Hallam 1971). If the sea-floor spreading rate should increase or decrease, then the mid-ocean ridges would become narrower or wider respectively, so causing global transgressions or regressions. This idea is supported by the finding that the rate of sea-floor spreading was uncommonly great, twice the normal rate, throughout the world in mid-Cretaceous times, with a peak at about 85 million years ago (Larson and Pitman 1972). If this increase in rate should have increased the volume of the mid-ocean ridge system appreciably, then a marine transgression would be expected to have occurred. A major transgression did occur in the Late Cretaceous that might have been caused by the elevated spreading rate of oceanic plates (Hays and Pitman 1973; Pitman 1978).

Other processes, including juvenile-water release at active ridge-edges, volume changes of the ocean basins owing to the differentiation of the lithosphere, variations in sedimentation, and crustal shortening, may lead to smaller relative changes of sea level.

FLOODS AND SUPERFLOODS: RATES OF HYDROSPHERIC CHANGE

The climate system may flip from one quasi-stable state to another with remarkable celerity. Such climatic flips affect the more responsive water stores of the hydrosphere. Lake levels are known to have changed abruptly during the last 14,000 years (Figure 5.9). Glacio-eustatic sea-level changes are slower than lake-level change, but early Holocene sea-level rise was fairly speedy (Figure 5.10; see also Lambeck 1990). Geological sea-level changes, as traced in Vail curves, show gradual rises and very rapid falls. The cause of the sudden plunges in sea level is unknown.

Not all hasty changes in the hydrosphere are linked to climatic change. Ultra-high-magnitude flooding events are caused by the rupturing of dammed lakes, by the passage of giant tsunami, and by oceanic impacts of asteroids and comets. These will be explored in detail.

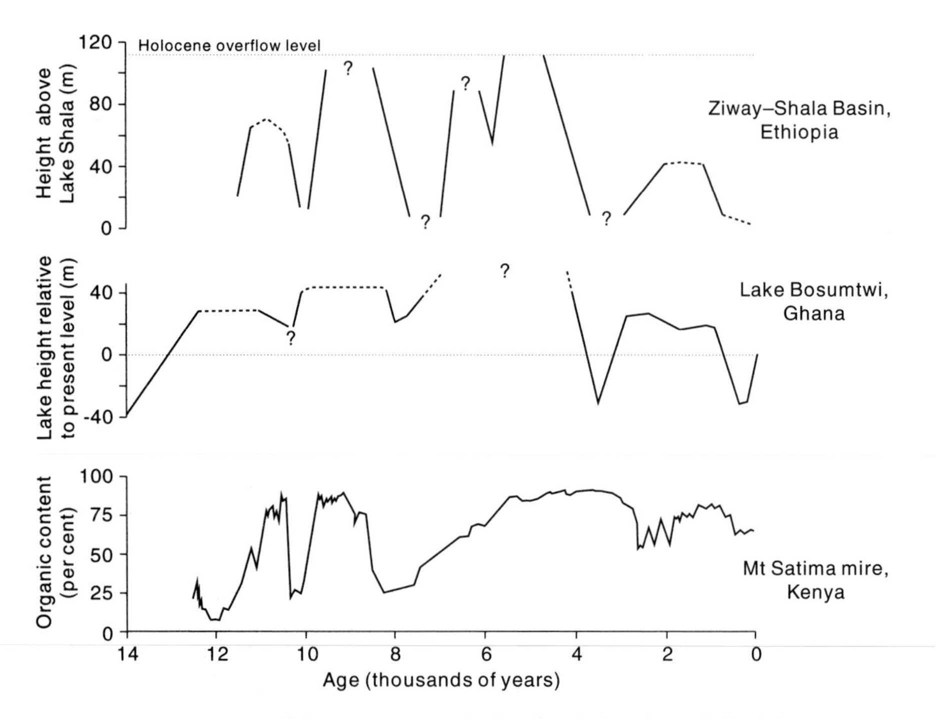

Figure 5.9 Rapid fluctuations in the levels of closed amplifier lakes
Source: After Street-Perrott and Perrott (1990)

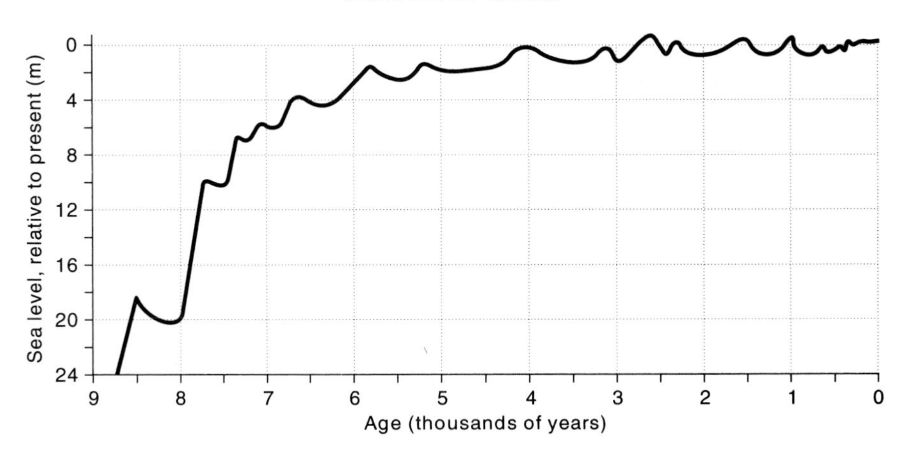

Figure 5.10 Holocene sea-level rise in north-western Europe. Note the rapid rise between about 9,000 and 7,000 years ago; this amounted to roughly 20m, an average increase of 1m per century
Source: After Mörner (1980)

Lake bursts

Large bodies of water impounded by ice or sediment cause catastrophic floods if suddenly freed. Such speculative events were controversial when first proposed, but, in the face of overwhelming evidence, they are widely accepted today (see Huggett 1989b: 149–59). Several cases of such dam breakage are well documented, the prime example being the Spokane Flood (Bretz 1923). This enormous event took place between 13,000 and 18,000 years ago in south-eastern Washington State. It involved two outbursts from glacial Lake Missoula following the failure of impounding dams of ice (V. R. Baker 1978a, 1978b). The floods filled normal valleys to the brim, and then spilt over the former divides, eroding the summits to complete the network of drainage ways. In doing so, it created the Channeled Scablands. This is a landscape of

> abandoned erosional waterways, many of them streamless canyons (coulees) with former cataract cliffs and plunge basins, potholes and deep rock basins, all eroded in the underlying basalt of the gently southwestward dipping slope of that part of the Columbia Plateau.
>
> (Bretz 1978: 1)

It also contains giant ripple marks, the size of small hills – an investigator, standing in a ripple trough, would be unable to see over the ripple humps! The flood water ran on down the Columbia River as far as Portland, Oregon, adding a 500-km^2 delta in the Willamette Valley. The discharge reached an estimated 21.3 million m^3/sec, and in some channels the flood-flow velocity touched 30 m/sec; but even at that extraordinary discharge, it would have taken a day to empty the lake of its 2.0 trillion m^3 of water (V. R. Baker 1973). Further studies have shown that, during Quaternary times, at least five major cataclysmic floods occurred in the general vicinity of the Channeled Scablands, of which the Spokane Flood was the last.

Other catastrophic lake bursts include the Lake Bonneville Flood (Malde 1968; Jarrett and Malde 1987), which took place about 15,000 years ago, and the catastrophic drainage of glacial Lake Agassiz through a north-western outlet following the incision of a drainage divide about 9,900 years ago (D. G. Smith and Fisher 1993). Pleistocene Lake Bonneville overtopped its rim at Red Rock Pass in south-eastern Idaho and rapidly lowered, decanting about 4,700 km^3 of water down the Snake River (Malde 1968). This debacle rushed down the Snake River Plain of southern Idaho to Hell's Canyon, causing extensive erosion and deposition. Today, the valley displays impressive abandoned channels, areas of scabland, and gravel bars composed of sand and angular and rounded boulders up to 3 m in diameter. The peak discharge, calculated using a step-backwater computational technique for the constricted reach of the Snake River Canyon at the mouth of Sinker Creek, was 793,000–1,020,000 m^3/sec (Jarrett and Malde 1987). At this rate of discharge, the shear stress for the flood would have been 2,500 N/m^2 (Newtons per square metre), and the unit stream power would have been 75,000 N/m.sec. This compares with shear stress and unit stream power for recent floods of the Mississippi and Amazon Rivers of 6–10 N/m^2 and 12 N/m.sec.

Catastrophic release of impounded water in glacial lakes creates smaller features. The Watrous spillway, Saskatchewan, Canada, was rapidly incised during a short-lived outburst from glacial Lake Elstow (Kehew and Teller 1994). In its outlet area, the bed of Lake Elstow was composed of stagnant ice. The 40-km-long spillway is incised across a divide, and ends in the glacial Last Mountain Lake Basin, where a coarse-grained fan was deposited. Large clasts are concentrated on the fan surface, and probably represent deposition at peak discharge.

Massive floods, though not so powerful as lake bursts, may be produced by extreme precipitation events. These superfloods have puzzled hydrologists because they cannot be understood in terms of drainage basin hydrology. Rather, they result from atmospheric circulation anomalies on an almost hemispherical scale (V. R. Baker 1983; Hirschboeck 1987). Although they are short-lived, these floods have long-term effects on the landscape (V. R. Baker 1977). Their discharge is limited by the physical constraints imposed by their drainage basins. This is the case for events with a recurrence interval of up to 2,000 years on the Salt River watershed, Arizona (J. Partridge and Baker 1987).

Giant tsunamis

Landforms and sediments in at least two coastal regions register what appear to be the effects of massive tsunamis propagated by submarine displacements. The first region is the Hawaiian Islands and the second is westernmost Washington State.

Several deposits on islands of the Hawaiian group suggest transport by giant waves. Gravel deposits on Lanai and Molokai were apparently laid down by such waves some 100,000 years ago (G. W. Moore and Moore 1984; J. G. Moore, Bryan, and Ludwig 1994). Giant landslides on the submarine flanks of the Hawaiian Ridge, which attain lengths of 200 km, probably generated the giant waves (J. G. Moore *et al.* 1989; J. G. Moore, Normark, and Holcomb 1994). The tsunami train would have moved out radially across the Pacific Ocean, eventually reaching continents around the Pacific Rim. Evidence of its passage is found along the eastern seaboard of Australia (Young and Bryant 1992). Sand barriers along the coast of southern New South Wales were almost utterly destroyed by a catastrophic tsunami, and vestiges of catastrophic wave erosion on coastal abrasion ramps are evident at least 15 m above present sea level. The barriers, which date from the last interglacial, appear to have been destroyed about 105,000 years ago, probably by the tsunamis generated near Hawaii.

In westernmost Washington State, extensive, well-vegetated lowlands (represented by peaty layers in estuarine sediments) have been buried beneath intertidal mud at least six times in the last 7,000 years. In three cases, the lowlands were also buried beneath sheets of sand. Tsunamis created by rapid tectonic subsidence (in the range 0.5–2 m) along the outer coast of Washington State may have caused these burials (Atwater 1987; Atwater *et al.* 1991). The subsidence was associated with large earthquakes (magnitude 8 or 9) emanating from the Cascadia subduction zone, where the Juan de Fuca plate slips below the North American plate (Figure 5.11).

Smaller tsunamis cause run-up and backwash, and leave traces in coastal regions (see Dawson 1994). For instance, tsunamis generated by the Lisbon earthquake of 1755 left traces in estuarine deposits in the Scilly Isles (Foster *et al.* 1991), and those generated by submarine slides off Iceland left vestiges in the Scottish and Norwegian landscapes (Dawson *et al.* 1988; Dawson *et al.* 1993).

Impact superfloods

All the above superfloods would be dwarfed by the scale of flooding hypothesized to be unleashed by an asteroid or comet crashing into the ocean. The gigantic waves thrown up by bombardment would produce floods that could truly be called super (Huggett 1989a).

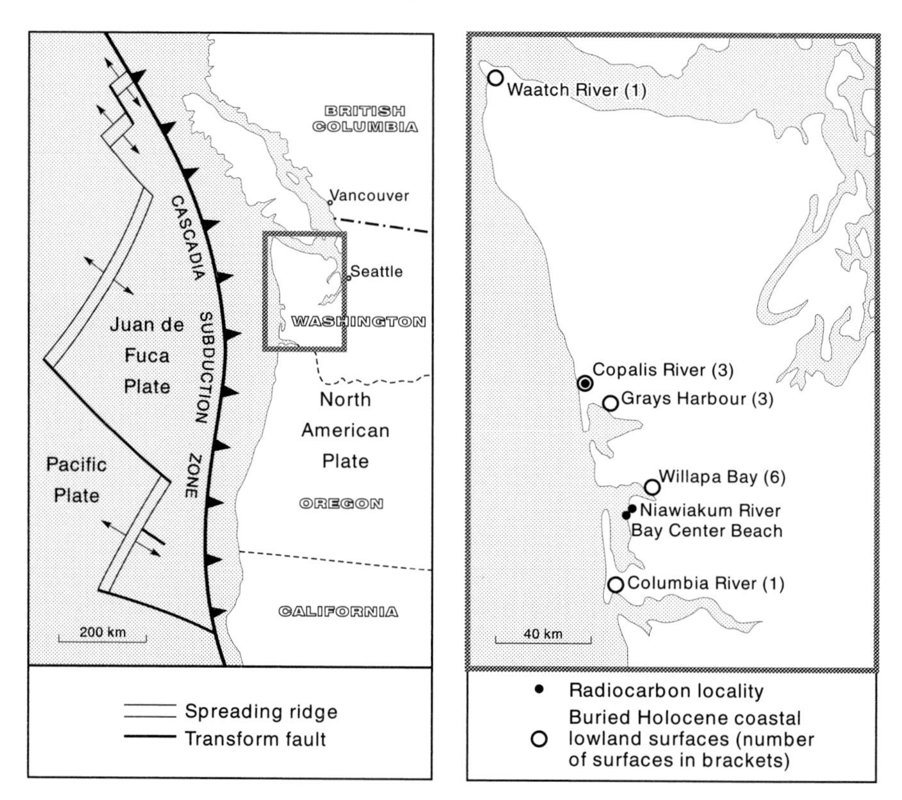

Figure 5.11 Location of the Juan de Fuca plate, rapid tectonic subsidence along which occasions giant tsunamis, and sites of buried Holocene coastal lowlands in Washington State
Source: After Atwater *et al.* (1991)

Earth-crossing asteroids and comets possess enormous kinetic energy. Should they crash into the ocean, they would create an enormous wave system that would flood continental lowlands (Figure 5.12; see also Hills *et al.* 1994). Fairly firm sedimentary evidence exists for the production of superwaves by asteroidal or cometary impacts. The Chicxulub structure on the Yucatán Peninsula, Mexico, is interpreted by some as an impact crater dating from the Cretaceous–Tertiary boundary (Figure 5.13). A plausible scenario is that an extraterrestrial object about 10 km in diameter smote the Earth, triggering massive earthquakes and the collapse of soft sediments down nearby continental slopes (Hildebrand *et al.* 1991). Giant tsunamis would have radiated from the Yucatán Peninsula, scouring sediments from the sea floor and coursing over surrounding lowlands, depositing a jumble of fine and coarse sediments. An outcrop at Mimbral, which lies across the

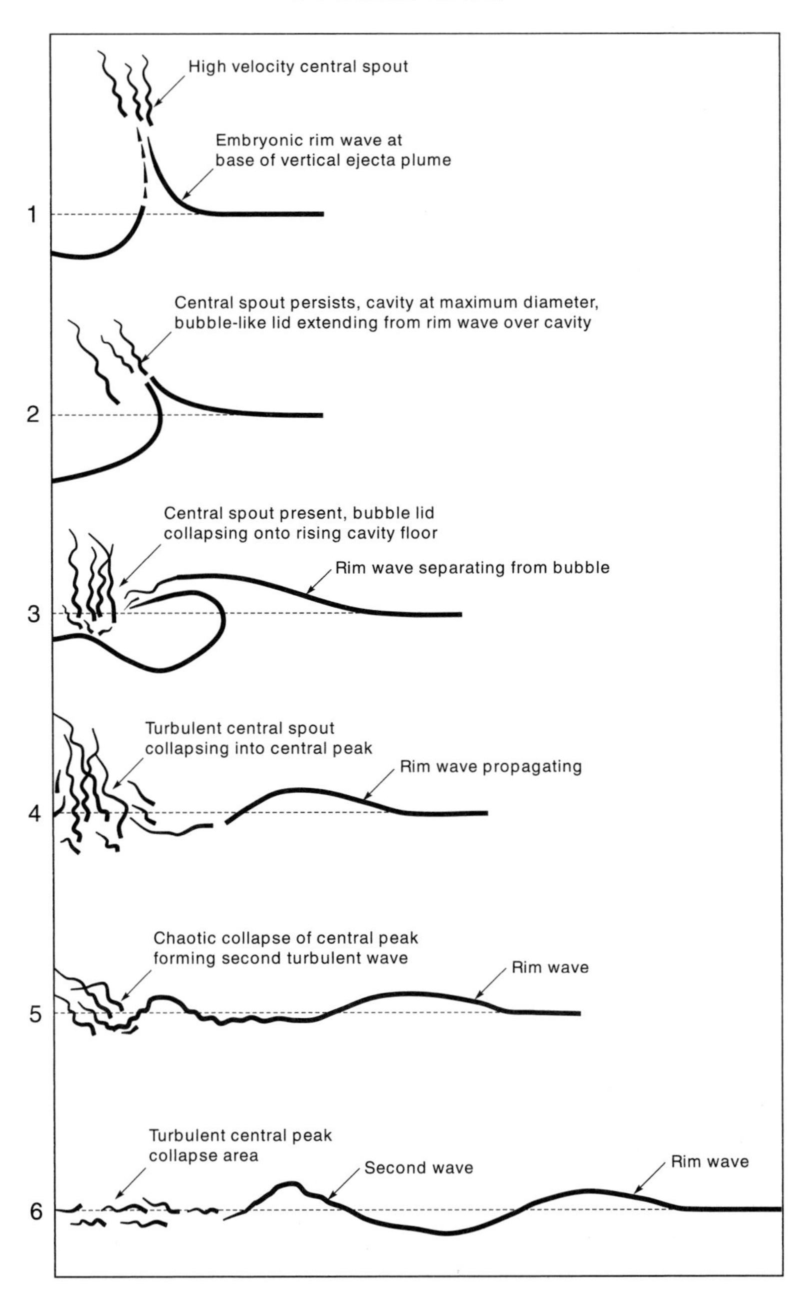

Figure 5.12 'Superwaves' produced in scale experiments. This is the sequence of waves generated by an aluminium sphere with a diameter of 0.317 cm impacting in water at 5.64 km/s. The time elapsed from start to finish is 1 second
Source: After Gault and Sonett (1982)

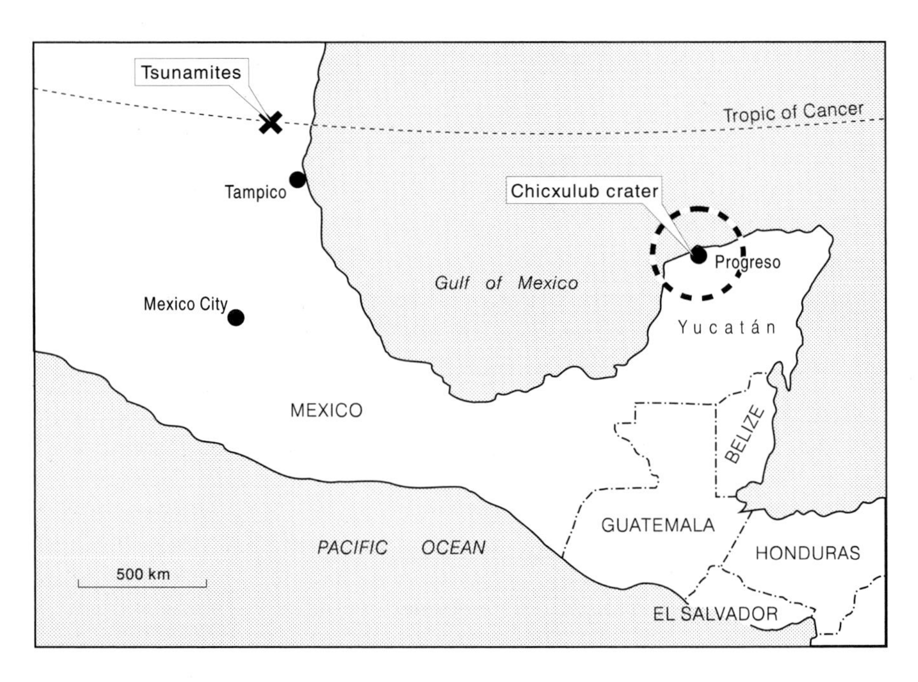

Figure 5.13 The Chicxulub crater, Yucatán, Mexico. Tsunamites containing marine sediments, glass spherules (produced by the impact), and bits of tree are found across the Gulf of Mexico

Gulf of Mexico from Chicxulub, records the sequence of events during the impact (Swinburne 1993). A spherule bed overlies Cretaceous deep-water sediments. These spherules would have been produced by molten droplets of rock thrown out of the impact crater and then cooled. On top of the spherule bed lies a wedge of sediments deposited by the train of tsunamis that would have rocked back and forth over the site. On top of the tsunami bed are ripples in fine-grained sediments that represent the last stages of tsunami dissipation. Similar tsunami deposits are found in Texas, where a wave amplitude of 50–100 m is indicated (Bourgeois *et al.* 1988).

Superflooding following oceanic impacts is not a process confined to the remote geological past. Several researchers believe that it occurred at the transition of the Pleistocene and Holocene epochs (e.g. Spedicato 1990; Kristan-Tollmann and Tollmann 1992). This might explain the flood myths found in nearly all cultures. It has been suggested that Noah's Flood occurred around 9,545 years ago when the Earth collided with a comet several kilometres in diameter that had broken into seven large pieces and several smaller bits. The cometary fragments generated truly gigantic waves, the waters from which gushed out from the sites of impact, streamed over mountain chains, and poured deep into continents (Kristan-Tollmann and Tollmann 1992). It has also been posited that a comet or asteroid might have struck the Laurentide ice sheet 11,000 years ago, creating enough water to submerge Canada and the north-eastern and northern midwestern United States to depths of 1–2 km, and to produce the Alberta erratics train and many drumlin fields (C. W. Hunt 1990: 169). That is certainly a novel explanation of rapid deglaciation!

WET AND DRY: DIRECTIONS OF HYDROSPHERIC CHANGE

The hydrosphere has changed significantly through Earth history. Its volume might have slowly increased as the rigid rocks of the lithosphere were converted into magma. This deep-seated mantle magmatism might have led to a contraction of the tectosphere and a concomitant expansion of the hydrosphere (Belyy 1994). Other long-term changes, many of which have a cyclical character, have resulted from various factors – the secular decrease in rotation rate, mountain building, continental drift, volcanism, changes in ocean circulation, and glacio-eustatic and tectono-eustatic changes of sea level. As the positions of the continents, the level of the oceans, and many other factors have changed over geological time, so the climate system, including the hydrosphere, has altered.

Many facets of long-term change in the hydrosphere could be discussed. Three will be selected: the origin of the hydrosphere, the late Precambrian water cycle, and cycles of sea-level change.

Watery beginnings

Opinions vary as to when the hydrosphere was created. Some people say it was nearly all formed early in Earth's history (Fanale 1971). Most say it accumulated slowly through geological time (A. T. Anderson 1975; Rubey 1951). At least one believes that most of the water was formed as modern ocean floors were created over the past 200 million years (Carey 1988).

If the Earth should have degassed very rapidly, within 50 million years of accretion, then the oceans could have appeared very early in Earth history (Condie 1989a: 342). Present volcanoes erupt volatiles that are 70 per cent water. Current production rates could replace the entire hydrosphere in about 55 million years. So creating a hydrosphere presents no difficulties. The puzzling question is why the hydrosphere does not contain more water (Ollier 1996). It is doubtful whether a hydrosphere could have existed before the early Archaean aeon as surface temperatures would have exceeded 100°C. Continued cooling led eventually to condensation of water vapour as rain. The first rains would have begun the first water cycle. The water cycle would have led to weathering, erosion, and transport of weathered products, and the deposition of sediments. Some of the oldest known rocks – greenstones from Godthaab, Greenland – are reliably dated at 3.8 billion years old (Moorbath 1977). They are shallow-water sedimentary rocks, so a water cycle must have been up and running by that time. The crust was still relatively thin then, so the oceans would have been shallower than at present, perhaps 2 km deep, and the continents not so elevated. The ocean covered much of the globe. Its surface was broken by volcanoes, on the flanks of which volcanic sediment aprons accumulated.

The composition of the early secondary atmosphere would have determined the composition of the early hydrosphere. This atmosphere, which was produced by volcanic gases, was reducing. Rainwater at the time would have been charged with copious quantities of hydrochloric and carbonic acids. Weathering would have continued apace, leaching carbonates, sulphates, halides, borates, and other salts from exposed rocks. Sea water would have been dominated by chloride and bicarbonate anions. Much iron would have been present in the highly soluble ferrous form (Fe^{2+}), as no oxygen was available to oxidize it to the poorly soluble ferric iron (Fe^{3+}). A higher partial pressure of carbon dioxide than at present would have given a pH of 6.7 (Garrels and Perry 1974). Salinity would have been higher than present salinity by up to 45 per cent (Mackenzie 1975). Little change in ocean composition would have occurred until the atmosphere, owing to photosynthetic organisms, became oxidizing. This change would have lowered the partial pressure of carbon dioxide (as more of it became incorporated either in organic material that was not re-oxidized or else in carbonate sediments), and the pH rose towards its present value. Iron 'rusted' out of the oceans, and sulphate became the main dissolved form of sulphur and the dominant anion in the sea. The balance between production and consumption of oxygen appears to have attained a steady state before the Phanerozoic aeon, by which time sea water was similar to modern sea water (Rubey 1951; Mackenzie 1975). Little overall change in seawater composition subsequently occurred (e.g. Horita *et al.* 1991), though it did fluctuate at times (see Clemens *et al.* 1993; Berner 1994). For example, the oceans have on occasions become anoxic over large areas (e.g. Wetzel 1991).

The late Precambrian water cycle

Little is known about Precambrian climates. The Precambrian Earth spun faster than the present Earth. An increased rotation rate would have significantly influenced climate. General circulation model simulations of the climate system with the rotation rate upped to five times its present value, as would have applied during Precambrian times, predicted distinct changes in the present climatic belts (B. G. Hunt 1979). The conventionally defined subtropical arid region of the globe was confined to a narrower latitudinal belt and was shifted towards the equator. Most of the globe poleward of latitude 45° became semiarid as the increased coldness of the troposphere led to a general decline in evaporation and precipitation in this region.

The implications of these simulation experiments for Precambrian Earth climates are very interesting. The smaller scale and less organized nature of meteorological systems in Precambrian times would have led to precipitation being less intense and more localized, which suggests that runoff and erosion would have operated at lower rates. Wind stress may have been reduced by a factor of two. In the oceans, the lower wind stress would have caused far less vertical mixing, a shallower mixed layer, a lower overall heat capacity, and a diminishment in the lateral transport of heat. All these changes would have caused the tropical atmosphere and oceans to have been warmer, and the polar regions to have been colder, than they are today. Lower wind stress would have influenced atmospheric composition: transfer of gases between the Earth's surface and the atmosphere would have been restricted, as would vertical transport in the troposphere. The colder and drier polar regions would have been biologically far less active than they are today, and this could have affected the production of minor, but very important, atmospheric gases, such as nitrogen oxides, ammonia, and methane.

Plainly, the faster Precambrian rotation rate strongly affected climate (see also p. 132). However, it would be interesting to model the combined effect of faster rotation rate and a larger (70°) obliquity (cf. Figure 2.6).

Cycles of sea-level change

Sea level is ultimately controlled largely by tectonics. In the case of tectono-eustasy, the control is direct. In the case of glacio-eustasy, the control is indirect: tectonics alter climates and climates alter sea level.

Sea level fluctuates over all time-scales. Medium-term and long-term changes are recorded in sedimentary rocks and revealed by the technique of seismic stratigraphy (e.g. Vail *et al.* 1977, 1991). This technique offers a precise means of subdividing, correlating, and mapping sedimentary rocks. It uses primary seismic reflections. The reflections come from geological discontinuities between stratigraphic units that result from relative changes of sea level. The discontinuities are lithological transitions caused by abrupt changes in sediment delivery, and they can be correlated world-wide. They display six superimposed orders of cyclical sea-level change during the Phanerozoic aeon (Table 5.8). Each cycle has a distinct signature. First-order cycles reflect major continental flooding. Second-order cycles register facies changes associated with major transgression–regression cycles. The lower-order cycles record stratigraphic deposition sequences, systems tracts (sets of linked, contemporaneous depositional systems), and parasequences (the building blocks of systems tracts).

Table 5.8 Quasi-periodic sea-level cycles revealed by seismic stratigraphy

Cycle order	*Duration (years)*	*Eustatic expression*
First	More than 50,000,000	Major continental flooding cycle
Second	3,000,000–50,000,000	Transgressive–regressive cycles
Third	500,000–3,000,000	Sequence cycles
Fourth	80,000–500,000	Systems tracts
Fifth	30,000–80,000	Episodic parasequences
Sixth	10,000–30,000	Episodic parasequences

Source: After Vail *et al.* (1991)

The secular decrease in the Earth's internal energy has caused changes in rates of tectonic processes. Concomitant changes have occurred in palaeogeography and sea level. First-order eustatic sea-level cycles, with periods in the range 46–90 million years, probably result from cyclic changes in mantle-upwelling rates and conductive heat transmission through the lithosphere (Jansa 1991). Interestingly, sea-level low-stands in first-order cycles correspond to the boundaries of geological periods, 'a circumstance that may be fundamental to our understanding of revolutions in the history of life' (G. E. Williams 1981: 218). These first-order cycles may themselves ride upon zeroth-order cycles lasting about 300 million years (Figure 5.14). This very protracted cycle is seen as overall high-stands during the early to middle Palaeozoic and the middle to late Mesozoic.

Geotectonic processes may be the dominant causes of long-term sea-level

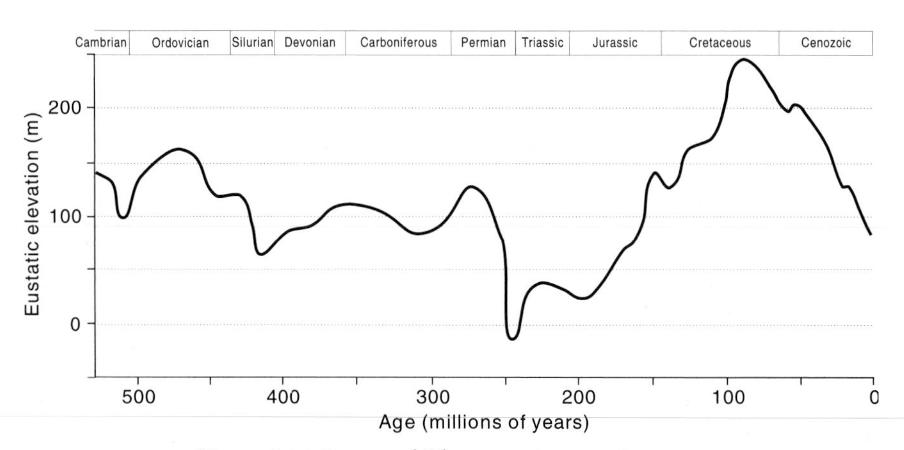

Figure 5.14 Proposed Phanerozoic eustatic curve

Sources: The Palaeozoic portion is an average estimated by Algeo and Seslavinsky (1995) and the Mesozoic and Cenozoic portion is from Haq *et al.* (1987)

change (Vail *et al.* 1977: 94). Sea-floor spreading rates and orogeny may determine the first-order cycle of sea-level change, and orogenic movements and volcanism may determine the second-order high-stands of sea level. All these long-term changes of sea level are thus tectono-eustatic in origin, and not glacio-eustatic. A plausible geotectonic explanation of very long-term sea-level change throughout Earth history lies in the supercontinent cycle (p. 88).

One of the many effects of the supercontinent cycle would be a systematic change of sea level resulting from thermal upheaving of continents and the creation and destruction of ocean basins (Figure 5.15). Sea level would be relatively low while a supercontinent existed. It would rise during breakup because the continental fragments would stretch and subside thermally, and because the breakup would replace old, Pacific-type ocean floor with young, Atlantic-type ocean floor. It would carry on rising for about 80 million years, as the proportion of younger oceans increased. Once the Atlantic-type oceans had aged and expanded, sea level would decline for another 80 million years or so, until the Atlantic-type oceans started to be subducted. The continents would begin to come together again and sea level would rise for another 80 million years, while older Atlantic-type crust was subducted. Sea level would drop for another 80 million years until continents collided and the growing supercontinent was uplifted thermally. It would stay static for another 120 million years once the supercontinent had been formed and before it broke up again, renewing the cycle.

A connection between the degree of continental fragmentation and sea level does exist. It is seen in the approximate correlation between the Phanerozoic sea-level curve (Figure 5.14) and the number of continents existing at various times during the Phanerozoic (Figure 3.16).

SUMMARY

The hydrosphere is all the Earth's waters. These waters circulate globally. They move round the water cycle, going from the oceans to the air, to the land, thence back to the oceans. Ocean waters circulate in surface gyres and in deep-water currents. Water stores and fluxes vary. Water-storage fluctuations register as changing lake levels, sea levels, water-table heights, ice cover, and so on. Water-flux fluctuations include changing rates of evaporation, advection, condensation, precipitation, and runoff. Hydrological change is a part of climatic change. The changes are expressed as phases of aridification and humification, the growth and decay of ice sheets, monsoon strength, droughts and floods, and so forth. Many of these changes are partly dictated by solar, lunar, and orbital forcings. Sea level varies according to the ice volume in ice sheets and glaciers (glacio-eustatic change), as well as ocean-basin volume (tectono-eustatic change). Water

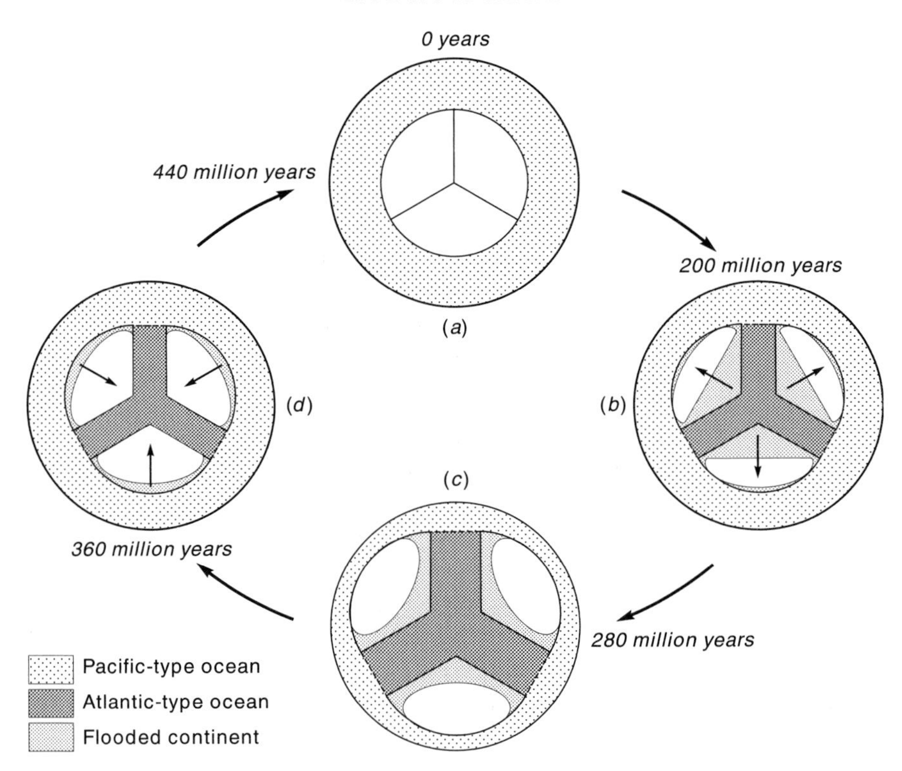

Figure 5.15 Eustatic sea-level change during a supercontinent cycle
Source: After Nance *et al.* (1988)

fluxes sometimes alter catastrophically. Massive floods result from the rupturing of dammed lakes, from the passage of giant tsunami, and from oceanic impacts of asteroids and comets. The hydrosphere originated by volcanic degassing of the solid Earth and, possibly, by cometary contributions. The volume of the Earth's waters might have slowly increased through geological time. Long-term cycles occur in sea level, the occurrence of ice ages, and phases of aridity and humidity.

FURTHER READING

The books suggested as further reading for the previous chapter are equally useful for this chapter. Readers may profitably consult *The World Water Balance* (Baumgartner and Reichel 1975), *Global Continental Palaeohydrology* (Gregory *et al.* 1995), and *Contemporary Hydrology* (Wilby 1996).

Plate 1 Mount Taranaki (formerly Mount Egmont), a dormant volcano rising to 2,158 m in North Island, New Zealand
Photograph by Brian S. Kear

Plate 2 A sedimentary sequence. This is a view of the 'goosenecks' in south-east Utah showing Desmoinesian (about 300-million-year-old) cyclic shelf-basin strata of the Paradox Basin. The strata record superimposed fourth-order and fifth-order cyclicity
Photograph by David W. Hunt

Plate 3a Mount Ngauruhoe, an active volcano rising to 2,291 m in North Island, New Zealand, emitting smoke
Photograph by Brian S. Kear

Plate 3b Occasional violent eruptions on Mount Ngauruhoe produce molten lava flows, seen as the dark ribbon emerging from beneath the white sinter in the lower part of the picture
Photograph by Brian S. Kear

Plate 4 Pohutu geyser, Rotorua, North Island, New Zealand. The solid Earth loses internal heat from such hydrothermal sites
Photograph by Brian S. Kear

Plate 5 Mountains: Mount Cook and the Hooker Glacier, New Zealand
Photograph by Brian S. Kear

Plate 6 A silver-studded blue butterfly
Photograph by N.R. Webb

Plate 7a Great Escarpment at Baker's Gorge, east of Armidale, New South Wales, Australia. The relief is about 200 m
Photograph by Cliff D. Ollier

Plate 7b Suicide Point (or Mrs Lamb's Leap), near Kodaikanal, Tamil Nadu, India. This is part of the Indian Great Escarpment on the east-facing slope of the Kerala–Tamil Nadu Plateau. The relief is over 400 m
Photograph by Cliff D. Ollier

6

PEDOSPHERE

SCRATCHING THE SURFACE: SOILS AND SEDIMENTS

Soils

Edaphosphere and debrisphere

Soils are defined in two ways. Geologists and engineers see soils as soft, unconsolidated rocks. The entire profile of weathered rock and unconsolidated rock material, of whatever origin, is then soil material. By this definition, soil is the same as regolith, that is, 'all surficial material above fresh bedrock' (Ollier and Pain 1996: 2). It includes *in situ* weathered rock (saprolite), disturbed weathered rock (residuum), transported surficial sediments, chemical products, topsoil, and a miscellany of other products, including volcanic ash.

Most pedologists confine soils to the portion of the regolith supporting plant life and dominated by soil-forming processes. This definition poses problems. Some saline soils and laterite surfaces cannot support plants – are they true soils? Is a bare rock surface encrusted with lichens a soil? Pedologists cannot agree on these vexatious matters. A possible way of skirting the problem is to define exposed hard rocks as soils (Jenny 1980: 47). This suggestion is not as harebrained as it appears on first acquaintance. Exposed rocks, like soils, are influenced by climate. Like some soils, they will support little or no plant life. Pursuing this idea, soil may be defined as 'rock that has encountered the ecosphere' (Huggett 1995: 12). This definition eschews the somewhat arbitrary distinctions between soil and regolith, and between soil processes and geomorphological processes (cf. Dan and Yaalon 1968; Brewer *et al.* 1970; B. E. Butler 1982). It means that the pedosphere is the portion of the lithosphere affected by living things, and that 'the soil' includes sedimentary material that is affected by physical and chemical processes, and, to far lesser degree, by biological processes. A definition along these lines is favoured by palaeopedologists (e.g. Nikiforoff 1959; Retallack 1990: 9).

Pedologists may feel uncomfortable with my geological definition of soil. However, they can find solace in a term of their own – solum. The solum is the genetic soil developed by soil-building forces (Soil Survey Staff 1975), and normally comprises the A and B horizons of a soil profile, that is, the topsoil and the subsoil. It is the 'soil proper' or edaphosphere (Huggett 1995: 13). The other portion of the pedosphere, the bit lying beneath the edaphosphere but above the limit of the ecosphere's influence, is the debrisphere (Huggett 1995: 13). It includes all weathered materials at the Earth's surface and at the bottom of rivers, lakes, and oceans that are not materially affected by animals, plants, and micro-organisms. It is approximately equivalent to the decomposition sphere (Büdel 1982), but includes detritus created by mechanical disintegration, as well as the productions of chemical weathering (cf. Nikiforoff 1935). The debrisphere is characterized by epimorphic processes (weathering, leaching, new mineral formation, and inheritance), while the edaphosphere is characterized by the 'traditional' soil-forming processes (cf. Paton *et al.* 1995: 110). Interestingly, some pedologists are broadening the scope of their purview to include the entire regolith (e.g. Creemans *et al.* 1994).

A soil profile is pedogenetically altered material plus the deep layers (the substrata) that influence pedogenesis. The 'deep layers' are called parent material or parent rock. They are also called lithospheric material (Paton *et al.* 1995: 108) Parent material is material from the lithosphere (igneous, metamorphic, and sedimentary rocks) or biosphere (peat and other organic debris) lying within the influence of, and subject to alteration by, the ecosphere. It is the material in the debrisphere. Parent materials derived from the lithosphere exist in a virtually unaltered state only in those parts of the lithosphere that the biosphere cannot reach. Such little-altered material is grandparent material or bedrock. It is not part of the debrisphere. However, the ecosphere's influence runs deep indeed. Bacterial populations exist more than 500 m below the floor of the Pacific Ocean (Parkes *et al.* 1994), and microbial life is widespread down to 4,200 m in the continental lithosphere, where it may be involved in subterranean geochemical processes (Pedersen 1993). That's one monumental soil profile!

Soil horizons

A pit dug into the ground reveals a series of roughly parallel, horizontal, and fairly distinct layers known as soil horizons (Plate 6.1). These horizons are labelled by a system of capital letters, which signify main divisions, and lower-case letters, which signify subdivisions. Unfortunately, no lettering system is internationally agreed. Table 6.1 lists some common horizon designations. As a rule, A and E horizons are depleted in solutes, colloids, and fine particles by the processes of eluviation and pervection (mechanical eluviation of silt and clay); whereas B horizons are enriched in the solutes,

Plate 6.1 A soil profile with distinct soil horizons. The podzol is formed in the Te Kopuru Sand, Dargaville, North Auckland, New Zealand. The eluvial (E) horizon, which lies behind the spade, is bleached and ash colour. Its top end is stained by organic matter from the surface organic (O) horizons. A very dark illuvial horizon of organic matter accumulation (a Bh horizon) lies beneath the E horizon. It changes abruptly into a paler illuvial horizon of iron accumulation (a Bfe horizon). This horizon is stained orange and brown by the iron, though is not evident in a black-and-white picture. At the base of the profile is weathered parent material (a C horizon). The soil originally evolved under Kauri pine (*Agathis australis*) forest
Photograph by Brian S. Kear

Table 6.1 Soil horizons

Horizons: type and symbol	*Description*
Organic horizons on mineral soil surfaces	
O	Horizon formed from organic litter derived from animals and plant (mainly mosses, rushes, and woody materials)
L	Largely undecomposed litter (mainly leaves, spent fruits, twigs, and woody materials)
F	Partly decomposed litter (mainly leaves, spent fruits, twigs, and woody materials)
H	Well-decomposed humus layer, low in mineral matter. Organic structures indiscernible
Mineral topsoil horizons	
A	Mineral horizons that are rich in organic matter, or that have lost clay, iron, and aluminium, or both
A1	Mineral–organic horizon formed at, or immediately below, the surface
A2 or E	Light-coloured eluvial horizon (clay or sesquioxides of iron and aluminium, or both, removed)
A3	Horizon transitional between A and B horizons. Dominated by properties of overlying A1 or E horizons, but with some B horizon properties
AB or A/B	Horizon of transitional character between A and B, with an upper part dominated by A horizon properties and a lower part dominated by B horizon properties, although the two parts cannot be separated into A3 and B1 horizons
AC or A/C	Horizon transitional between A and C and having subordinate properties of both
Mineral subsurface horizons	
B	Altered horizons distinguished from superjacent A or E horizons and subjacent C horizon by colour, structure, and illuvial concentrations of silicate clay, iron, aluminium, or humus
B1	Transitional horizon between B and A1 or E. Properties of the underlying B2 horizon dominate, while properties of overlying A1 or E horizon are subordinate
B2	The part of the subsoil where properties are clearly expressed without subordinate properties suggestive of a transitional layer
B3 or B/C	Horizon transitional between B and C horizons. Dominated by properties of an overlying B2 horizon, but with some C or R horizon properties. Only defined if a B2 horizon exists
Mineral substrata horizons	
C	Lowest mineral horizon, excluding bedrock. Little altered save by epimorphic processes, irreversible and reversible cementation, gleying, and the accumulation of calcium and magnesium carbonates or more soluble salts. May or may not be the same as the overlying material
R	Underlying consolidated bedrock
Subhorizon symbols	
b	Buried soil horizon
c	Irreversibly cemented horizon
ca	Accumulations of secondary carbonates, commonly calcium carbonate
cs	Accumulations of calcium sulphate

Table 6.1 (continued)

Horizons: type and symbol	*Description*
cn	Accumulations of concretions or hard, non-concretionary nodules enriched in iron and aluminium sesquioxides, with or without phosphorus
e	Eluvial horizon (eluviation of clay, iron, aluminium, and organic matter, alone or in combination)
f	Permanently frozen horizon
g	Strongly gleyed (mottled) horizon formed under reducing conditions
h	Illuvial humus accumulation
fe (or ir)	Illuvial iron accumulation
m	Strong cementation and induration
p	Disturbance by ploughing, cultivation, or other human activities
sa	Accumulations of secondary salts more soluble than calcium and magnesium carbonates
si	Nodular or continuous cementation by siliceous material, soluble in alkali
t	Illuvial silicate clay accumulation
u	Marked disruption by pedoturbation (other than cryoturbation)
x	Fragipan character (firm, brittle, high-density, low-organic matter)

colloids, and fine particles lost from the overlying A horizons. C horizons are weathered parent material. O horizons are organic material that accumulates on the soil surface.

The A, E, B, and C horizon labels were devised to describe mid-latitude soils. Their applicability to tropical and subtropical soils is debatable. Tropical and subtropical soil horizons are sometimes denoted by the letters M, S, and W (e.g. Watson 1961; M. A. J. Williams 1968). They describe the tripartite profile of many tropical and subtropical soils – a mineral layer (M) overlies a stony layer (S) that sits upon weathered rock (W). It now seems likely that mid-latitude A, E, and B–C horizons are equivalent to tropical M, S, and W horizons (Johnson 1994).

Soil types

Soil profiles differ from one another in varying degrees and may be classified accordingly. Soil classification is essentially a matter of comparing horizon sequences, as well as chemical and physical horizon properties.

Soil classification schemes are multifarious, nationalistic, and use confusingly different nomenclature. Old systems were based on geography and genesis. They designated soil orders as zonal, intrazonal, and azonal; divided these into suborders; and then subdivided the suborders in Great Soil Groups such as tundra soils, desert soils, and prairie soils. Newer systems give more emphasis to measurable soil properties that either reflect the genesis of the soil or else affect its evolution. The most detailed and comprehensive new classification was prepared by the Soil Conservation Service of the US Department of Agriculture and published, after many approximations, in 1975. To ease communication between soil surveyors, the nomenclature eschews the early genetic terms and, for units above the series level, uses names derived mainly from Greek and Latin. The taxonomy is based on class distinction according to precisely defined diagnostic horizons, soil moisture regimes, and soil temperature regimes. Eleven orders are distinguished (Table 6.2 and Figure 6.1). The orders are successively subdivided into suborders, great groups, subgroups, families, and series.

Oxisols, Ultisols, and Vertisols dominate huge tracts of low-latitude landscapes. They have a well-developed horizon structure and display signs of prolonged and intense weathering. Oxisols are red, yellow, and yellowish brown in colour. They are exceptionally well weathered, mainly because of their extreme old age. They lack distinct horizons, apart from darkened

Table 6.2 Soil orders

Order	*Diagnostic character(s)*	*Older names of soils included*
Entisols	Very slightly developed with no diagnostic horizons	Recently formed soils – hydromorphic soils, alluvial soils, regosols, lithosols
Vertisols	Contain swelling clays	Shrinking and swelling dark clay soils – grumusols
Inceptisols	Rapidly forming umbic or cambic (B) horizons	Embryonic soils – pseudogleys, rankers, plaggen soils, some brown soils, acid brown soils
Aridisols	Soils of arid climates	Desert soils – sierozems, brown sierozems, solonchak, some brown and reddish brown soils
Mollisols	Mollic A1 horizon	Temperate grassland soils – calcareous soils (rendzinas), planosols, solonetz, gleyed humic soils, chernozems, brunizems (prairie soils), reddish chestnut soils, chestnut soils, some brown soils
Spodosols	Spodic B horizon	Podzols – gleyed humic podzols, humic podzols, podzolic soils, iron–humus podzols, iron podzols
Alfisols	Argillic B horizon, not very weathered (high base status)	Lessived soils – grey-brown podzolic soils, gray wooded soils, noncalcic brown soils, planosols, ferruginous and fersiallitic soils
Ultisols	Argillic B horizon, weathered and very unsaturated (low base status)	Ferruginous soils or ferrisols – red-yellow podzolic soils, reddish-brown lateritic soils, associated planosols
Oxisols	Oxic horizon rich in sesquioxides	Ferrallitic soils of tropical regions – latosols
Histosols	Histic horizon	Organic soils – peats
Andisols	More than half parent material is volcanic ash	Soils formed on volcanic ash – andosols

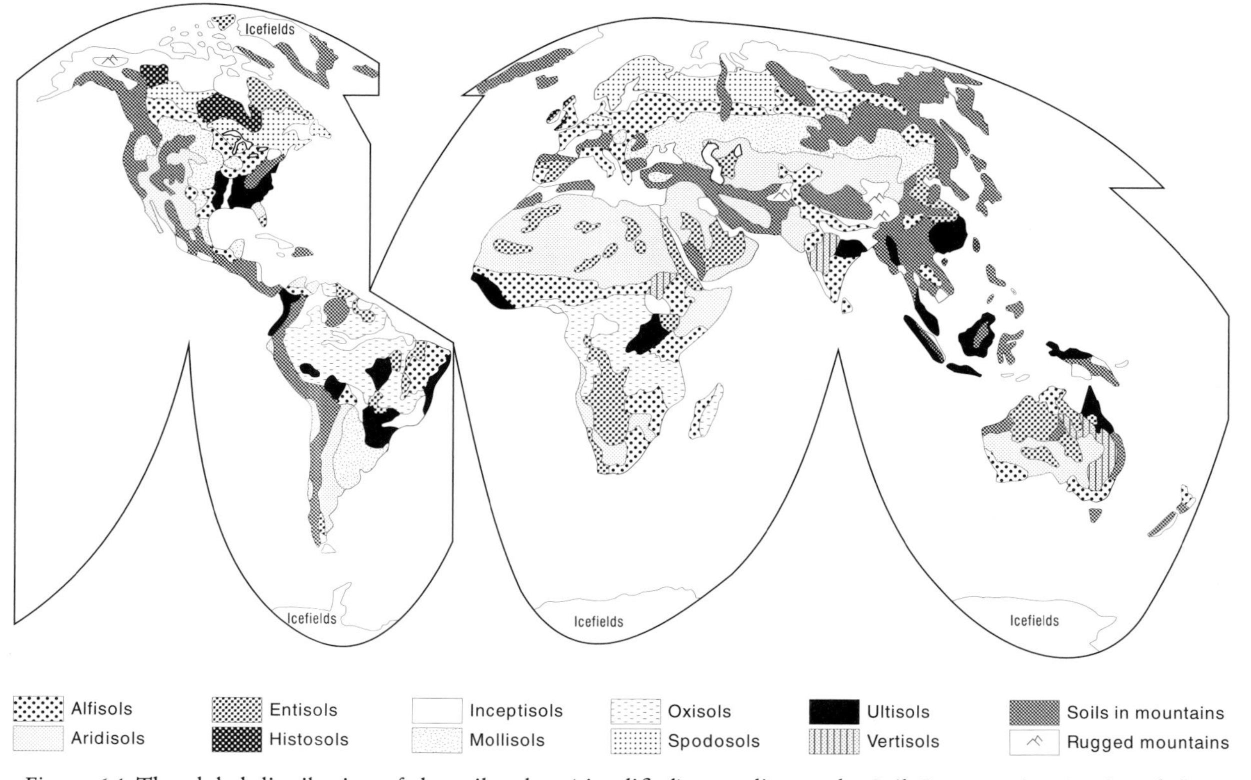

Figure 6.1 The global distribution of the soil orders (simplified) according to the Soil Conservation Service of the United States Department of Agriculture

surface layers and, in places, a subsurface sesquioxide horizon (called plinthite). Ultisols are closely related to Oxisols. They are reddish to yellowish in colour and possess an argillic (rich in illuvial clay) B horizon. Vertisols are black soils with a high clay content. They contain a large proportion of expanding clays that swell and contract with wet and dry seasonal cycles.

Alfisols, Spodosols, Mollisols, and Aridisols all have well-developed horizons and fully weathered minerals. Mostly, they have evolved over thousands of years or more and reflect temperature and moisture conditions. Alfisols are associated with humid and subhumid climates. They have a subsurface horizon of clay accumulation and high base status. Spodosols are characteristic of moist, cold climates. They have well-developed illuvial B horizons. Mollisols are associated with semiarid and subhumid grasslands of middle latitudes. They have a dark epipedon, rich in humus and of high base status. Aridisols are associated with dry climates. They have little organic matter and subsurface horizons with calcium carbonate accumulations and accumulations of soluble salts.

Histosols have a deep upper layer very rich in organic matter. Entisols, Inceptisols, and Andisols are relatively little evolved soils. Entisols form on recently deposited material and display no horizons. Inceptisols show weak horizon development and possess unweathered or partially weathered minerals. Andisols are formed on material with a high percentage of glassy volcanic fragments. They have weakly developed horizons.

Sediments

Sediments are a stage in the rock cycle. They are material temporarily resting at or near the Earth's surface. Sedimentary material is derived from weathering, from volcanic activity, from the impact of cosmic bodies, and from biological processes. Nearly all sediments accumulate in neat layers that obligingly record their own history of deposition. In the fullness of Earth history, deposition has produced the geological or stratigraphic column. If the maximum known sedimentary thickness for each Phanerozoic period is summed, about 140,000 m of sediment results (A. Holmes 1965: 157).

Clastic sediments

Clastic sediments form through rock weathering. Weathering attacks rocks chemically and physically and so softens, weakens, and breaks them. The process releases fragments or particles of rock, which range from clay to large boulders. These particles may accumulate *in situ* to form a regolith. Once transported by a fluid medium (air, water, or ice) they become clastic sediments.

Clastic sediments are normally grouped according to their size. Loose sediments and their cemented or compacted equivalents have different names. The coarsest loose fragments (2 mm or more in diameter) are gravels, pebbles, cobbles, and boulders. When indurated, these deposits form rudaceous sedimentary rocks. Examples are conglomerate, which consists largely of rounded fragments held together by a cement, and breccia, which consists largely of angular fragments cemented together. Loose fragments in the size range 2–0.02 mm (the lower size limit varies a little between different systems) are sands. Indurated sands are known as arenaceous sedimentary rocks. They include sandstone, arkose, greywacke, and flags. Loose fragments smaller than about 0.02 mm are silts and clays. Silt is loose particles with a diameter in the range 0.02–0.002 mm. Clay is loose and colloidal material smaller than 0.002 mm in diameter. Indurated equivalents are termed argillaceous rocks (which embrace silts and clays). Examples are claystone, siltstone, mudstone, and shale. Clay-sized particles are often made of clay minerals, but they may also be made of other mineral fragments.

Chemical sediments

The materials in chemical sediments are derived mainly from weathering, which releases mineral matter in solution and in solid form. Under suitable conditions, the soluble material is precipitated chemically. The precipitation usually takes place *in situ* within soils, sediments, or water bodies (oceans, seas, lakes, and, less commonly, rivers). Iron oxides and hydroxides precipitate on the sea-floor as chamosite, a green iron silicate. On land, iron released by weathering goes into solution and, under suitable conditions, precipitates to form various minerals, including siderite, limonite (bog-iron), and vivianite. Calcium carbonate carried in groundwater precipitates in caves and grottoes as sheets of flowstone (travertine) or as stalagmites, stalactites, and columns of dripstone. It sometimes precipitates around springs, where it encrusts plants to produce tufa. Evaporites form by soluble-salt precipitation in low-lying land areas and inland seas. They include halite or rock-salt (sodium chloride), gypsum (hydrated calcium sulphate), anhydrite (calcium sulphate), carnallite (hydrated chloride of potassium and magnesium), and sylvite (potassium chloride). Evaporite deposits occur where clastic additions are low and evaporation high. At present, evaporites are forming in the Arabian Gulf, in salt flats or sabkhas, and around the margins of inland lakes, such as Salt Lake, Utah. Salt-flat deposits are known in the geological record, but the massive evaporite accumulations, which include the Permian Zechstein Basin of northern Europe and the North Sea, may be deep-water deposits, at least in part.

Chemicals precipitated in soils and sediments often form hard layers called duricrusts. These occur as hard nodules or crusts, or simply as hard

Plate 6.2 Calcrete horizon developed in loess, Tajikistan, central Asia
Photograph by Andrew S. Goudie

layers (Plate 6.2). The chief types are ferricrete (rich in iron), calcrete (rich in calcium carbonate), silcrete (rich in silica), alcrete (rich in aluminium), gypcrete (rich in gypsum), and manganocrete (rich in manganese). Hardpans and plinthite also occur. These are hard layers but, unlike duricrusts, are not enriched in a specific element.

Biogenic sediments

Ultimately, the chemicals in biogenic sediments and mineral fuels are derived from rock, water, and air. They are incorporated into organic bodies and may accumulate after the organisms die. Limestone is a common biogenic rock. It is formed by the shells of organisms that extract calcium carbonate from sea water. Chalk is a fine-grained and generally friable variety of limestone. Some organisms extract a little magnesium as well as calcium to construct their shells – these produce magnesian limestones. Dolomite is a calcium–magnesium carbonate. Other organisms, including diatoms, radiolarians, and sponges, utilize silica. These are sources of siliceous deposits such as chert and flint and siliceous ooze.

The organic parts of dead organisms may accumulate to form a variety of biogenic sediments. The chief varieties are organic muds (consisting of finely divided plant detritus) and peats (called coal when lithified). Traditionally, organic materials are divided into sedimentary (transported) and sedentary (residual). Sedimentary organic materials are called dy, gyttja, and alluvial peat. Dy and gyttja are Swedish words that have no English equivalent. Dy is a gelatinous, acidic sediment formed in humic lakes and pools by the flocculation and precipitation of dissolved humic materials. Gyttja comprises several biologically produced sedimentary oozes. It is commonly subdivided into organic, calcareous, and siliceous types. Sedentary organic materials are peats, of which there are many types (see Shotyk 1992).

Sedimentary environments

Weathering renders sediment susceptible to transport by wind and water. Solid and soluble materials are carried from one place to another, normally from upland to lowland, to the sea. Where transporting capacity is insufficient to carry the solid sediment load, or where the chemical environment leads to the precipitation of the solute load, deposition of sediment occurs. Sedimentary bodies occur where deposition outpaces erosion, and where chemical precipitation exceeds solutional loss. Sediments accumulate in all terrestrial and marine environments to produce depositional landforms. As a rule, the land is a sediment source and the ocean is a sediment sink. None the less, there are extensive bodies of sediments on land and many erosional features on the ocean floor.

Three main sedimentary environments exist – terrestrial, shallow marine, and deep marine. Each of these is dominated by a single sedimentary process: gravity-driven flows (dry and wet) in terrestrial environments; fluid flows (tidal movements and wave-induced currents) in shallow marine environments; and suspension settling and unidirectional flow created by density currents in deep marine environments (Fraser 1989). Transition zones separate the three main sedimentary environments. The coastal transition zone separates the terrestrial and shallow marine environments; the shelf-edge–upper-slope transition zone separates the shallow and the deep marine environments. Sediments found in terrestrial and marine environments will be considered in a little detail.

Terrestrial deposits

On land, the most pervasive 'sedimentary body' is the weathered mantle or regolith. The thickness of the regolith depends upon the rate at which the weathering front advances into fresh bedrock and the net rate of erosional loss (the difference between sediment carried in and sediment carried out by water and wind). At sites where thick bodies of terrestrial sediments accumulate, as in some alluvial plains, the materials would normally be called sediments rather than regolith. But regolith and thick sedimentary bodies are both the product of exogenic processes. They are thus distinct from the underlying bedrock, which is a production of lithospheric processes.

Unconsolidated weathered material in the regolith is transported by gravity, water, and wind across hillslopes and down river valleys. Local accumulations form stores of sediment. Sediment stored on slopes is called talus, colluvium, and talluvium. Talus is made of large rock fragments, colluvium of finer material, and talluvium of a fine and coarse material mix. Sediment stored in valleys is called alluvium. It occurs in alluvial fans and in floodplains. All these slope and valley stores, except for talus, are fluvial deposits (transported by flowing water).

Marine deposits

Marine clastic sediments form in shallow water and in deep water. Shallow-water clastic sediments are laid down in seas 10–200 m deep. Two types of sea fall into this depth category: a marginal or pericontinental type, which covers the modern continental shelves; and an epeiric or epicontinental type, which lies upon a continental area (e.g. the North Sea). Clastic rocks in these shallow-water environments are chiefly sandstones, siltstones, and mudstones. They often display signs of bioturbation (mixing by organisms).

Chemical sediments are sometimes found in these shallow-marine environments. The predominant chemical deposit is calcium or magnesium carbonate, which forms limestone and dolostone. These may form marine carbonate platforms. Carbonates form most rapidly in warm seas, such as the Bahamas Banks. Some carbonates accumulate as reefs built by metazoans. The reef structures, or bioherms (which mainly occur as atolls, barrier reefs, and fringing reefs), normally show a vertical succession of colonization and growth. Authigenic minerals occur in places. Phosphate deposits, for example, are laid down in areas of low clastic sedimentation. They result from the direct precipitation of calcium phosphate as nodules or laminae from sea water rich in phosphorus-bearing phytoplankton. Glauconite is produced mainly by the alteration of faecal pellets. Colloidal silica is sometimes deposited. It aggregates during diagenesis and later hardens to create chert or flint.

Deep-water marine sediments accumulate in abyssal plains lying beyond the edge of continental shelves. They are divided into three groups: non-pelagic, hemipelagic, and pelagic (Table 6.3). Non-pelagic sediments contain much land-derived (terrigenous) material created by chemical and physical weathering. They are found in the shallower, peripheral parts of the oceans. Pelagic sediments form from material in the deep oceans and contain no significant components derived from the land. They occur on the deep-

Table 6.3 Deep-ocean sediments

Type	*Comment*
Non-pelagic	
Terrigenous muds (silty clays)	Contain more than 30 per cent silt and sand derived from land. They include black, red, green, blue, and white varieties
Turbidites	Largely graded sands produced by turbidity currents from land areas or from submarine hills. They also occur in deep ocean trenches
Slide deposits	Carried to deep water by slumping
Glacial marine	Contain a significant percentage of allochthonous (foreign) particles derived from ice-rafting. Include dropstones
Volcanic debris	Consists of pyroclastic ash, tuff, and volcanic glass
Aeolian dust	Terrigenous sediment carried by wind
Cosmic dust	Tiny meteoroids
Hemipelagic	
Pelagic–non-pelagic sediment mix	Turbidites and olistromes
Pelagic	
Brown or red clay	Contains less than 30 per cent biogenous material
Diagenetic deposits	Contain mainly minerals crystallized in sea water, such as zeolites and manganese nodules
Biogenous deposits	Contain more than 30 per cent material derived from organisms. Include oozes formed from the calcareous tests of foraminifers, coccoliths, discoasters; from the aragonite shells of molluscs (pteropods); and from the opaline silica shells of radiolarians and frustules of diatoms
Planktonic debris	Dinoflagellates, pollen, spores, silicoflagellates, and others
Benthic debris	Sponge spicules, fish fragments, fish teeth, and others
Coral reef debris	Coral sands and white muds derived from slumping around reefs

Source: Partly after Shepard (1963)

ocean floor. Hemipelagic sediments are an intermediate group. They are confined to continental margins and the adjacent abyssal plain, while pelagic sediments occur in deep-ocean floor. They are sometimes laid down in large basins. They consist mainly of turbidite deposits and olistromes, which are mixtures of very ill-sorted sediments derived largely from submarine fault scarps.

Chemical and biogenic sediments are common in deep-marine environments. Carbonates are precipitated chemically from ocean water. Tests and other inorganic parts of small marine organisms – including foraminifers, coccoliths, discoasters, radiolarians, and diatoms – rain out of the oceans and accumulate on the ocean floor as oozes.

INS AND OUTS, TRANSFERS AND TRANSFORMATIONS: PEDOLOGICAL CHANGE

Soil and sedimentary systems

Soils (the edaphosphere) and sediments (the debrisphere) may be viewed as stores of materials at, and immediately below, the Earth's surface. The stores change owing to material gains and losses, transfers and transformations (see Simonson 1959, 1978). Edaphic and sedimentary systems are thus akin to a lithospheric slab: they are all open systems engaging processes of creation, destruction, movement, and change.

For all practical purposes, the edaphosphere changes independently of an almost unchanging debrisphere over short time intervals. For this reason, soil scientists studying annual, decennial, and centennial edaphic change can safely take parent material as a fixed quantity (unless, of course, soil erosion should be rife). This assumption works well when the relationships between edaphic change and agriculture and forestry are being considered. From the longer perspective of geology and geomorphology, parent material in many parts of the world is in a state of flux. Soil change thus involves short-term and long-term components, or what are sometimes styled short and long 'cycles' (Duchaufour 1982: 110). As a generalization, short-term changes involve the edaphosphere, while long-term changes affect the entire pedosphere (debrisphere and edaphosphere). Edaphic change includes several fast-running processes of physical and chemical transfer and transformation. These speedy processes allow quick adjustments of steady state in such properties as pH and nitrogen content. Sedimentary change generally involves slower-running processes of denudation (weathering, erosion, transport) and deposition that take longer to attain steady states. But the edaphic and sedimentary changes seldom act in isolation, and it is usually unrealistic to separate them where change over thousands of years is under consideration.

Change in soils

Soil genesis is explained by two chief theories – soil formation theory and dynamic denudation theory.

Soil formation

Soil formation theory, which arose in the closing decades of the nineteenth century, is the traditional paradigm. It originally postulated that soil genesis is the product of downwards-acting processes that lead to two sets of interrelated layers – the A horizons and the B horizons, which together constitute the solum. Eluviation washes solutes and fine-grained materials out of the A horizons and deposits them in subjacent, illuvial B horizons. Continued eluviation and illuviation produce coarse-textured residual A and E horizons over more heavily textured B horizons. Under some conditions, organic matter accumulates as distinct O horizons that lie on top of the uppermost A horizon. Later, it was realized that some soil processes mix soil materials and in doing so tend to destroy soil horizons. Soil formation theory was then modified to include the effects of horizon creation (horizonation) and horizon destruction (haploidization) (Hole 1961). The main pedogenic processes are listed and explained in Table 6.4.

According to soil formation theory, the nature and rate of pedogenic processes are regulated by 'factors of soils formation' – climate, organisms, relief, parent material, and time (Jenny 1941). This factorial–functional approach to soil genesis, which had a far-reaching impact in many environmental sciences, was the ruling theory until recently (see Johnson and Hole 1994).

Dynamic denudation

The dynamic denudation theory of pedogenesis emerged in the 1990s as a new model of soil–landscape evolution (Johnson 1993a, 1993b; Paton *et al.* 1995). It sees lithospheric material, topography, and life (through its role in biomechanical soil processes) as the prime determinative factors of pedogenesis. Its chief tenet is that A horizons are a biomantle (Johnson 1990) created by biomechanical processes, while B horizons are created by epimorphic processes (weathering, leaching, and new mineral formation) acting upon lithospheric material. This means that the textural contrast between A and B horizons is not primarily due to the eluviation and illuviation of fine materials.

The role played by biomechanical processes in soil genesis was recognized by Charles Darwin (1881) in his disquisition on earthworms, and by later workers (see Johnson 1993b). Not until recently was the action of animals and plants seen as a major factor in soil (and landscape) evolution (e.g. Johnson 1993b; D. R. Butler 1995). It is now accepted that the topmost portion of the weathered mantle is subject to mixing by organisms that live in the soil. This mixing is referred to as bioturbation. It is mainly caused by the activities of animals (faunal turbation). It is equally important in shallow-water marine sediments (e.g. Sepkoski *et al.* 1991). In soils, earthworms are the most effective bioturbator, followed in order by ants and termites, small burrowing mammals, rodents, and invertebrates. Bioturbation is vigorous enough in almost all environments for the near-surface soil to be described as a bioturbated mantle or biomantle. Some bioturbatory processes produce mounds of bare soil on the ground surface. These little piles of sediment are susceptible of erosion by rain splash and

Table 6.4 Soil processes

Process	*Type of process*	*Definition*
Littering	Gain	Accumulation of organic litter on the soil surface
Humification	Transformation	The conversion of raw organic matter into humus
Melanization	Translocation	The darkening of light-coloured mineral matter by organic matter (as in dark A1 horizons)
Leucinization	Translocation	The paling of soil horizons by the loss of dark organic matter
Paludinization	Transformation	The accumulation of organic matter (> 30 cm) to form peat
Ripening	Transformation	Chemical, biological, and physical changes in organic soil when air enters previously waterlogged soil
Mineralization	Transformation	The release of oxide solids through the decomposition of organic matter
Eluviation	Translocation	The washing out of material from a portion of a soil profile
Illuviation	Translocation	The washing in of material into a portion of a soil profile
Leaching (depletion)	Loss	The eluviation of material from the entire solum
Enrichment	Gain	The addition of material to a soil body
Calcification	Translocation	The accumulation of calcium carbonate
Decalcification	Translocation	The removal of calcium carbonate
Salinization	Translocation	The accumulation of soluble salts (e.g. sulphates and chlorides of calcium and sodium) in salty horizons
Desalinization	Translocation	The removal of soluble salts from salic soils horizons
Alkalization (solonization)	Translocation	The accumulation of sodium ions on the exchange sites in a soil
Dealkalization (solodization)	Translocation	The leaching of sodium ions and salts from natric horizons
Lessivage (pervection)	Translocation	The mechanical eluviation of silts and clays
Pedoturbation	Translocation	Biological, chemical, and physical churning of the soil
Podzolization (silication)	Translocation and transformation	The chemical migration of aluminium, iron, and/or organic matter resulting in a relative accumulation of silica in the eluviated layer
Desilication (ferrallitization, ferritization, allitization)	Translocation and transformation	The chemical migration of silica from the solum leading to a concentration of iron and aluminium sesquioxides in the solum
Decomposition (weathering)	Transformation	The breakdown of mineral and organic materials
Synthesis (neoformation)	Transformation	The formation of new particles of mineral and organic species
Brunification (rubification, ferrugination)	Translocation and transformation	Release of iron from primary minerals and the dispersion of iron oxide, the progressive oxidation or hydration of which gives the soils brown or red coloration respectively
Gleization	Translocation and transformation	The reduction of iron under anaerobic conditions to produce bluish to greenish grey matrix colours, with or without yellowish brown, brown, and black mottles and ferric and manganiferous concretions
Loosening	Transformation	Increase in the volume of soil voids
Hardening	Transformation	Decrease in the volume of soil voids

Source: After Buol *et al.* (1980)

wash processes. The cumulative effect of wash on surface material is a gradual winnowing of fines from the biomantle and a concomitant coarsening of texture.

A new view of pedogenesis emerges by viewing epimorphism and bioturbation together in a three-dimensional landscape (Figure 6.2). In brief, epimorphism produces saprolite, the upper and finer portion of which is mined by mesofauna to form a topsoil (biomantle). The topsoil is further sorted and moved downslope by rain wash. The result is a soil profile comprising a mobile biomantle, commonly lying on a stone layer, that rests upon subsoil saprolite. The contrast between topsoil and subsoil is often seen in soil texture – the biomantle is dominated by residual quartz and displays few features of the bedrock. Rarely, the contrast is seen in fabric. Aeolian processes add to the development of the texture contrast by winnowing fines and leaving behind coarser, well-sorted topsoils. The residue may be capable of forming mobile sand dunes. Thus the wind is responsible for the accumulation of quartz as the final residual product of pedogenesis. This is evident in Africa and Australia at present. It was far more common a process in the past.

Soil chronosequences

Current change in the soil may be measured in the field using appropriate equipment. Long-term changes are more difficult to assess. Reconstruction of past changes in present soils is possible using either mass balance or chronosequence techniques. A mass balance calculates the relative accumulation and depletion of materials using a stable index mineral, such as zircon, as a reference point. Such calculations permit an assessment of rates of pedogenic change. For example, secondary clay constitutes one-third the weight of a 240,000-year-old Alfisol evolved in a beach sand on a northern Californian marine terrace (Chadwick *et al.* 1990). Mass balance computations show that desilication has removed 29 per cent of the silica originally present in the beach sand at a rate of 2.1 $t/km^2.yr$.

Soil chronosequence construction substitutes spatial soil differences for temporal soil differences. It establishes the rate and direction of pedogenic change. There are two methods for constructing soil chronosequences. The first method assumes that soil evolution is a progressive, or developmental, process. Different stages of a developmental sequence of soils are assumed to exist in modern soil–landscapes. Identifying the different stages and placing them in chronological order allows a chronosequence to be inferred. A well-known sequence of this kind is the development of a lithosol formed in marly limestone, through a rendzina, brunified rendzina, calcareous brown soil, and calcic brown soil, to a lessived brown soil (Figure 6.3). The same sequence on a pure limestone stops at the rendzina stage (Plate 6.3). This method rests on shaky foundations. The age of different soils in the

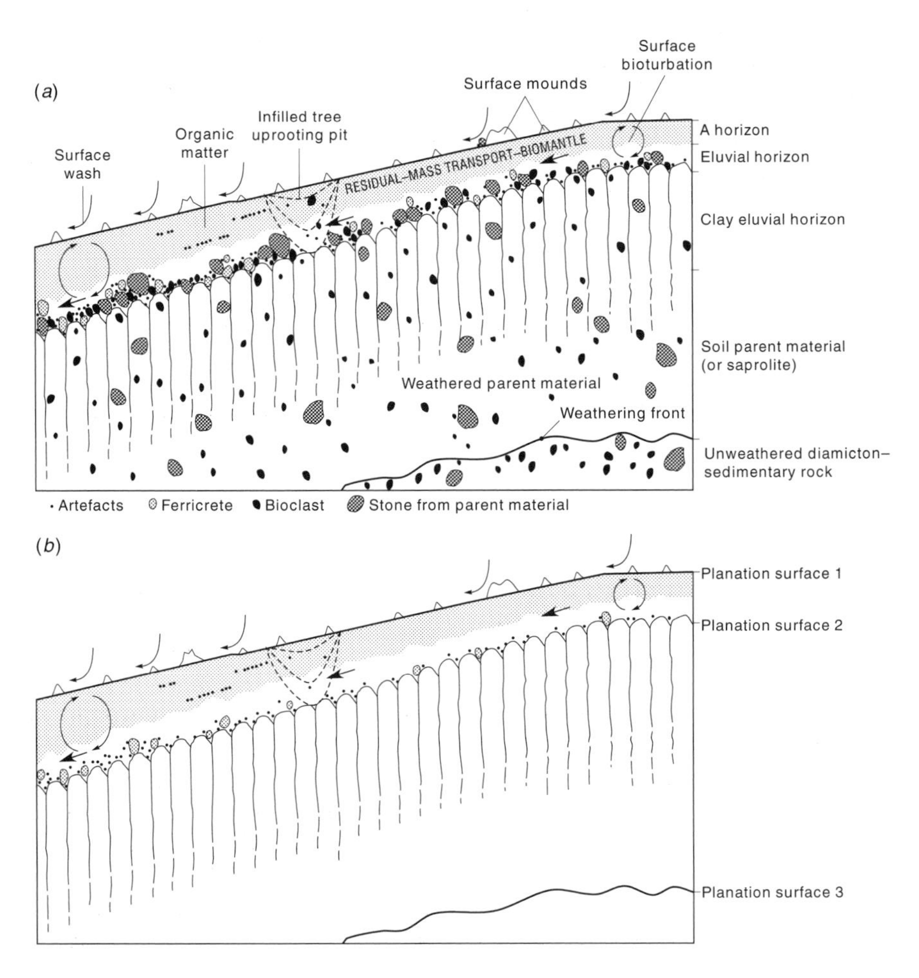

Figure 6.2 The interaction of soils and landform evolution – dynamic denudation. (a) Dynamic denudation in landscapes underlain by stony sediments. (b) Dynamic denudation in landscapes underlain by stone-free sedimentary rocks. In both cases, three plantation surfaces are present. The first is the chemical weathering or dissolution front that migrates downward with time. Material released by dissolution is carried away laterally by groundwater and throughflow. The second is the dominantly wash surface (the ground surface). Disturbance by animals and plants is great. Rain and wind carry fine materials downslope, the finest particles moving the furthest. The third is the boundary between the topsoil (A and E horizons) and the subsoil (B horizons). It separates the biomantle from largely *in situ* parent material and is commonly marked by a stone-line or metal nodules. Much of the soluble material washed out of the biomantle is carried downslope by throughflow just above this boundary

Source: After Johnson (1993a)

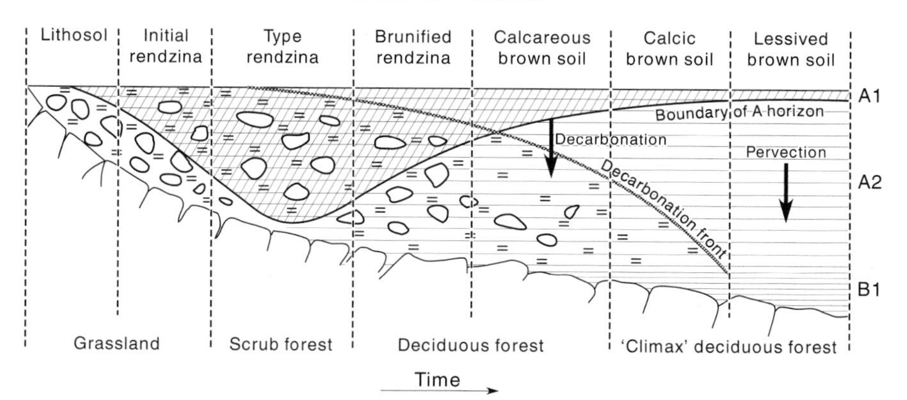

Figure 6.3 Classic developmental sequence of soils on marly limestone
Source: After Duchaufour (1982)

assumed sequence is at best educated guesswork. And there is no reason to suppose that, for example, rendzinas will evolve into brown soils under all circumstances. Those on steep slopes, for example, may be roughly in balance with local soil-forming factors and 'progress' no further.

The second method for constructing soil chronosequences is far more trustworthy. It involves building true soil chronosequences in situations where soils have developed on surfaces of known age. Sequences of coastal sand dunes, river terraces, and lava flows have proved particularly fruitful sites for this purpose; earthflows, abandoned pasture, strandlines, and other datable landscape features are also usable. Classic examples of soil chronosequences are those developed on sand dunes (e.g. Salisbury 1925) and those formed in the wake of a retreating glacier (e.g. Crocker and Major 1955; Olson 1958). Neoglacial ('Little Ice Age') moraine ridges in southern Norway provide several suitable sites for chronsequence investigations (e.g. Mellor 1987). Horizon thickness chronofunctions for four sites (Haugabreen, Austerdalsbreen, Storbreen, and Vestre Memurubreen) are shown in Figure 6.4.

Soil chronosequences are not common, but the number of well-documented and reliably dated examples is mounting fast. They consider changes over years to centuries, millennia, and hundreds of thousands of years (Table 6.5). Most of them show an exponential increase or decrease of soil properties with time, as predicted by the laws of chemical kinetics. An example of this is a study of net nitrogen mineralization and net nitrification along a tropical forest-to-pasture chronosequence that used present forest and pasture of three different ages – 3, 9, and 20 years (Piccolo *et al.* 1994). But most studies consider Holocene and Quaternary changes.

Plate 6.3 A humic rendzina (Icknield Series) in the Yorkshire Wolds, England
Photograph by Brian S. Kear

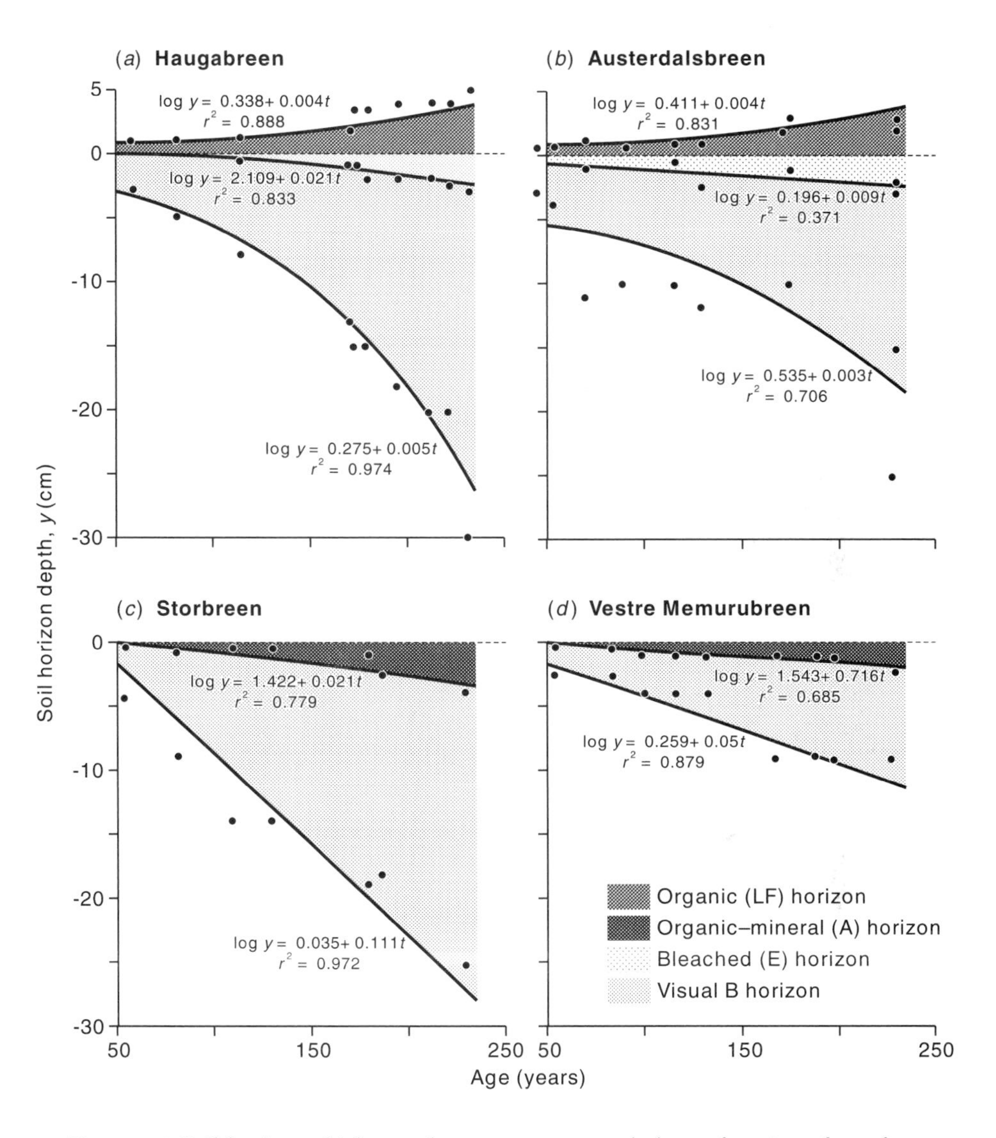

Figure 6.4 Soil horizon thickness chronosequences and chronofunctions from four moraine sequences near the west coast of southern Norway. Haugabreen and Austerdalsbreen are near the Josterdalsbreen ice cap; Storbreen and Vestre Memurubreen are in the Jotunheim Mountains
Source: After Mellor (1987)

Table 6.5 Some recent soil chronosequence studies

Site characteristics	*Location*	*Main soil changes studied*	*Reference*
Years to centuries			
Forest and old pasture	Rondonia, western Brazilia	Soil nitrogen	Piccolo *et al.* (1994)
Current and old pasture	Sevillet National Wildlife Refuge, New Mexico, United States	Soil carbon and respiration	Kieft (1994)
Forelands of Athabasca glacier	Canada	Soil nitrogen	Kohls *et al.* (1994)
Sand dunes	Westhoek Nature Reserve, Belgium	Horizon thickness, litter content, humus content	Ampe and Langohr (1993)
Holocene			
Andesitic ocean beach-ridge sediments	Costa Rica	Weathering and neoformation of minerals	Nieuwenhuyse *et al.* (1994)
Alluvial fans	Negev Desert, Israel	Gravel shattering by salts	Amit *et al.* (1993)
River terraces	Western Cairngorms, Scotland	Exchangeable calcium, magnesium, sodium, and potassium; base saturation; clay type	Bain *et al.* (1993)
Quaternary			
Alluvial terraces	Appalachian Highlands, south-eastern United States	Clay content, rubification index, iron oxide content, elemental ratios	Leigh (1996)
Alluvial surfaces	Inner coastal plain, central Virginia, United States	Soil type, duripans, incipient plinthite, ferricrete, weatherable minerals	Howard *et al.* (1993)
Coastal dunes	Southern Queensland, Australia	Various extraction of elemental carbon, aluminium, silicon, iron, manganese, phosphorus, titanium, and zircon	C. H. Thompson (1992); Skjemstad *et al.* 1992)

Change in sediments

Sediment sources and sinks

The weathered mantle of rock waste tends to move down hillslopes, down rivers, and down submarine slopes. This downslope mass wasting is caused by gravity and by fluid forces. Ice, water, and air are the transporting media. Moving air may erode and carry sediments in all subaerial environments. It is most effective where vegetation cover is scanty or absent. Winds may carry sediments up slopes and over large distances (see Simonson 1995). Dust-sized particles may travel around the globe.

Water and ice in the pedosphere may be regarded as liquid and solid components of the weathered mantle. Weathered products, along with water and ice, tend to flow downhill along lines of least resistance (which typically lie at right angles to the topographic contours). The flowlines run from summits to sea-floor. In moving down a flowline, the relative proportion of water to sediment alters. On hillslopes, there is little, if any, water in a large body of sediment. Mass movements prevail. These take place under the influence of gravity, without the aid of moving water, ice, or air. In glaciers, rivers, and seas, there is a large body of water that bears some suspended and dissolved sediment. Movement is achieved by glacial, fluvial, and marine transport.

Mass movements are multifarious. There are six basic types – creep, flow, slide, heave, fall, and subsidence (Varnes 1978). The chief kinds of mass movement, together with flows of ice and water, grouped according to water content, are given in Table 6.6. Notice that mass movements in high water-content materials grade into transport by flowing ice and water.

Eroded material is moved by gravitational and fluid forces. When these forces are no longer equal to the task of transporting the sediment, deposition occurs. The deposits so produced are usually named after the processes responsible for creating them. Wind produces aeolian deposits, rain and rivers produce fluvial deposits, lakes produce lacustrine deposits, and the sea produces marine deposits.

Sediment budgets

Sediment budgets may be calculated for a drainage basin, though it is common practice to confine the budget to alluvial or valley storage. The change in storage during a time interval is the difference between the sediment gains and the sediment losses. Where gains exceed losses, storage increases with a resulting aggradation of channels or floodplains or both. Where losses exceed gains, channels and floodplains are eroded (degraded). It is feasible that gains counterbalance losses and a steady state obtains. This condition is surprisingly rare, however. Usually, valley storage and fluxes

Table 6.6 Mass, glacial, and fluvial movements

	Water content					
Main mechanism	*Very low*	*Low*	*Moderate*	*High*	*Very high*	*Extremely high*
Creep		Rock creep Continuous creep				
Flow	Dry flow	Slow earthflow Debris avalanche (struzstrom) Snow avalanche		Solifluction Gelifluction Debris flow	Rapid earthflow Rainwash Sheet wash	Mudflow Slush avalanche Ice flow Rill wash River flow Lake currents Ocean currents Turbidity currents
Slide (translational)		Debris slide Earth slide Debris block slide Earth block slide Rock slide Rock block slide	Debris slide Earth slide Debris block slide Earth block slide		Rapids (in part) Ice sliding	
Slide (rotational)		Rock slump	Debris slump Earth slump			
Heave		Soil creep Talus creep				
Fall		Rock fall Debris fall (topple) Earth fall (topple)				Waterfall Ice fall
Subsidence		Cavity collapse Settlement				

Source: Partly after Varnes (1978)

conform to one of four common patterns under natural conditions (Trimble 1995): a quasi-steady-state typical of humid regions, vertical accretion of channels and aggradation of floodplains, valley trenching (arroyo cutting), episodic gains and losses in mountain and arid streams (Figure 6.5).

Sedimentary sequences

Sedimentary layers form a hierarchy of stratigraphic units (Table 6.7; Colour plate 2). The smallest units are laminae and the most inclusive units are sequences. The terms used are the subject of considerable confusion (see Posamentier and James 1993). In Table 6.7, the stratal units are defined. They would be identified in the field only by the physical relationships of the strata – thickness, time of formation, and interpretations of origin are

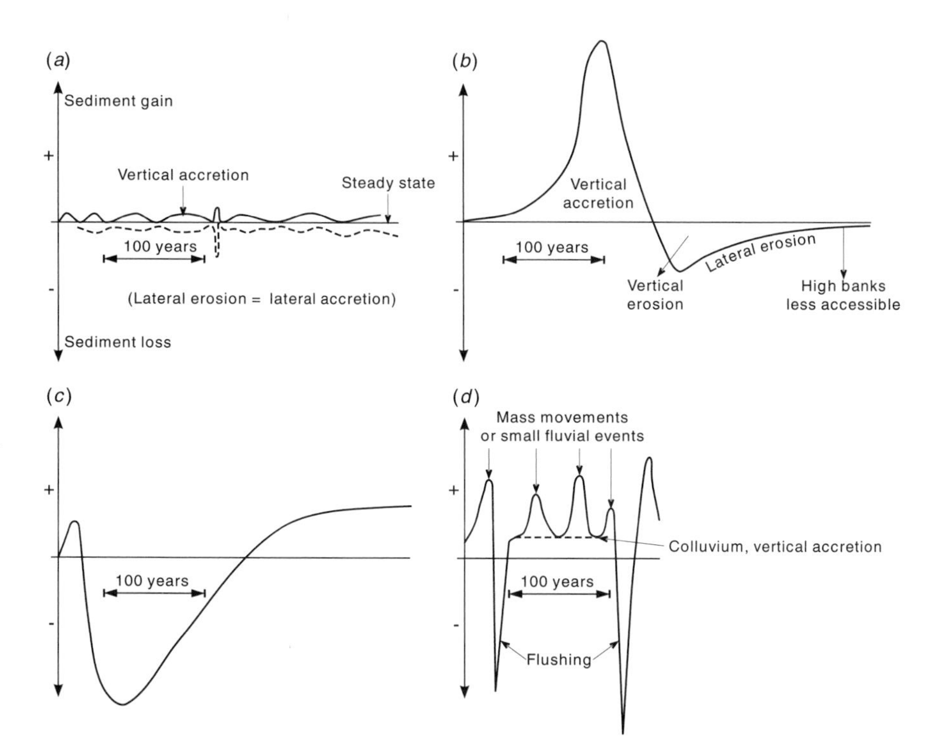

Figure 6.5 Four common patterns of valley sediment storage and flux under natural conditions. (a) Quasi-steady-state typical of humid regions. (b) Great sediment influx with later amelioration producing vertical accretion of channels and aggradation of floodplains. (c) Valley trenching (arroyo cutting). (d) High-energy instability seen as episodic gains and losses in mountain and arid streams
Source: After Trimble (1995)

Table 6.7 A hierarchy of stratal units: definitions and characteristics

Stratigraphic unit	*Definition*	*Approximate formation time (years)*
Sequence	A relatively conformable succession of genetically related strata bounded by unconformities and their correlative conformities	2,000,000–100,000
Parasequence set	A succession of genetically related parasequences forming a distinctive stacking pattern bounded by major marine-flooding surfaces and their correlative surfaces on coastal plains and shelves	200,000–5,000
Parasequence	A relatively conformable succession of genetically related beds or bed sets bounded by marine flooding surfaces and their correlative surfaces	20,000–100
Bed set	A relatively conformable succession of genetically related beds bounded by surfaces (called bed-set surfaces) of erosion, non-deposition, or their correlative conformities	1,000–1
Bed	A relatively conformable succession of genetically related laminae or lamina sets bounded by surfaces (called bedding surfaces or planes) of erosion, non-deposition, or their correlative conformities	Centuries to minutes
Lamina set	A relatively conformable succession of genetically related laminae bounded by surfaces (called lamina-set surfaces) of erosion, non-deposition, or their correlative conformities	Days to minutes
Lamina	The smallest megascopic layer	Hours to minutes

Source: After Van Wagoner *et al.* (1990)

not used to define stratal units nor to place them in the hierarchy. Bounding surfaces are critical in making such identification (Figures 6.6 and 6.7). They include marine flooding-surfaces, unconformities, and conformities:

1 Marine flooding-surfaces separate younger strata from older strata and display evidence of an abrupt increase in water depth. This deepening of water is commonly accompanied by a little submarine erosion or non-deposition, and a minor hiatus in sedimentation is normally indicated. Marine flooding-surfaces correlate with surfaces in coastal plains and on submarine shelves.
2 An unconformity is a surface separating younger from older strata that contains evidence of subaerial erosional truncation, that, in some cases, correlates with submarine erosion or subaerial exposure, and that implies a significant hiatus in sedimentation. Two types of unconformity are recognized, imaginatively dubbed type 1 and type 2 unconformities. A type 1 unconformity develops when there is a relative fall of sea level. It is associated with an abrupt basinward shift of coastal onlap characterized by forced regression and, in some cases, fluvial incision. Type 1 unconformities are readily identified in rock outcrops. A type 2

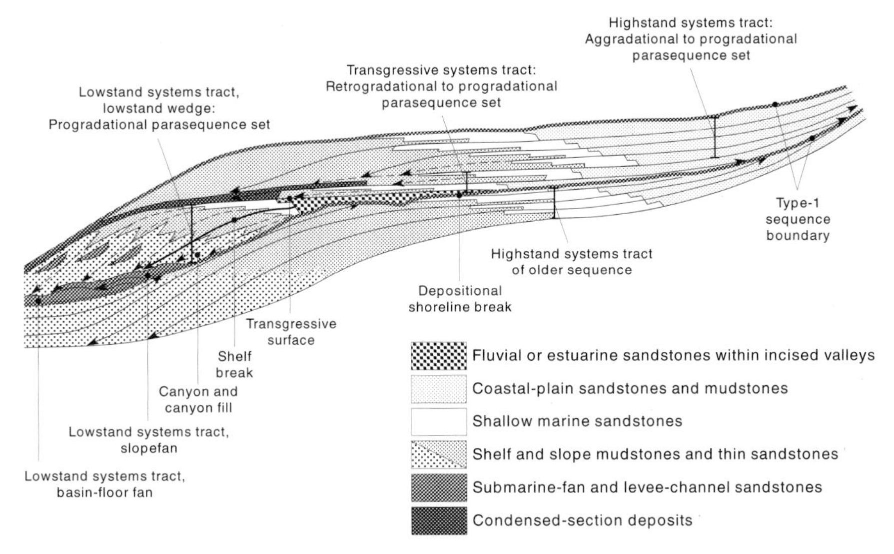

Figure 6.6 Stratal patterns in a type-1 sequence deposited in a basin with a shelf break
Source: After Van Wagoner *et al.* (1990)

unconformity results from a decelerating, and then accelerating, relative rise of sea level. It is characterized by an abrupt basinward shift of coastal onlap without forced regression and significant fluvial incision. It is almost impossible to see in a rock outcrop because, in the marginal marine setting, no relative sea-level fall occurs to punctuate the succession (as the maximum rate of eustatic fall never quite attains the rate of subsidence). Rather, there is a change from an increasingly progradational to decreasingly progradational and subsequently aggradational stacking pattern.

3 A conformity is a surface separating younger from older strata that contains no evidence of erosion (either subaerial or submarine) or of non-deposition, and that indicates no hiatus in sedimentation. It includes surfaces upon which there is very slow deposition or low rates of sediment accumulation, and long periods of geological time are represented by very thin deposits.

Systems tracts are sometimes defined interchangeably with parasequence sets. However, systems tracts refer to a linked assemblage of coeval depositional systems. In turn, depositional systems are three-dimensional assemblages of lithofacies. Unlike parasequence sets, their identification specially involves inferences about sea-level changes, stratal geometries, temporal and spatial relationships between facies tracts, and the nature and significance of bounding surfaces (Posamentier and James 1993).

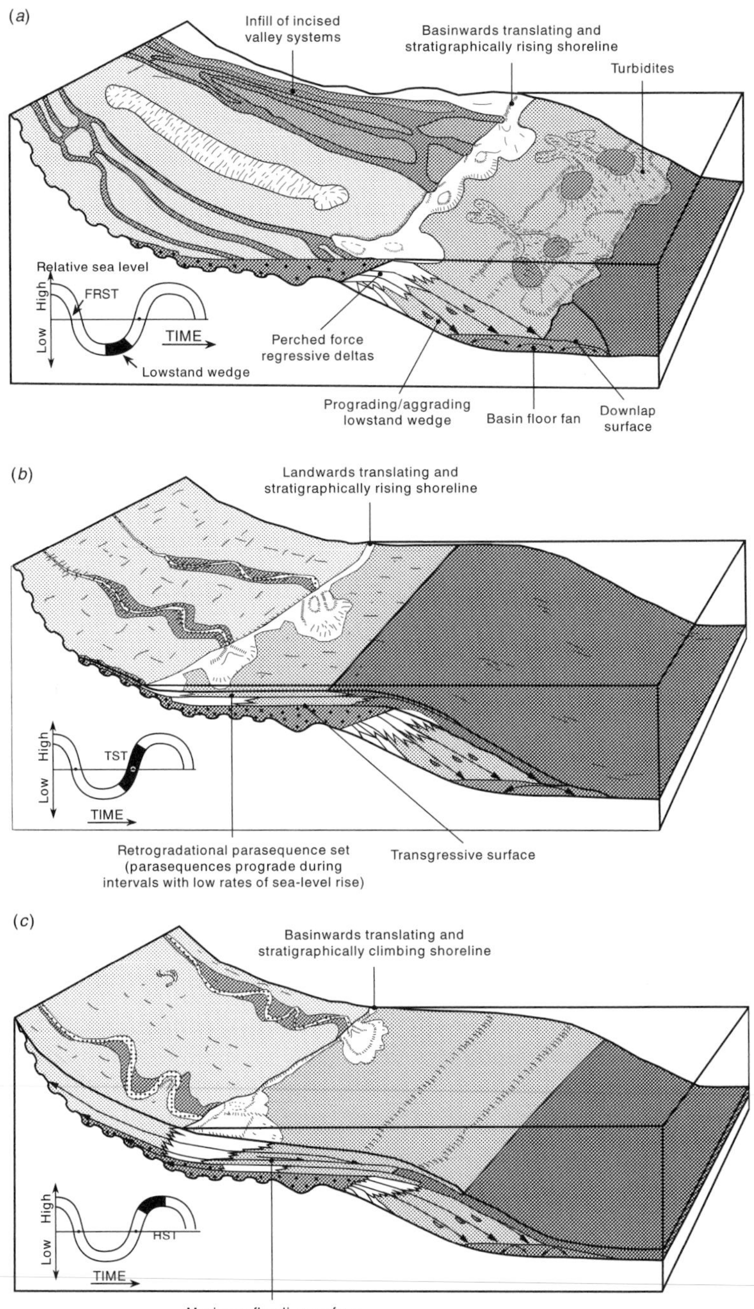
(a)
Infill of incised valley systems
Basinwards translating and stratigraphically rising shoreline
Turbidites
Relative sea level
High
Low
FRST
TIME
Lowstand wedge
Perched force regressive deltas
Prograding/aggrading lowstand wedge
Basin floor fan
Downlap surface
(b)
Landwards translating and stratigraphically rising shoreline
High
Low
TST
TIME
Retrogradational parasequence set (parasequences prograde during intervals with low rates of sea-level rise)
Transgressive surface
(c)
Basinwards translating and stratigraphically climbing shoreline
High
Low
HST
TIME
Maximum flooding surface (downlap surface)

Figure 6.7

CHANGING HORIZONS: CAUSES OF PEDOLOGICAL CHANGE

What drives changes in soils and sediments? Is it environmental factors? Or is it internal dynamics of soil and sediment systems?

Internal mechanisms

Some changes in soils and sediments undoubtedly result from the inner workings of soil and sediment systems. Such changes will occur without environmental change. In sedimentology, internally driven changes are called autogenic (originating inside a sedimentary basin), and are contrasted with allogenic change that is induced by environmental fluctuations (originating outside a sedimentary basin). These terms are equally applicable to pedogenic changes.

In sediment systems, external and internal agents often have a cyclical character that imparts rhythm to sedimentation (e.g. Einsele *et al.* 1991b: 7). Autocyclic processes include mud slumping, storms, and turbidity currents. Allocyclic processes include global sea-level fluctuations, climatic change, tectonic change, and changing biological productivity. It is sometimes difficult, but not usually impossible, to disentangle the effects of autocyclic and allocyclic processes. Orbital signals in small-scale stratigraphic sequences (in the millimetre to decametre range) are plainly allogenic in origin. However, a large number of sediment sequences lack clear-cut orbital signals. This may be due to the cloaking effect of autocyclic perturbations. Some autocyclic processes produce a chaotic pattern of sedimentation rates and camouflage the orbital signals (Peper and Cloetingh 1995).

In soils, autogenic changes often occur when internal thresholds are crossed and may be distinguished from allogenic changes. Under Mediterranean climates, a pedological threshold is crossed in moving from Alfisols on summits and backslopes to Vertisols on toeslopes (Muhs 1982). The Vertisols evolve when sufficient smectite (an expanding clay) accumulates. The smectite comes from *in situ* weathering and from downslope movement of clay suspended particles and base cations. There

Figure 6.7 Sedimentary sequence evolution on a shelf due to sea-level cycles. (a) Forced regressive and lowstand systems tracts associated with sea-level fall and early rise. A forced regressive systems tract (not illustrated separately) comprises incised valleys and canyons that produce perched forced-regressive deltas at the coast and fans on the basin floor. During the lowstand of sea level, prograding or aggrading wedges of sediment accumulate on the shelf, together with turbidites, to form the lowstand systems tract. (b) Transgressive systems tract associated with a rising sea level. (c) Highstand system tract associated with a high sea level
Sources: After Van Wagoner *et al.* (1990) and Gawthorpe *et al.* (1994)

are many other examples of intrinsic thresholds (e.g. Muhs 1984). The leaching of carbonates is a prerequisite for clay migration in soils. This is because divalent cations of calcium and magnesium, among others, are strong flocculants that inhibit pervection of clay. Conversely, monovalent sodium ions tend to disperse clays. In coastal or playa-margin environments, clay migration occurs rapidly once enough sodium has accumulated to cause clay dispersal. Laminar calcretes only form in arid and semiarid regions when the pores in growing K horizons become plugged with calcium carbonate. After that time, a laminar zone of calcium carbonate accumulation develops and grows upwards.

External forces

Soils and environmental factors

The traditional state-factor model of pedogenesis sees soil as the result of external soil-forming factors – climate, organisms, relief, parent material, and time. In the dynamic-denudation model of pedogenesis, soil-forming factors exert a strong influence on soil evolution, but their relative importance is seen in a different light. Thus lithospheric (parent) material and topography are held as the primary determinants of global soil distribution (Paton *et al.* 1995: 111). In turn, lithospheric materials and topography are largely the products of plate tectonic processes. Continental lithospheric plates comprise three segments, each of which is associated with characteristic lithospheric materials and topography (Table 6.8). Most soils are formed in plate centres. These centres are very old. Pedogenic processes have operated since at least the breakup of Pangaea. In this long time-span, pedogenesis could have reached its endpoint several times, and this is borne out by the accumulations of inert end products (such as fine-grained kaolins, iron and aluminium oxides, and quartz sand) found on all continental fragments (Paton *et al.* 1995: 111). However, the pedological clock has been reset in temperate and high latitudes by glaciation: ice sheets and glaciers stripped off the old covers of soil and laid down a fresh supply of unweathered rock (Chesworth 1982).

Table 6.8 The main types of continental plate and associated rocks and topography

	Continental plate segments		
Environmental factors	*Divergent margins*	*Plate centre*	*Convergent margins*
Lithospheric material	Basaltic	Granitic	Mixed
Topography	Steep with plateaux	Gentle	Steep

Source: After Paton *et al.* (1995)

Climate does exert an influence on soil and weathering processes at a global scale. Weathering reactions are usefully defined within an Eh–pH framework (Figure 6.8). Three extreme conditions are recognized – acid, alkaline, and reduced – and one non-extreme condition (Chesworth 1992). Each of these is associated with a particular set of soil processes. A strong climatic control is mirrored in the world distribution of these four soil and weathering conditions (Figure 6.8).

External, or environmental, thresholds must be crossed for some pedogenic processes to operate. For instance, sufficient clay with iron bound to it must be present in parent material before brunification can occur (Duchaufour 1982: 270). More generally, most soil properties possess a degree of inertia and will remain stable within a range of environmental change (cf. Bryan and Teakle 1949). Furthermore, when considered from a longer-term perspective than a few thousand years, many soils display cycles of change, similar to those exhibited by sediments, that are caused by environmental change. In the case of soils, a cycle normally consists of a phase of soil evolution, which occurs while the landscape is geomorphically calm, broken in upon by a phase of soil erosion or burial, which occurs when the landscape is geomorphically very active. The soil–landscape evolves through a succession of cycles (K-cycles), each cycle comprising a stable phase, during which soils develop forming a 'groundsurface', and an unstable phase, during which erosion and deposition occur (B. E. Butler 1959, 1967).

The cyclical nature of soil–landscape evolution was particularly pronounced in middle and high latitudes during the glacial–interglacial cycles (Starkel 1987). During warm and wet interglacials, deep soils and regoliths evolved in a strong regime of predominantly chemical weathering. During cold and dry glacials, permafrost, ice sheets, and cold deserts developed and chemical weathering was greatly subdued. Equivalent changes occurred in arid and semiarid environments. For instance, talus deposits formed during prolonged mildly arid to semiarid pluvial climatic modes were eroded by gullying, leaving talus flatiron relicts during arid to extremely arid interpluvial climatic modes (Gerson and Grossman 1987). In north-western Texas and eastern New Mexico, a vast sheet of Quaternary loess covering more than 100,000 km^2 and up to 27 m thick, known as the Blackwater Draw Formation, records more than 1.4 million years of aeolian sedimentation (Holliday 1988, 1989). Six buried soils in the formation reveal that stable landscapes obtained under subhumid to semiarid conditions, similar to those of the past several tens of thousands of years, whereas regional wind deflation and aeolian deposition prevailed during periods of prolonged drought.

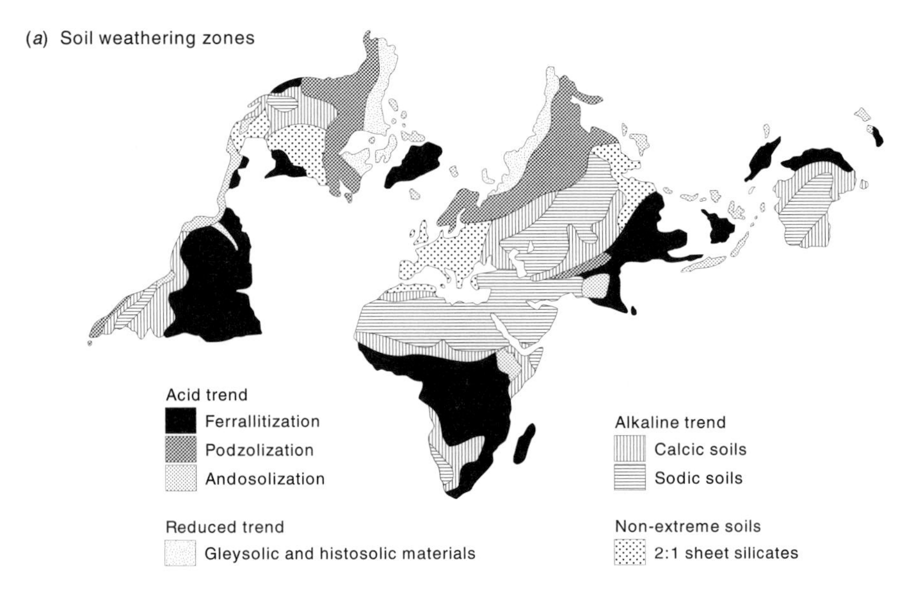

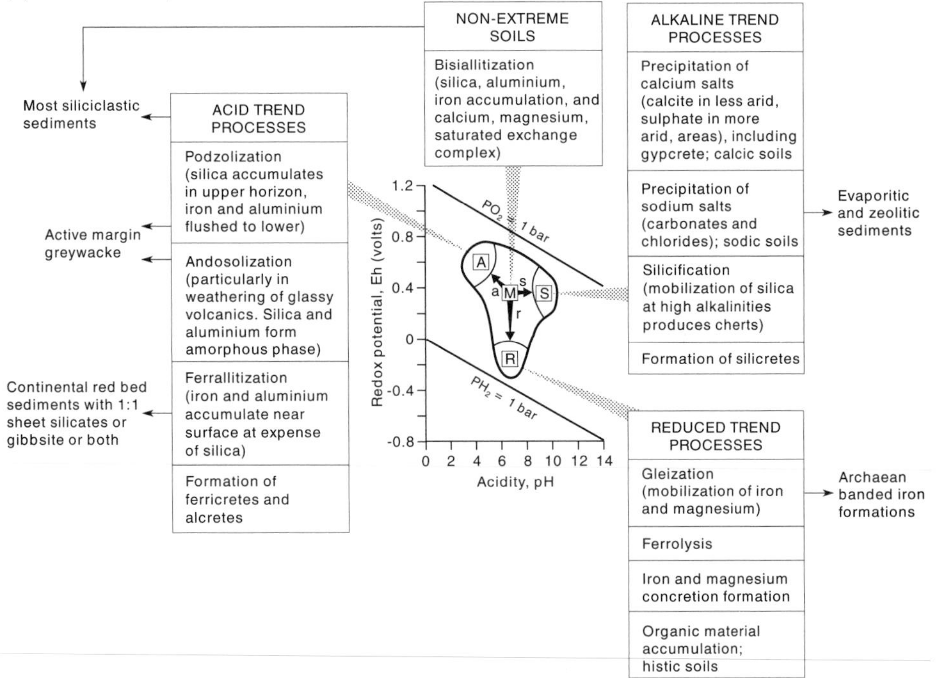

Figure 6.8 Soil weathering zones and soil chemical processes
Source: After Chesworth (1992)

Allocycles in sediments

Sedimentary sequences are replete with rhythmic changes, many of which appear to result from allogenic cycles. Soli-lunar cycles and Croll–Milankovitch cycles are found in a range of sedimentary environments. Solar, and to a lesser degree lunar, cycles are found in lake varves (Table 6.9). They are also encoded in some riverine varves (e.g. Sonett *et al.* 1992). Varve thickness provides a rough guide to precipitation and temperature. The solar and lunar signals are filtered by atmosphere–ocean processes, such as the quasi-biennial oscillation and El Niño–Southern Oscillation, and by geomorphological process of solution, entrainment, transport, and deposition (R. Y. Anderson 1991). Nevertheless, they can be teased out of the geological and historical record. The oldest known evidence of soli-lunar

Table 6.9 Rhythmic sedimentation in lakes

Site	*Origin*	*Laminae*		*Periodicity (years)*	*Reference*
		Light	*Dark*		
Recent					
Lake Skilak, Alaska	Distal glacial	Silt	Clay	1, 11, 22	Sonett and Williams (1985)
Lake Turkana, Kenya	Tropical	Clastic, carbonate	Clastic, carbonate	4, 25, 31, 44, 78, 100, 165, 200, 270	Halfman and Johnson (1988)
		Carbonate	Carbonate	11, 18.6, 22, 32, 76, >1,000	Halfman *et al.* (1994)
Pleistocene					
Lake Barlow–Ojibway, Ontario, Canada	Proglacial	Silt	Clay	11, (14), 22, 200	Agterberg and Banerjee (1969)
Rita Blanca Lake, Texas, United States	Non-glacial, meromictic, hard-water lake	Thick; carbonate, clay	Thin; clay, organic	1, 10–12?, 22, 200	R. Y. Anderson (1986); R. Y. Anderson and Dean (1988)
Pre-Pleistocene					
Todilto Formation, New Mexico, United States (Jurassic)	Non-glacial	Clastic, carbonate	Organic	1, 10–13, 60, 85, 170, 180	R. Y. Anderson and Kirkland (1960)
Green River Formation, Rocky Mountains, United States (Eocene)	Lake–playa complex	Thick	Thin	4.8–5.6 (ENSO events), 10.4–14.7	Ripepe *et al.* (1991)

Source: Partly after Glenn and Kelts (1991)

forcing comes from sedimentary rhythmites in the Big Cottonwood Formation, central Utah (Chan *et al.* 1994). These rocks are about 800–1,000 million years old. An excellent modern example of solar forcing is the wind record in Elk Lake, Minnesota (R. Y. Anderson 1992). A 10,400-year time series of varve thickness for this lake was collected. Varve thickness within a 2,000-year, mid-Holocene window defines changes in palaeowinds. Spectral density for varve thickness is high at periods of 200 years, 40–50 years, and 22 years (Figure 6.9a). Virtually the same periods occur in carbon isotope ratios from tree rings (Figure 6.9b), which provide a proxy for changes in solar activity (as they are linked to the solar cosmic-ray flux).

Many sediments contain orbital signals in the Croll–Milankovitch frequency band. Signs of orbital forcing are found in a wide spectrum of terrestrial and marine sedimentary environments – caves (speleothems),

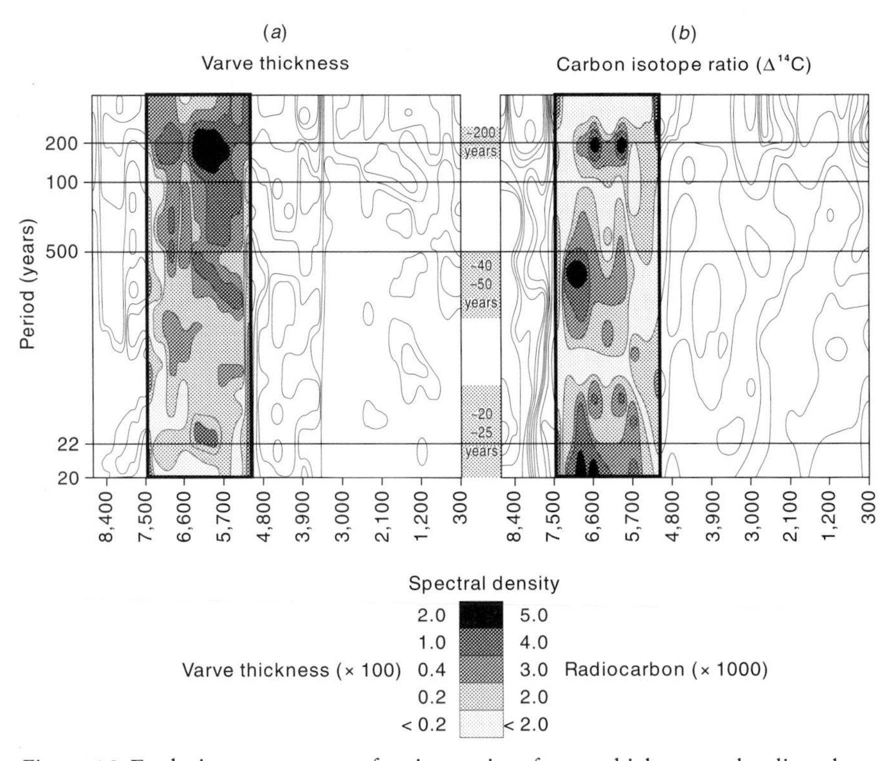

Figure 6.9 Evolutionary spectrum for time series of varve thickness and radiocarbon. (a) Varve thickness. Increased spectral density occurs at periods of ~200 years, 40–50 years, and ~22 years. (b) Carbon isotope ratio. Increased spectral density occurs at periods of ~200 years, 40–50 years, and 20–25 years. These increased spectral densities apply only to the shaded 2,000-year time-windows
Source: After R. Y. Anderson (1992)

plains (loesses), epicratonic basins, marine carbonate platforms, and pelagic and hemipelagic systems (Table 6.10). Indeed, geology is currently gripped by 'cyclomania'.

Astronomical forcing is probably best expressed in lake sediments, which register variations in lake level, and in evaporite sediments, which record variations in salinity. Lake and evaporite deposits are laid down in essentially closed systems. Long sequences of varves commonly display strong precessional signals grouped into 100,000-year bundles corresponding to the short ellipticity cycle. The Lake Lockatong complex existed in the Newark rift valley, eastern North America, during Triassic times (van Houten 1964; Olsen 1984, 1986; Gore 1989; Olsen *et al.* 1996). Owing to climatic variations, the lakes fluctuated between expanses of water, over 7,000 km^2 in extent and more than 100 m deep, and restricted playas or completely dry lake basins. Lake varves show that short sedimentary cycles involving about 5 m of sediment were produced by lakes whose depth oscillated with a period of about 21,800 years. Longer cycles, involving about 25 m and 100 m of sediment, echo periodic changes in the magnitude of the 21,800-year lake-level cycles. They are characteristically 101,000, 400,000, and 418,000 years in length. During the Triassic period the Newark Basin lay at palaeolatitude 15° N, or thereabouts. At that latitude, solar insolation changes should reflect the precession and eccentricity changes, but not changes of tilt. This theory seems to hold, for there is no sign of a 41,000-year obliquity cycle in the Lockatong sediments.

A study of the varved Castile evaporites, which formed at the close of the Permian period in the Delaware Basin of western Texas and south-eastern New Mexico, came to similar conclusions (R. Y. Anderson 1984). The start and finish of deposition, the trends in annual thickness, and the symmetry of response were consistent with control by a 100,000-year eccentricity cycle. Nearly symmetrical changes in the calcium-sulphate deposition rate in the same deposits matched the ~20,000-year precessional cycle. There was no apparent sign of forcing by the tilt cycle, but, as with the Lockatong sediments, that may be due to the low palaeolatitude of the area.

NEW HORIZONS: RATES OF PEDOLOGICAL CHANGE

The pulse of pedogenesis

Soil chronosequences are exceedingly useful for they provide a means of gauging the rate of soil processes and the pace of soil evolution. Studies of chronosequences suggest that soil processes fall into three groups according to their rate of operation (Table 6.11). The fleetest are essentially reversible, self-regulating processes and the slowest are mostly irreversible, self-terminating reactions. An intermediate group includes processes that have

Table 6.10 Orbital signals in soils and sediments

Deposit	*Place*	*Age*	*Signals present (years)*	*Reference*
Terrestrial				
Speleothems	Britain	Late Pleistocene	Tilt and precession	Kashiwaya *et al.* (1991)
Loess	Czechoslovakia	Quaternary	Short eccentricity	Kukla (1968)
Loess	Halfway House, Alaska	Pleistocene	Short eccentricity, tilt, and precession	Begét and Hawkins (1989)
Loess	Baoji, Shaanxi Province, China	Quaternary	Long and short eccentricity, tilt, and precession (weak)	Y. Wang *et al.* (1990); Yu (1994)
Old Red Sandstone	North Sea	Devonian	Short eccentricity and precession	Kelly (1992)
Alluvial deposits	Ireland	Devonian	Long and short eccentricity	Kelly (1992)
Lake and evaporite				
Lockatong Formation	Eastern North America	Triassic–Jurassic	Short eccentricity, tilt (weak), and precession	Olsen (1984, 1986)
Lake Biwa	Japan	Pliocene to present	Short eccentricity, tilt, and precession	Kanari *et al.* (1984), Kashiwaya *et al.* (1987)
Epicratonic				
Cyclothems	Kansas	Middle and Upper Pennsylvanian, Permian	Long and short eccentricity	Crowell (1978); Fischer (1986)
Calcrete–coal cyclothems	Sydney Basin, Nova Scotia, Canada	Late Carboniferous	Short eccentricity and precession	Tandon and Gibling (1994)
Marine carbonate				
Chalk–marl sequences	Italy North American interior	Albian Late Cretaceous	Long and short eccentricity and precession	Fischer (1993)
Peritidal carbonates	Southern Spain	Uppermost Portlandian to Berriasian	Short eccentricity, tilt, and precession	Jimenez de Cisneros and Vera (1993)
Pelagic and hemipelagic				
Carbonate-rich and carbonate-poor couplets	Southern Alps, northern Italy	Barremian–Cenomanian	Short eccentricity, tilt, and precession	Claps *et al.* (1991); Claps and Masetti (1994)
Biogenous sediments	Equatorial Atlantic Ocean	Late Pleistocene	Short eccentricity, tilt, and precession	Verardo and McIntyre (1994)
Sapropel and carbonate cycles	Mediterranean Sea	Pliocene and earliest Pleistocene	Tilt and precession	Hilgen *et al.* (1993)
Microfossils (calcareous nannofossils, dinocysts, and foraminifers)	Atlantic Ocean	Latest Jurassic through Lower Cretaceous	Short eccentricity, tilt (strong), and precession	Gradstein *et al.* (1993)
Planktonic foraminifera	Mediterranean Sea	Pliocene	Tilt and precession	Sprovieri (1992)

reaction rates sluggish enough to make the detection of change in the short term difficult. Reversible, self-regulating processes adjust to changing conditions very rapidly, and produce characteristic soil horizons within centuries or millennia. Soil nitrogen, for example, responds rapidly to changing climate. Other examples include the evolution of mollic horizons (rich in organic matter), salic horizons (rich in soluble salt), mottles, gilgai features (microscale mounds and depressions), and some spodic horizons (rich in illuvial iron and aluminium sesquioxides and organic matter). Long-term processes take several hundred thousand years to act and to create soil horizons. Many of these long-term reactions are essentially irreversible and produce very persistent features in soil–landscapes that may survive hundreds of megayears in palaeosols. Examples of such features include oxic horizons (rich in kaolinite and sesquioxides) in ferrallitic soils, laterite, natric horizons (clayey and rich in sodium), and duricrusts. Productions of intermediate-rate processes include some spodic horizons, cambic horizons (loamy fine sand or finer texture with weak signs of argillic or spodic horizon properties), fragipans (dense and hard), argillic horizons, histic horizons (very rich in organic matter), and gypsic horizons (rich in calcium sulphate). These evolve over several millennia, but they will change in response to environmental change and are not necessarily persistent features.

Sedimentation rates

One of the oldest controversies in geology concerns the nature of sedimentation. Does it happen little by little, year by year? Or is most of it condensed into short, sharp bouts? Arguments about this issue hinge upon the catastrophist–gradualist debate, though this will not be pursued here (but see Huggett 1990; Ager 1993b). Current research shows that sedimentation involves slow *and* fast changes in rate. Many stratigraphic sequences display signs of slow and gradual rhythmic sedimentation, the

Table 6.11 Features and horizons of soils grouped according to formation rate

Rate of evolution	*Characteristic origin*	*Examples of features*	*Examples of horizons*
Fast (less than 1,000 yr)	Created by reversible, self-regulating processes	Gilgai, mottles, slickenslides	Cambic, gypsic, mollic, salic, spodic
Intermediate (1,000–10,000 yr)	Created by slow-acting processes	Fragipans, mottles	Argillic, calcic, cambic, gypsic, histic, natric, spodic, umbric
Very slow (more than 10,000 yr)	Created by irreversible, self-terminating processes	Duricrusts, fragipans, gypsic crusts, plinthite	Albic, argillic, gypsic, histic, natric, oxic, petrocalcic, placic

Source: After Yaalon (1971)

rhythms reflecting periodic environmental fluctuations. Many other sequences record episodes of abrupt changes in sedimentation or depositional events. Some of these are due to extraordinarily energetic events of regional extent – explosive volcanic eruptions, large body impacts, giant mass failures, catastrophic floods, huge earthquakes, large violent storms, and enormous tsunamis (see Clifton 1988). Slow and gradual rhythmic sedimentation was described in previous sections of this chapter. A few points will be made here about episodic sedimentation.

Floods, storms, turbidity currents, mass flows, and earthquakes may all produce event deposits, that is, sediments resulting from abrupt changes in deposition rate (Seilacher 1991). Floods produce flash flood conglomerates on land and inundites in protected bays and other coastal sites. Storm waves produce tempestites, while tsunamis produce tsunamites. Turbidity currents, commonly associated with continental slopes, produce turbidites. Mass-flow deposits are similar to turbidites, but the particles are supported and tossed by other grains and pressurized pore water, and are not fully suspended. Earthquakes may create seismites, sediments that display several features of seismic shock. They may also trigger tsunamis and thus lead to tsunamite deposits. Asteroidal and cometary impacts produce impact deposits, including tektites.

Giant tsunami deposits

The Hawaiian island of Lanai carries sedimentary evidence of enormous tsunamis. Coastal slopes are blanketed by a limestone-bearing gravel – the Hulopoe Gravel (G. W. Moore and Moore 1984). The deposit reaches a maximum altitude of 326 m (Figure 6.10). It was originally believed to be the remnant of several different marine strandlines. Dated submerged coral reefs and tide-gauge measurements show that the south-eastern Hawaiian Islands sink so fast that former world-wide high-stands of the sea now lie beneath local sea level. Other recent evidence suggests that the Hulopoe Gravel and similar deposits on nearby islands were deposited about 100,000 years ago, during the Pleistocene epoch, by a giant ocean wave that swept several hundred metres up the flanks of Lanai and its neighbours. Marine material in the deposits would have been torn from the littoral and sublittoral zone and mixed with basaltic debris as the wave travelled inland. It is significant that the deposit mantles the surface and is well preserved where topographic ‘traps’, such as the Kaluakapo Crater, would have blocked backflow from the wave. The deposit thins and becomes finer grained with increasing distance from the shoreline. Its lower layers contain subrounded to rounded clasts of basalt and limestone, ranging in size from 0.3 to 1 m, which may have been laid down by the upsurging wave. Its upper layers contain subangular to angular basaltic clasts, ranging from 0.2 to 1.5 m, possibly dropped by backflow from the same wave. The boulders in both

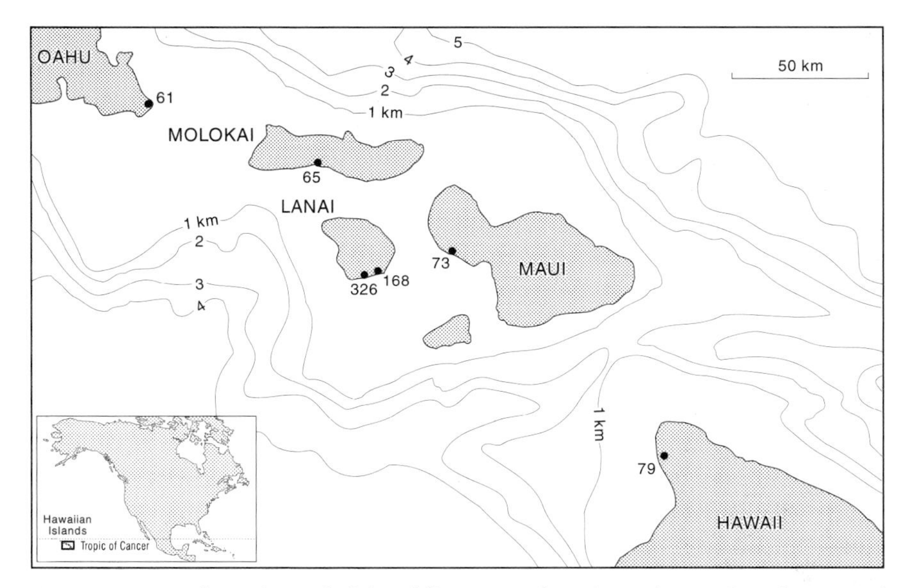

Figure 6.10 Local maximum height of limestone clasts in Hulopoe Gravel on Lanai and other tentatively correlated high-level gravel deposits on other south-eastern Hawaiian Islands

Source: After G. W. Moore and Moore (1984)

layers are supported by clasts, rather than by a matrix, which indicates that they were deposited by flowing water and not by a submarine debris flow.

Similar evidence for catastrophic deposition from a giant wave comes from Molokai, also in the Hawaiian Islands (J. G. Moore, Bryan, and Ludwig 1994). A marine conglomerate, containing carbonate (mainly coral) and basalt clasts, is exposed more than 60 m above present sea level, and nearly 2 km inland of the present shoreline, on the south-west side of East Molokai Volcano. 'Onwash' and 'offwash' facies are present. It may have been laid down by a giant wave breaking over an outer reef, rushing forwards as a turbulent bore over the back-reef flat, picking up a slurry of debris rich in carbonate and dropping it from the onwash on inland slopes. The offwash picked up loose basalt debris on its return flow. The deposit, which is 0.5–3 m thick, is dated at 240,000–200,000 years old. It was therefore not caused by the same wave that produced the Hulopoe Gravel on Lanai.

Impact ejecta

Tektite fields are strong evidence of impact events. However, the terrestrial record of impact cratering suggests that more than 3 km of impact-crater ejecta deposits should have been produced during the last 2 billion years.

The question is: where has all the ejecta gone? A possible answer is that it remains as diamictites. Diamictites are poorly sorted mixtures of sediment, commonly with boulder-sized clasts in a fine-grained matrix, whose textural characteristics are similar to those of impact crater deposits predicted by an impact model (Oberbeck *et al.* 1993; see also Rampino 1994). According to the model, even modest impacts in shallow seas, forming craters with diameters as small as 5 km, would have ejected large amounts of material and produced characteristic sequences of sediments. The sedimentary productions would include thick bodies of sediments traditionally recognized as tillites (a variety of diamictite), on the ocean floor, in deltas, and on land (Figures 6.11 and 6.12). At present, all tillites, even those ten times thicker than Pleistocene glacial deposits, are construed as glacial deposits. If they should be of glacial origin, then a difficulty arises: what process leads to the preservation of ancient glacial deposits but removes all traces of impact crater deposits that, in theory, should total at least 3 km? If the reinterpretation of tillites and diamictites as impact-crater deposits laid down within a few hours should be correct, then the implications are truly astounding. It would mean, for instance, that there might not have been so many glaciations in the past, a fact that would demand a fresh look at theories of geological climates. And it would explain some palaeoclimatic anomalies, to wit, the occurrence of Lower Proterozoic tillites when the global was warm, and the low-latitude distribution of some Upper Proterozoic glacial deposits (Rampino 1994). It would also lend credibility to the diluvial hypothesis of landscape development. There are many Miocene breccias, thousands of metres thick, with clasts that are tens of metres in diameter, that appear to have formed violently, or at least chaotically (J. R. Marshall, personal communication 1993). These deposits are enigmatic and do not appear to be impact ejecta. The possibility that they are impact-induced superflood deposits seems worth considering.

THE PAST IN PROFILE: DIRECTIONS OF PEDOLOGICAL CHANGE

Trends or cycles?

Soil and sediments have evolved over geological time. However, superimposed upon evolutionary trends of soils and sediments are what appear to be cyclical fluctuations. The geological record reveals at least three major cycles of weathering (Valeton 1994). The cycles began in the early Precambrian, the late Precambrian–Palaeozoic, and the Mesozoic–Cenozoic. They started with special world-wide conditions of flat relief, a hothouse atmosphere, and a deep weathering (lateritic) sequence that later became more differentiated. However, rather than being true cycles, these weathering episodes may represent extreme cratonic events (Ollier and Pain

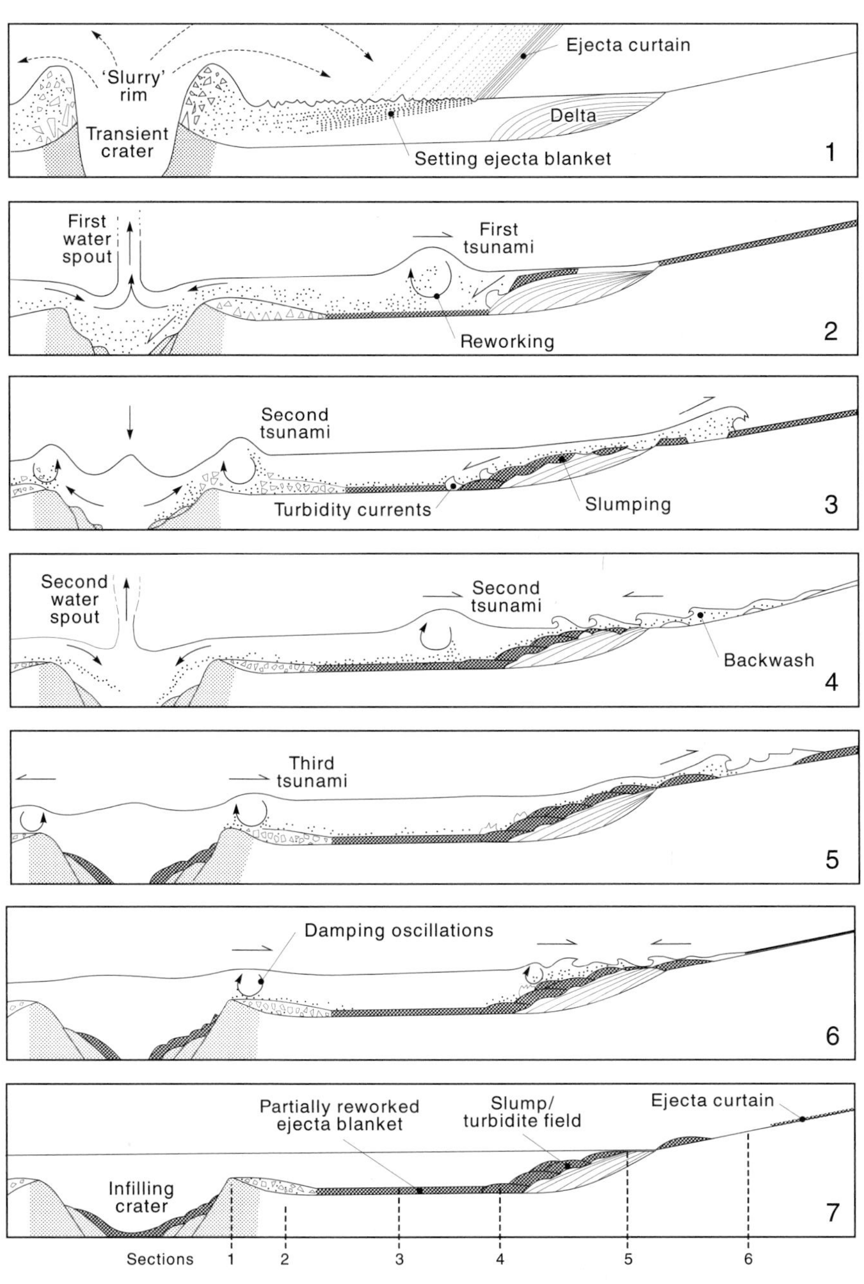

Figure 6.11 Hypothetical sequence of events following the impact of a mantle-penetrating bolide in a shallow sea. The full sequence would take several hours to run. The vertical scale is exaggerated
Source: After Oberbeck *et al.* (1993)

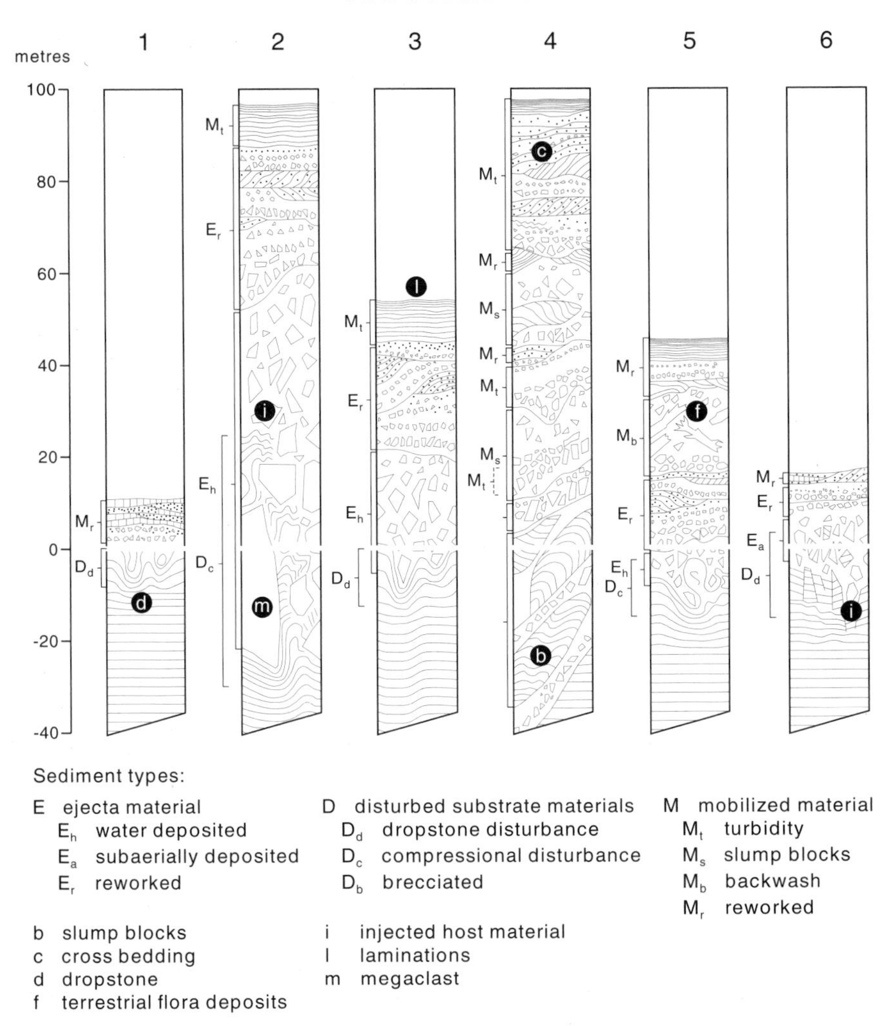

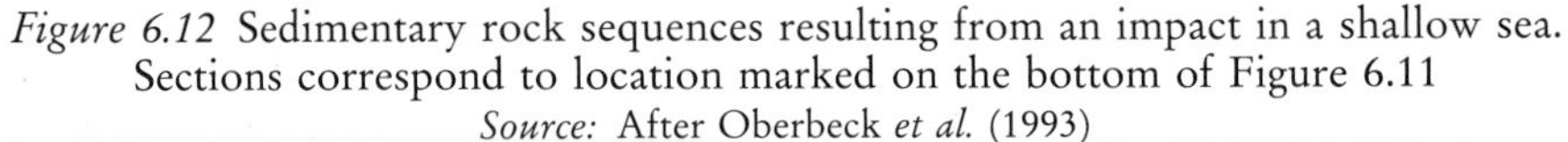

Figure 6.12 Sedimentary rock sequences resulting from an impact in a shallow sea. Sections correspond to location marked on the bottom of Figure 6.11
Source: After Oberbeck *et al.* (1993)

1996: 283). And, although exogenic processes may track the long-running climatic cycles, directional changes in soils and sediments are unmistakable in the rock record.

Evolving sediments

The long-term history of weathering, denudation, and deposition is intimately linked with the changing states of the atmosphere and oceans.

In turn, the atmosphere, oceans, and exogenic processes were radically affected by biological evolution. Signal evolutionary events were oxygen-liberating photosynthesis, various chemolithotrophic activities of bacteria (nitrogen fixation, sulphate reduction, and the like), the rise of metazoans, the invention of burrowing, the evolution of calcareous and siliceous skeletons, the colonization of land, the appearance of calcareous skeletons in plankton, the appearance of siliceous skeletons in phytoplankton, the invention of flowers, and the rise of humans (Fischer 1965, 1972, 1984b).

A unique event was the transition from an anoxic to an oxic atmosphere resulting from the evolution of photosynthesis that released oxygen as a waste product. This oxygen 'watershed' is plainly recorded in palaeosols and sediments (Figure 6.13). For instance, all palaeosols formed after about 2.0 billion years ago indicate that the atmosphere contained sufficient oxygen to oxidize all the iron in soils developed on igneous rocks. The Flin Flon palaeosol formed 1.8 billion years ago in the Amisk Group volcanics (Holland *et al.* 1989). It shows many characteristic features of modern well-drained soils. Corestones of spheroidally weathered pillow lavas occur at the base of the C horizon, and decrease in size upwards, eventually to disappear in the haematite-rich horizon at the top of the palaeosol. Oxides of calcium and magnesium and ferrous iron decrease up the profile while oxides of aluminium and titanium increase in the same direction. The ferrous iron was apparently oxidized to ferric iron and held within the palaeosol profile during weathering, indicating the presence of oxygen in the atmosphere, albeit at a much lower partial pressure than at present. In the 2.2 billion-year-old Hekpoort palaeosol of the Transvaal, South Africa, the profile is depleted in iron (Retallack 1986a). The difference between the Flin Flon and Hekpoort palaeosols suggests that the oxygen content of the atmosphere rose appreciably from 2.2 to 1.8 billion years ago (but see Retallack and Krinsley 1993).

Other biological events profoundly altered soils and sediments. The evolution of infauna (animals living in, as opposed to on, the sea bed) during the Phanerozoic aeon, for instance, left a biological overprint on marine sediments, and especially on shallow platform facies (Sepkoski *et al.* 1991). Trends parallel major changes in the diversity and composition of marine benthos, and an increase in bioturbation by burrowing. Cambrian and Early Ordovician rocks, when faunal diversity was low and bioturbation slight, contain sequences with abundant thin (less than 1 cm) storm beds and flat-pebble conglomerates. After the Ordovician radiations, very thin tempestites and flat-pebble conglomerates become rare. In addition, tempestites are commonly separated by gritty shales, representing mud and sand winnowed from tempestites by organisms. After the Triassic period, storm stratification is greatly obscured in many shelf environments. This bioturbation mirrors the increase in faunal diversity and the expansion of the deep infauna.

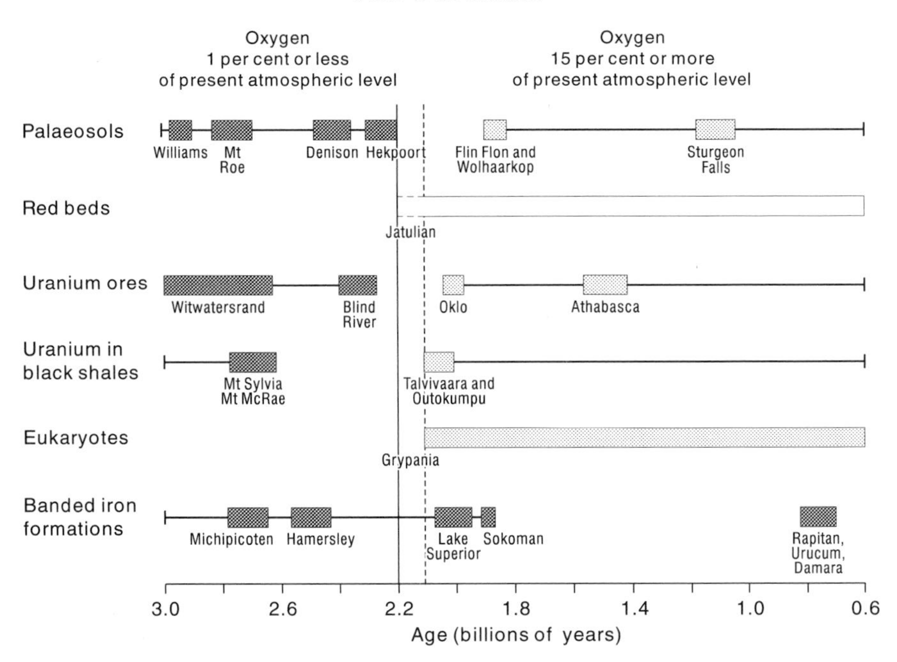

Figure 6.13 A summary of geochemical and palaeobiological data pertaining to the 'oxygen watershed' during the early Proterozoic aeon
Source: After Knoll and Holland (1995)

The colonization of the land in the Late Silurian period, and the evolution and spread of grasses during the Cretaceous period, were key events in landscape history (Ollier 1981: 308). The establishment of the first terrestrial ecosystems and, later, the addition of grass to the ecosphere, would have created new exogenic process regimes (cf. Retallack 1992a).

Environmental changes wrought by biological innovations have created at least five non-actualistic divisions of Earth surface history (Table 6.12). In each of these divisions, the parameters of weathering, erosion, transport, and deposition would have altered (cf. Jong 1976; Cocks and Parker 1981: 59). The evolving exogenic process regimes produced several secular trends. These trends are seen in the present-day preserved area, thickness, and volume of sedimentary rocks (Figure 6.14), in the relative proportions of different sediments in successive geological periods (Figure 6.15), and in rates of denudation and sedimentation (Figure 6.16).

Not all the changes in sediments are driven by biological evolution. On occasions, unique combinations of geography and climate created very unusual conditions. This happened in the Miocene epoch and resulted in a disproportionately high number of middle-latitude phosphate deposits (Parrish 1990). Given this environmental evolution, current exogenic

Table 6.12 A guide to non-actualistic divisions of ecospheric history

Age (billions of years)	Presence and absence of characteristics				
	Water	*Life in water*	*Oxygen*	*Life on land*	*Grass*
0.1–0	Yes	Yes	Yes	Yes	Yes
0.4–0.1	Yes	Yes	Yes	Yes	No
2.0–0.4	Yes	Yes	Yes	No	No
4.0–2.0	Yes	Yes	No	No	No
4.6–4.0	No	No	No	No	No

Source: After Huggett (1990: 148)

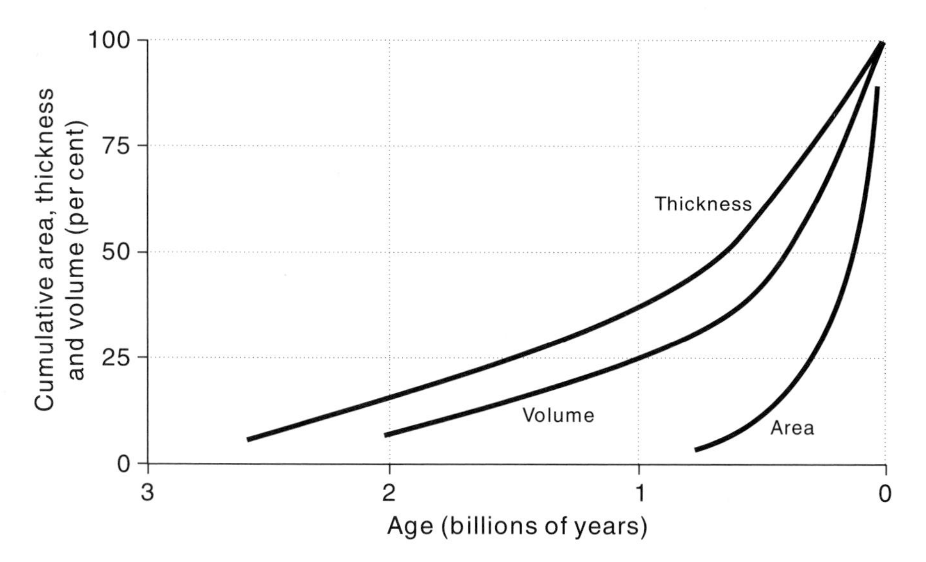

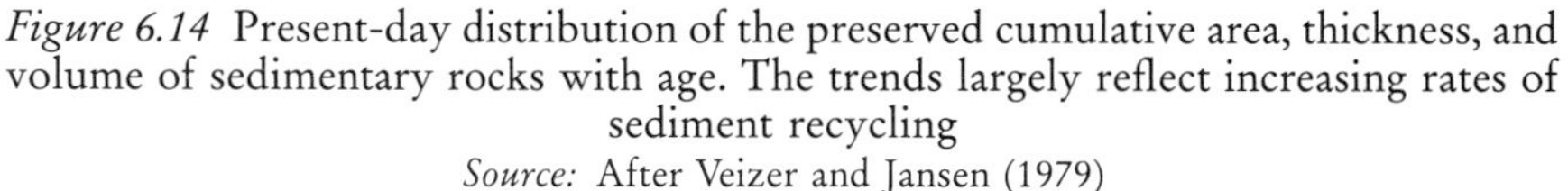

Figure 6.14 Present-day distribution of the preserved cumulative area, thickness, and volume of sedimentary rocks with age. The trends largely reflect increasing rates of sediment recycling

Source: After Veizer and Jansen (1979)

processes cannot be used as a key to all past exogenic phenomena, and modern environments cannot be held up as perfect analogues for all past environments. However, in the same way that modern endogenic processes may be used to aid our interpreting ancient crustal phenomena, modern exogenic processes can guide the deciphering of past surface environments.

Evolving soils

Soils and weathering features have evolved over geological time. Most of this evolution involved the appearance of new soil types and novel

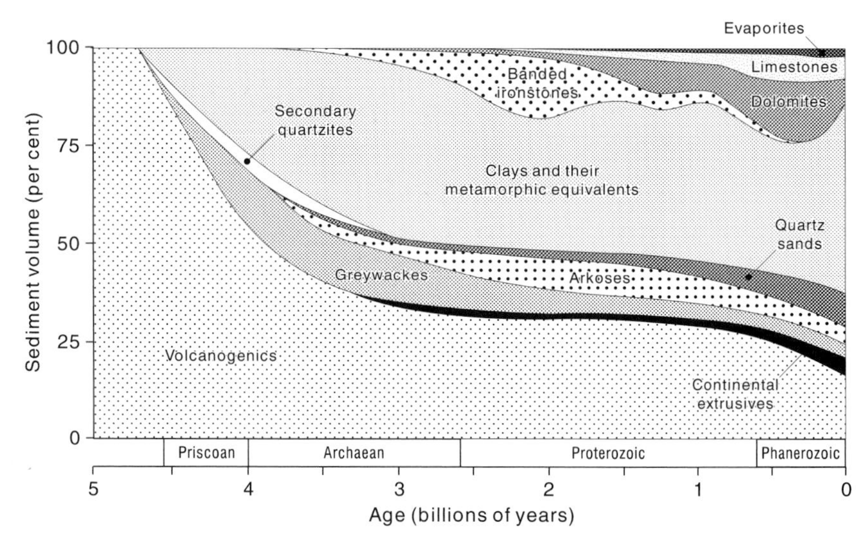

Figure 6.15 The relative proportions of major sediment types over geological time
Source: After Ronov (1964)

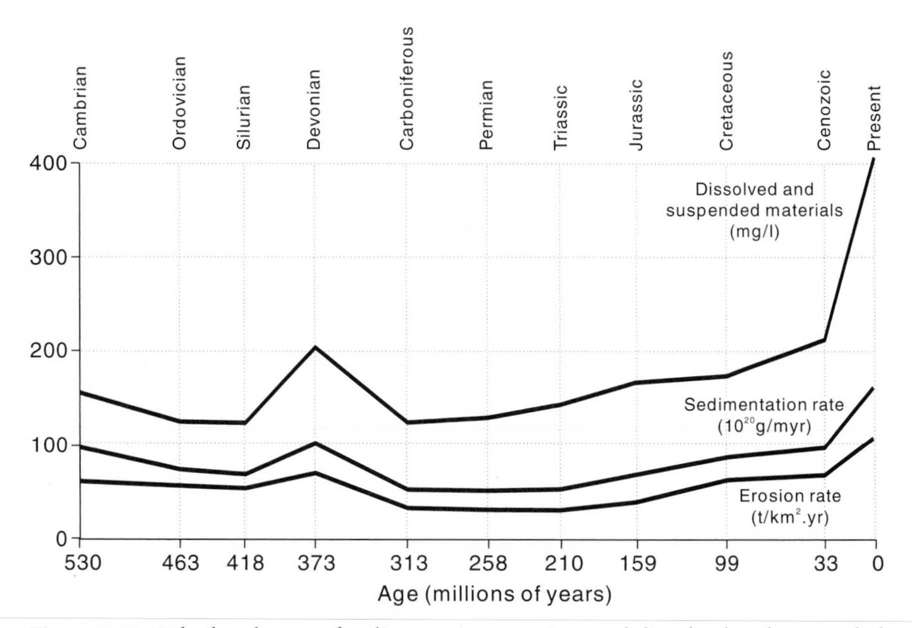

Figure 6.16 Calculated rates of sedimentation, erosion, and dissolved and suspended loss from continents during the Phanerozoic aeon. The times are the mean ages of geological periods
Source: After Tardy *et al.* (1989)

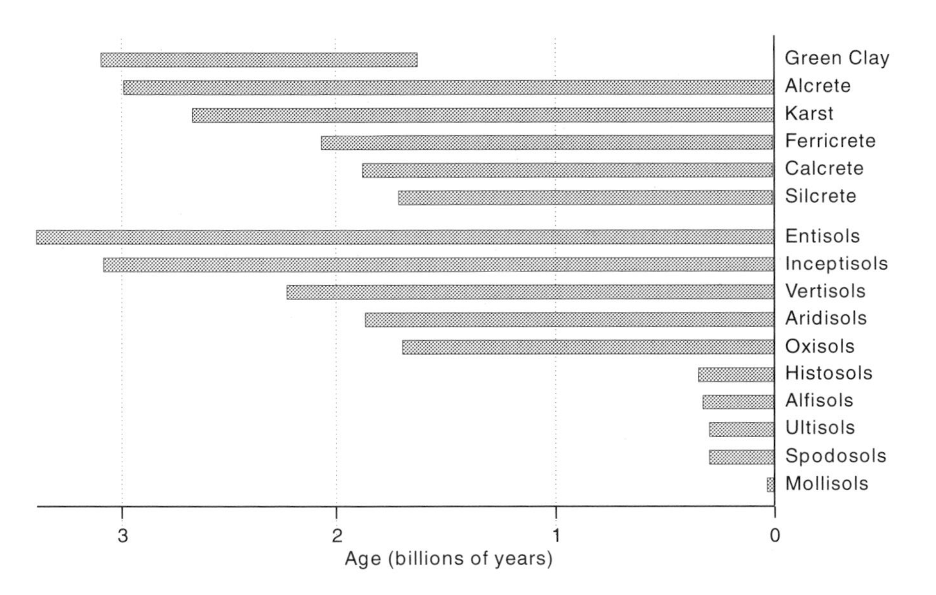

Figure 6.17 Geological time ranges of soil orders and weathering features
Source: After Retallack (1986b)

weathering features (Figure 6.17). The new soils evolved in new environments and new ecosystems. Ancient soils have persisted alongside newly evolved soils throughout Earth history (Retallack 1986b, 1992a). In consequence, soil diversity has increased in parallel with mounting environmental diversity, though progenitors of all modern soil orders (except for Mollisols, which are unknown since the Eocene epoch) existed during the Palaeozoic era (Figure 6.17). In theory, soils could go 'extinct', but in practice only the Green Clay palaeosols from the early Precambrian no longer exist (Retallack 1990: 293). Green Clay palaeosols formed in iron-rich parent materials such as basalt. They are, as their name suggests, green and clayey. They are also rich in alumina and poor in bases. Basalts currently weather to iron-rich red soils. The Green Clay palaeosols evolved under a weathering regime with very low atmospheric oxygen levels. Extinct soils appear to be a rarity. This could be because it is easier to recognize a palaeosol that has a modern counterpart, even when it has changed during burial, than it is a unique palaeosol.

SUMMARY

The pedosphere is soil (edaphosphere) and sediment (debrisphere). Soils and sediments are layered. Soils consist of soil horizons. Sediments consist of beds (strata). Soils almost defy classification, but eleven 'soil orders' are recognized (Alfisols, Andisols, Aridisols, Entisols, Histosols, Inceptisols, Mollisols, Oxisols, Spodosols, Ultisols, and Vertisols). Sediments are clastic, chemical, or biogenic. They form in three basic environments, each of which is dominated by a single sedimentary process – terrestrial (gravity-driven flows), shallow marine (tidal movements and wave-induced currents), and deep marine (suspension settling and unidirectional density currents). Soils and sediments change owing to material gains, losses, transfers, and transformations. According to traditional soil-formation theory, soil-changing processes are regulated by 'factors of soil formation' (climate, organisms, relief, parent material, and time). According to dynamic denudation theory, soil changes owing to biomechanical processes in A horizons, which form a biomantle, and epimorphic processes (weathering, leaching, and new mineral formation) in B horizons. Soil chronosequences reveal pedogenic changes. Sediment eroded from continents ends up in oceans. Marine deposits display characteristic sequences, depending on the relative rise or fall of sea level. These sequences correlate globally. Change in soils and sediments is either autogenic (internal) or allogenic (external). Autogenic changes in soils stem largely from internal soil thresholds. Allogenic changes arise from the environmental 'state factors'. Parent material and topography are important in explaining global soil distribution. Allocycles in sediments are locked into solar, lunar, and planetary rhythms. Soil chronosequences permit pedogenic rates to be measured. Fast, reversible, self-regulating processes are distinguished from slow, mostly irreversible, self-terminating reactions. Sedimentation engages slow and fast changes. Floods, storms, turbidity currents, mass flows, earthquakes, and asteroid impacts produce event deposits (sediments resulting from abrupt changes in deposition rate). Soils and sediments have evolved through geological time. New environmental conditions, largely created by evolving organisms, have engendered new sediments and new soils.

FURTHER READING

Erosional and depositional landforms are described in the excellent *Global Geomorphology: An Introduction to the Study of Landforms* (Summerfield 1991a). Marine sedimentary environments are dealt with in *Clastic Depositional Sequences: Processes of Evolution and Principles of Interpretation* (Fraser 1989). Of the many recent books on soils, *Soil Geomorphology: An Integration of Pedology and Geomorphology* (Gerrard 1992), *Soil Geomorphology* (Daniels and Hammer 1992), and *Soils: A New Global View*

(Paton *et al.* 1995) make rewarding reading. *Regolith, Soils and Landforms* (Ollier and Pain 1996) is highly recommended. Sedimentary cycles are mostly tackled in highly expensive, but useful, technical books. *Cycles and Events in Stratigraphy* (Einsele *et al.* 1991a) is an excellent example of this genre. *Climate, Earth Processes and Earth History* (Huggett 1991) has a chapter on sediments that discusses Milankovitch cycles. Palaeosols are discussed in the superb *Soils of the Past: An Introduction to Paleopedology* (Retallack 1990), which, incidentally, also provides a first-rate introduction to pedology. Catastrophes in Earth history are discussed in *Catastrophism: Systems of Earth History* (Huggett 1990) and *The New Catastrophism: The Importance of the Rare Event in Geological History* (Ager 1993b).

7

TOPOSPHERE

UPS AND DOWNS: EARTH'S RELIEF

Continents and oceans

The continents, extended to the edge of the presently submerged continental shelves, and the oceans are the major topographical features of the Earth (Figure 7.1). This dissimilarity is of immense significance to the biosphere. The distribution of land and sea affects climate, ocean currents, the degree of continentality, and the access between different oceans and different continents. Changing sea level and the redisposition of continents alters the number of islands and island-continents, and consequently the insularity of terrestrial habitats. This, in turn, influences climate and species diversity.

Large-scale topographical features of continents are ancient crystalline shields, old and young mountains, and upland and lowland plains. Upland plains may form plateaux (tablelands). Mountains, plateaux, and plains are illustrated in Plates 7.1 and 7.2 and Colour plate 5. The distribution of continental topographic forms mirrors the distribution of the crustal types (p. 61; Figure 7.2). Continental shields and platforms tend to have low to moderate elevation. The sediments on the platforms are either horizontal or undulate a little. The mountain belts produce distinct topographic features. Palaeozoic mountain belts are often expressed as curved belts of eroded upland. Mesozoic and Cenozoic mountain belts are seen as eroded uplands, or as eroded upland areas separated by basins, as in the Basin and Range Province in the Great Basin, western United States. Young mountain belts, or active orogens, are commonly expressed as ribbons of high elevation (over 3 km) with signs of intense folding and thrusting. In places, they contain high plateaux, which are relatively undeformed and uplifted regions. An example is the Colorado Plateau. Inland sea basins are not uncommon in continental lowlands. Smaller-scale topographical features are associated with folds, fault-blocks, rift systems (Plate 7.3), and other lineaments (see Summerfield 1991a).

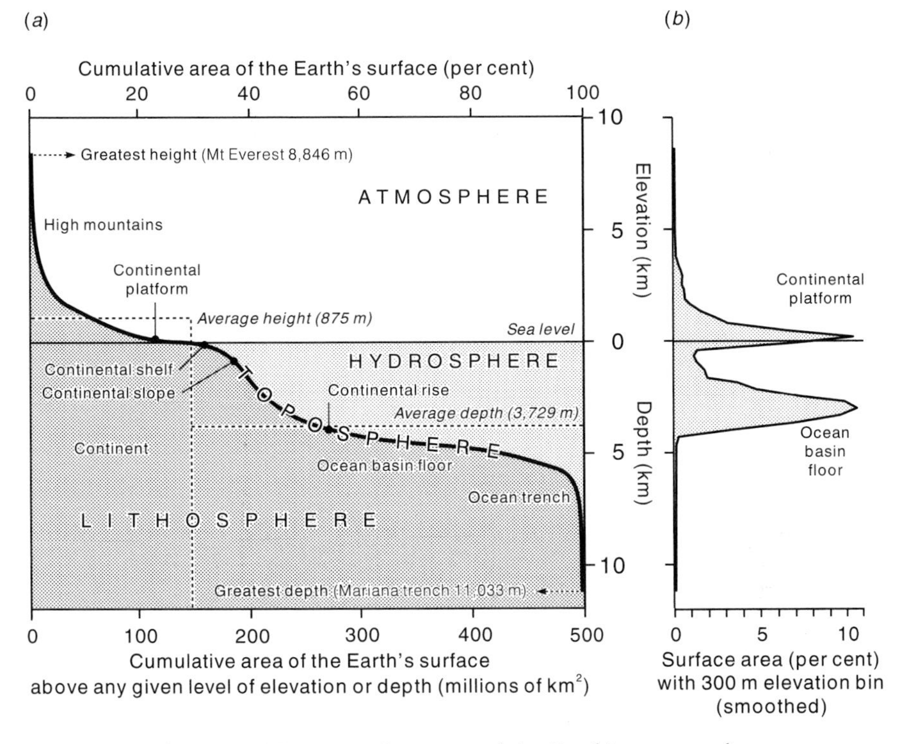

Figure 7.1 Hypsometric curves of the Earth's topography.
(a) Cumulative hypsometric curve with the chief features indicated.
(b) Differential hypsometric curve showing the area of the toposphere between successive levels using a 300 m elevation bin
Source: The hypsometric curves are after Rosenblatt *et al.* (1994)

The major topographical features of the oceans are ocean basins (the abyssal plain), ocean ridges, ocean trenches, volcanic islands, island arcs, and marginal-sea basins (Figure 7.1). These reflect the gross tectonic structure of the ocean floor and, in the case of some volcanic islands, the position of hot-spots.

Landforms

The toposphere contains a stupendous array of landforms. Regrettably, landforms are notoriously difficult to classify quantitatively. A fundamental distinction is made between erosional landforms (sculptured by the action of wind, water, and ice) and depositional landforms (built by sediment accumulation). Landforms also differ fundamentally in terrestrial, shallow marine, and deep marine environments (Tables 7.1–7.3). However, many

Plate 7.1a Plateaux: dissected plateau in Palaeozoic sediments, north-west of Alice Springs, Northern Territory, Australia
Photograph by C. R. Twidale and CSIRO

Plate 7.1b Plateaux: lateritized plain of Sturt Plateau, Northern Territory, Australia
Photograph by C. R. Twidale

Plate 7.2a Plains: an erosional plain, probably produced by etching into the underlying granite and gneiss, around Meekatharra, central Western Australia
Photograph by C. R. Twidale

Plate 7.2b Plains: a depositional plain (part of the Canterbury Plains), South Island, New Zealand. The mountains in the background are the Southern Alps
Photograph by the New Zealand Tourist Bureau

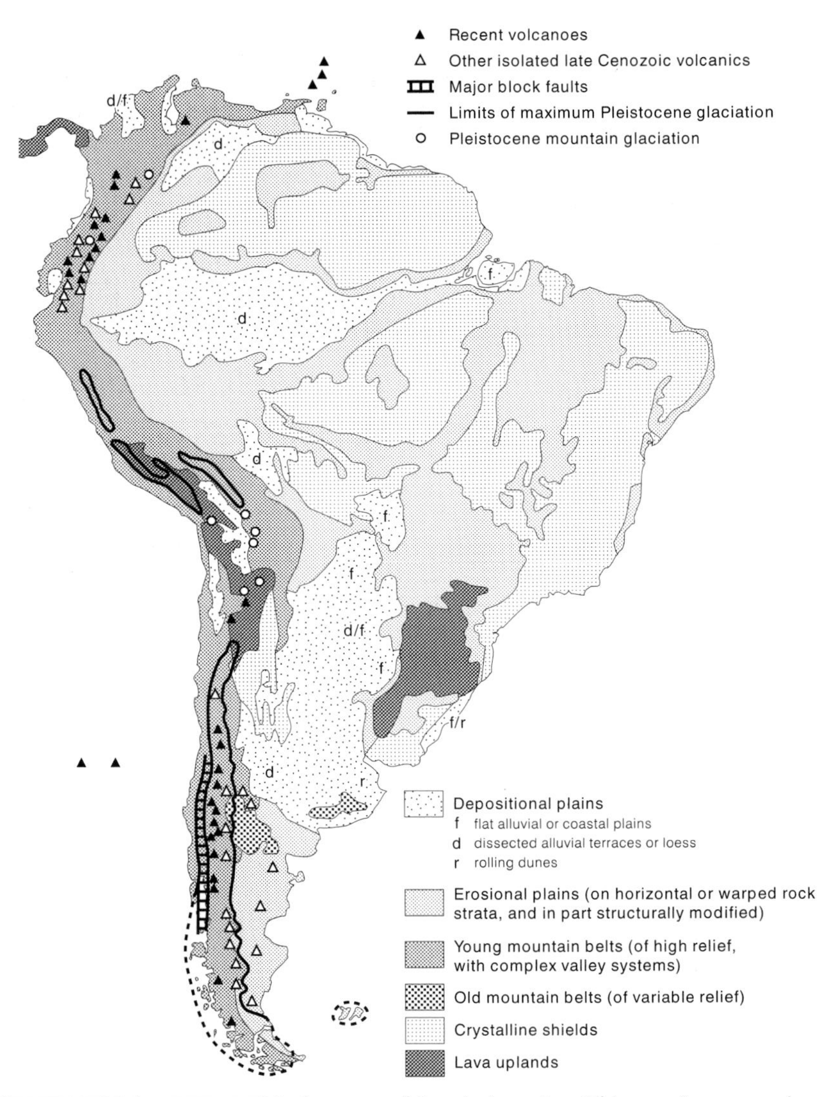

Figure 7.2 Major topographic features of South America. This continent consists of three main topographic provinces: the western cordillera (Andes); the continental lowlands; and the eastern uplands. The Andes consist of high fold mountains, low mountains, block-fault mountains, hilly coastal plains, intermontane basins, intermontane valleys, and volcanoes. The continental lowlands include the Llanos and Paraná–Chaco plainlands, the Amazonian lowlands, and the Patagonian Plateau. The eastern uplands, which are underlain by part of the Gondwana crystalline shield, are mantled with a patchy cover of sedimentary and volcanic rocks; they include the Guiana highlands, the interior plateau of Brazil (Mato Grosso), the northern highlands, the Serro del Mar (an upfaulted block of the crystalline shield), and the Paraná–Uruguay uplands
Source: After Butzer (1976)

Plate 7.3 A rift-valley wall on the west side of Lake Albert, Uganda, looking south from a position just south of Butiaba. The relief is about 300 m. (See Ollier 1990)
Photograph by Cliff D. Ollier

Table 7.1 Terrestrial environments and landforms

Exogenic regime	*Depositional landforms*	*Erosional landforms*
Arid	Alluvial fans Ripples, dunes, draa (megadunes) Playas or lakes	Ventrifacts, yardangs, deflation hollows, pans Lag deposits and desert (or stone) pavements Ephemeral streams
Humid	Floodplains Alluvial plains Alluvial fans Bajadas River terraces Loess plains Lakes	River valleys Bedrock channels Alluvial channels
Frigid	Till plains Fluvioglacial – kames, eskers, marginal ridges, sandar (singular sandur), deltas, terraces Periglacial – scree slopes and debris fans, permafrost horizons, ice wedges, pingos, palsas, thermokarst, loess, gelifluction features, block slopes, block streams, block sheets, rock glaciers	Glacial troughs, hanging valleys, Alpine and Icelandic troughs, fjords, open troughs, rock bars Whalebacks, rock drumlins, roches moutonnées, flyggberg, rock basins Plastically sculptured forms Meltwater channels Cryoplanation (altiplanation) terraces Cirques (cwms, corries), biscuit board topography, arêtes, horns, nunataks

landform classifications are based purely on topographic form, and ignore geomorphic process. For example, one scheme for large-scale landform classification uses three chief topographic characteristics (Hammond 1954). The first characteristic is the relative amount of gently sloping land (land with less than an 8 per cent slope). The second characteristic is the local relief (the difference between highest and lowest elevation in an area). The third characteristic is the 'generalized profile'. This defines the location of the gently sloping land – in valley bottoms or in uplands. In combination, these characteristics define the following landforms:

1 Plains with a predominance of gently sloping land combined with low relief.
2 Plains with some features of considerable relief. This group may be subdivided by the position of the gently sloping land into three types – plains with hills, mountains, and tablelands.

Table 7.2 Coastal environments and landforms

Exogenic regime[a]	*Depositional landforms*	*Erosional landforms*
Tide dominated	Salt marshes Tidal flats Sand waves, sand dunes, sand tongues, and shoals	Tidal creeks and channels
Muddy coastal	Chenier plain (muddy shoreline with long, thin sandy ridges) Marshes, mudflats, muddy shoals	Tidal creeks
Wave dominated (strandplains, barrier islands, lagoons)	Dunes, beaches, beach ridges and swales, sand waves and crescentic bars Berms, ridges, runnels, swash bars, longshore troughs, longshore bars, ripples	Tidal creeks Tidal inlets
Wave dominated (rocky shore)	–	Cliffs Shore platforms (intertidal platforms, wave-cut platforms, abrasion ramps, high-tide platforms)
Deltaic	Delta plain (marshes, lakes, shallow bays) Delta margin (channels flanked by levees, splays, and subdeltas; distributary mouth bars) Prodelta (linear tidal ridges)	Distributary channels
Estuarine	Tidal flat Tidal inlet Shoals	Channels
Coastal dunes	Sand dunes	Blowouts

Note: [a] Partly after Fraser (1989)

3 Hills with gently sloping land and low-to-moderate relief.
4 Mountains with little gently sloping land and high local relief.

There are many such schemes, all with their good and bad points.

FROM PEAKS TO PENEPLAINS: TOPOGRAPHIC CHANGE

The toposphere is a two-dimensional surface. At any place, its height may change owing to two types of processes – endogenic and exogenic. Endogenic processes involve tectonic and volcanic forces and take place in the solid Earth. They may cause uplift or subsidence of the ground surface. Exogenic processes include weathering, erosion, transport, and deposition. They occur at or near the Earth's surface. Where erosion exceeds deposition,

Table 7.3 Marine environments and landforms

Exogenic regime[a]	*Depositional landforms*	*Erosional landforms*
Shallow marine environment		
Marginal marine (continental shelves)	Ripples, dunes, sand waves (up to 10 m high and 500 m long), massifs (large, linear sand features – up to 100 km long and 20 km wide), linear sand ridges (on massifs), sand ribbons (up to 100 m wide and 15 km long), sand patches	Subsurface channels
Epeiric sea and epicontinental seaway	Shorelines (deltas, barrier islands, lagoons, strandplains) Shelves (sand ridges and sand waves)	Subsurface channels
Continental slope transition zone		
Canyon	Canyon walls Canyon axes	Submarine canyons Submarine gullies
Slope	Intercanyon areas Intraslope basins	Slump and slide scars Subsurface channels
Deep marine environment		
Active margin	Trench and basin plains Trench fans Turbidite wedges Levees	Canyons Axial channels
Passive margin	Continental rise (sediment wedges, contourites, turbidites, slumped sediments, giant bedforms, fans, levees) Abyssal plain	Distributary channels
Transform margin	Continental borderland basins	Submarine canyons

Note: [a] After Fraser (1989)

they cause a decrease of elevation; where erosion is less than deposition, they cause an increase of elevation. In simple terms, exogenic processes create a weathered mantle or regolith and remove material from topographic 'highs' and add it to topographic 'lows'. Very minor changes of topography are caused by expansion and contraction of soil and regolith, and by compaction of sediments. The balance between these various endogenic and exogenic forces of topographic creation and destruction fashions landscapes.

Processes of landform change

Landforms are the product of endogenic and exogenic agents. Endogenic or tectonic agents trigger earthquakes, deform rocks, and generate volcanic activity. In doing so, they determine geological structures and rock types that underlie the toposphere, and cause upheaval and subsidence of the land surface and sea floor. Landforms are also influenced by exogenic or gradational agents that constantly even out differences in relief. Endogenic agents create raw topography, but exogenic agents fashion the actual topography out of the raw form. This constant battle between endogenic and exogenic forces is called the 'principle of antagonism' (Scheidegger 1979).

Large-scale and medium-scale topographical features result primarily from endogenic processes, while smaller-scale landforms, as listed in Tables 7.1–7.3, are primarily the product of exogenic process. However, interpretations of landscape change depend on assumptions made about uplift and denudation.

Slope decline

The 'geographical cycle' was the first modern theory of landscape evolution (W. M. Davis 1889, 1899, 1909). It assumed that uplift takes place quickly. The raw topography is then gradually worn down by exogenic processes, without further complications from tectonic movements. Furthermore, slopes within landscapes decline through time – maximum slope angles slowly decrease (though few field studies have subtantiated this claim). So, topography is reduced, little by little, to an extensive flat region close to base level – a peneplain. The reduction process creates a time sequence of landforms that progresses through the stages of youth, maturity, and old age. However, these terms, borrowed from biology, are misleading and much censured (e.g. Ollier 1967; Ollier and Pain 1996: 204–5).

The Davisian system consists of two separate and distinct cyclical models, one for the progressive development of erosional stream valleys and another for the development of whole landscapes (Higgins 1975). Valleys are thought to be V-shaped in youth, flat-bottomed in maturity, after lateral erosion has become dominant, and to possess very shallow features of

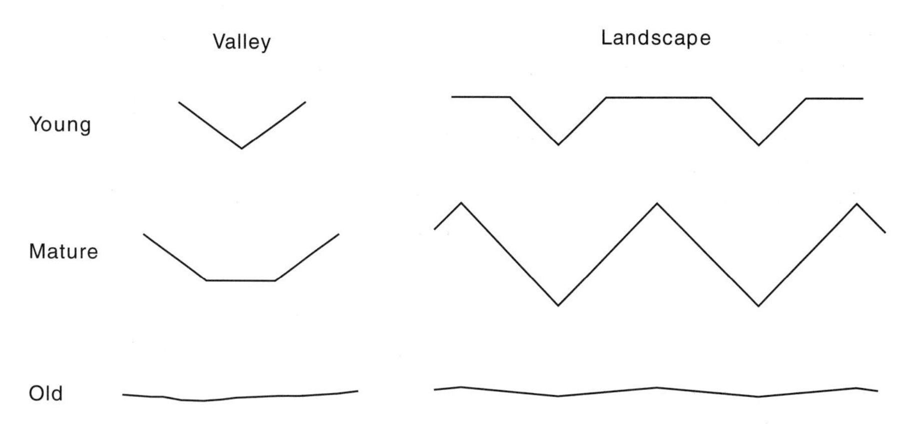

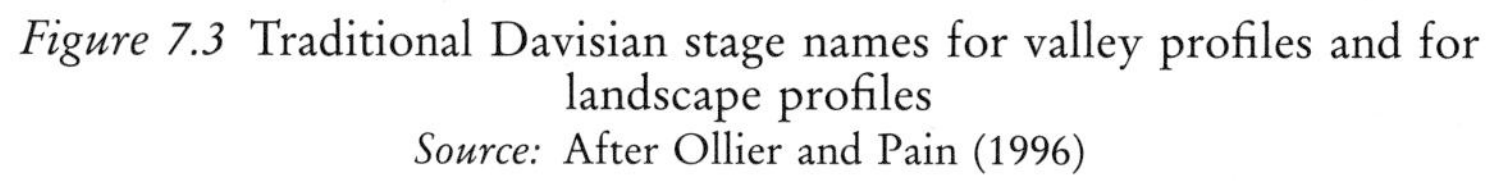

Figure 7.3 Traditional Davisian stage names for valley profiles and for landscape profiles
Source: After Ollier and Pain (1996)

extensive plains in old age, after lateral erosion has removed all hills (Figure 7.3). Young landscapes are characterized by much flat topography of the original uplifted peneplain. Mature landscapes have deeper and wider V-shaped valleys that have consumed much of the interfluves bearing remnants of the original land surface. Old landscapes are characterized by a peneplain, in which the interfluves are reduced to minor undulations (Figure 7.3).

The 'geographical cycle' was designed to account for the development of humid temperate landforms produced by prolonged wearing down of uplifted rocks offering uniform resistance to erosion. It was extended to other landforms, including arid landscapes, glacial landscapes, periglacial landscapes, to landforms produced by shore processes, and to karst landscapes (see Huggett 1985: 39).

Slope recession

According to the Davisian model, uplift and planation take place alternately. But, in many landscapes, uplift and denudation occur at the same time. The continuous and gradual interaction of tectonic processes and denudation leads to a different model of landscape evolution, in which the evolution of individual slopes determines the evolution of the entire landscape (Penck 1924, 1953). Three main slope forms evolve with different combinations of uplift and denudation rates. First, convex slope profiles, resulting from waxing development (*aufsteigende Entwicklung*), form when the uplift rate exceeds the denudation rate. Second, straight slopes, resulting from stationary (or steady-state) development (*gleichförmige Entwicklung*), form when uplift and denudation rates match one another. And third, concave slopes, resulting from waning development (*absteigende Entwicklung*), form when the uplift rate is less than the denudation rate.

A notable facet of the Penckian model is the omnipresence of slope recession. The argument runs that if the whole slope should weather equally, and if the weathered debris should be removed from the slope base, then a new slope should form parallel to the original slope. This view is contrary to the Davisian idea of slope decline (Figure 7.4). It means that maximum slope angles within landscapes will remain roughly the same until the very final stages of slope evolution. Field studies have confirmed that slope retreat is common in a wide range of situations. However, a slope that is actively eroded at its base (by a river or the sea) may decline if the basal erosion should stop.

Another model of landscape evolution in which slope recession is paramount involves cycles of pedimentation (King 1953, 1967, 1983). Each cycle starts with a sudden burst of cymatogenic diastrophism and passes into a period of diastrophic quiescence, during which subaerial processes reduce the relief to a pediplain (analogous to a peneplain). However, cymatogeny and pediplanation are interconnected. As a continent is denuded, so the eroded sediment is deposited offshore. With some sediment removed, the continental margins rise. At the same time, the weight of sediment in offshore regions causes depression. The outcome of this uplift and depression is the development of a major scarp near the coast that cuts back inland. As the scarp retreats, leaving a pediplain in its wake, it further unloads the continent and places an extra load of sediment offshore. Eventually, a fresh bout of uplift and depression will produce a new scarp. Thus, because of the cyclical relationship between continental unloading and the offshore loading, continental landscapes come to consist of a huge staircase of erosion surfaces (pediplains), the oldest steps of which occur well inland.

A variation on the slope-recession theme concerns the notion of unequal activity (Crickmay 1975). Davisian, Penckian, and Kingian models of

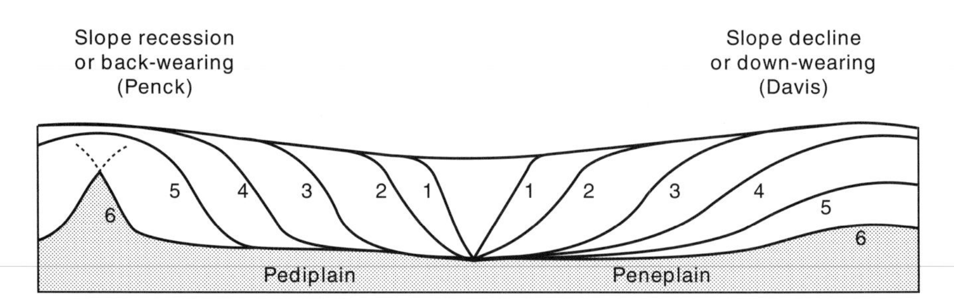

Figure 7.4 Slope recession and slope decline
Source: After Gossman (1970)

landscape evolution assume that slope processes act evenly on individual slopes. However, exogenic agents act unequally. For this reason, a slope may recede only where a stream (or the sea) erodes its base. If this should be so, then slope denudation is largely achieved by the lateral corrasion of rivers (or marine erosion at a cliff foot). This will mean that some parts of the landscape will stay virtually untouched by slope recession. Some evidence supports this contention (p. 256).

Slopes as catenae

All landforms share a basic feature: they consist of one or more slopes of variable inclination, orientation, length, and shape (Butzer 1976: 79). The tripartite classification of slopes into convex, straight, and concave elements appears to be fundamental – all landscapes consist of geomorphic catenae (Scheidegger 1986). A geomorphic catena consists of a flat eluvial region at the top, a colluvial region in the middle (with steep slopes and large mass flow rates), and a flat alluvial zone at the bottom. The several schemes devised to describe slope profiles recognize these three basic units, though subunits are also distinguished (Figure 7.5). One scheme recognizes four slope units: the waxing slope, also called the convex slope or upper wash slope; the free face, also called the gravity or derivation slope; the constant slope, also called the talus or debris slope where scree is present; and the waning slope, also called the pediment, valley-floor basement, and lower wash slope (A. Wood 1942). A widely used system has five slope units – summit, shoulder, backslope, footslope, and toeslope (Ruhe 1960). A similar system uses different names – upland flats, crest

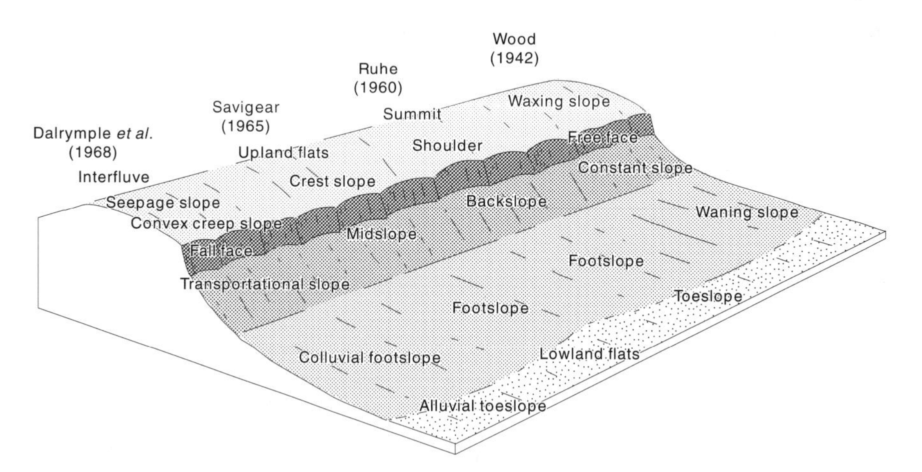

Figure 7.5 Systems for naming slope elements

slope, midslope, footslope, and lowland flats (Savigear 1965). The nine-unit land-surface model embraces and embellishes all these schemes and distinguishes nine units – interfluve, seepage slope, convex creep slope, fall face, transportational slope, colluvial footslope, alluvial toeslope, channel wall, and channel bed (Dalrymple *et al.* 1968).

The ubiquity of geomorphic catenae may result from increasing erosion rates with increasing slope gradients (Scheidegger 1991: 247). The steeper the slope, the faster it recedes. Flat regions at either end of the catena tend to remain intact while the steep middle section erodes and wears back, cutting into the summit plain. However, mathematical models of slope evolution suggest that the wearing back or wearing down of the midslope depends upon the mass movements in operation. As a generalization, surface wash processes lead to a back-wearing of slopes, whereas creep processes lead to a down-wearing of slopes (e.g. Scheidegger 1961; Hirano 1975; Gossman 1970, 1981; Nash 1981). But the pattern of slope retreat and slope decline is crucially dependent on conditions at the slope base, and especially the transport capacity of streams (e.g. Kirkby 1993).

Etching

Mechanical erosion was assumed to predominate in traditional models of landscape evolution. It was realized that chemical weathering reduces the mass of weathered material, but only on rocks especially vulnerable to solution (such as limestones) were chemical processes thought to have an overriding influence on landscape evolution. However, it now seems that forms of chemical weathering are important in the evolution of landscapes. Groundwater sapping, for instance, shapes the features of some drainage basins (e.g. Howard *et al.* 1988). And the solute load in three catchments in Australia comprised more than 80 per cent of the total load, except in one case where it comprised 54 per cent (Ollier and Pain 1996: 216). What makes these figures startling is that the catchments were underlain by igneous rocks. Information of this kind is making some geomorphologists suspect that chemical weathering plays a starring role in the evolution of nearly all landscapes.

In tropical and subtropical environments, chemical weathering produces a thick regolith that is then stripped by erosion (Wayland 1933; Jessen 1936; Thomas 1989a, 1989b). This process is called etchplanation. It creates an etched plain or etchplain. The etchplain is largely a production of chemical weathering. In places where the regolith is deeper, weakly acid water lowers the weathering front, in the same way that a sponge soaked in sulphuric acid would etch, and sink into, a metal surface. Some researchers contend that surface erosion lowers the land surface at the same rate that chemical etching lowers the weathering front (Figure 7.6). This is the theory of double planation. It envisages land surfaces of low relief being maintained during prolonged, slow uplift by the continuous lowering of double planation surfaces – the wash surface and the basal weathering surface (Büdel 1957, 1982; Thomas 1965). A rival view, depicted schematically in Figure 7.7, is that a period of deep chemical weathering precedes a phase of regolith stripping (e.g. Linton 1955; Ollier 1959, 1960; Hill *et al.* 1995).

Whatever the details of the etching process, it is very effective in creating landforms, even in regions lying beyond the present tropics. The Scottish Highlands experienced a major uplift in the early Tertiary. After 50 million

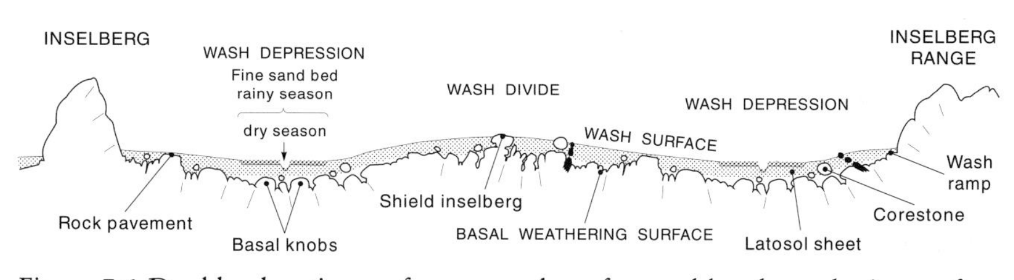

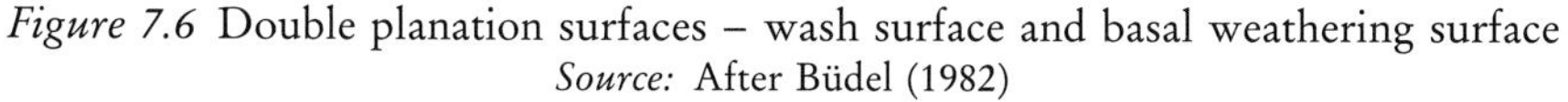

Figure 7.6 Double planation surfaces – wash surface and basal weathering surface
Source: After Büdel (1982)

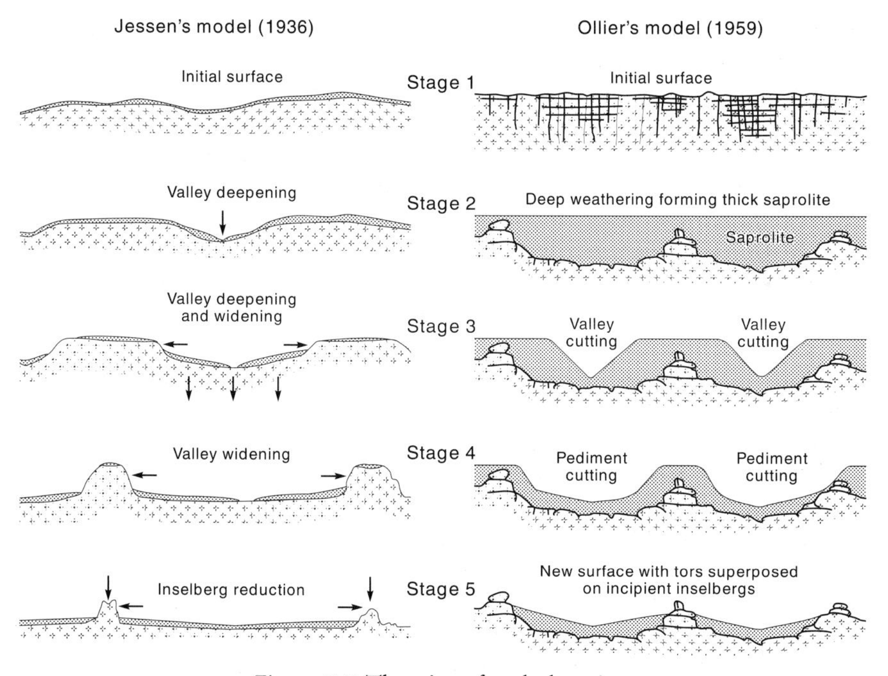

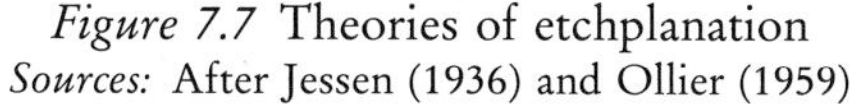

Figure 7.7 Theories of etchplanation
Sources: After Jessen (1936) and Ollier (1959)

years, the terrain evolved by dynamic etching with deep weathering of varied geology under a warm to temperate humid climate (A. M. Hall 1991). This etching led to a progressive differentiation of relief features, with the evolution of basins, valleys, scarps, and inselbergs.

Steady-state, non-linear complexities, and chaos

The landscape should attain a dynamic equilibrium or steady state where uplift is constant and a constant exogenic-process regime operates. In an erosional landscape, dynamic equilibrium prevails where all slopes, both hillside and river, are adjusted to one another (cf. Gilbert 1877: 123–4; Hack 1960: 81), and 'the forms and processes are in a steady state of balance and may be considered as time independent' (Hack 1960: 85). This view of landscape evolution may provide a more reasonable basis for interpreting topographical forms in an erosionally graded system, such as certain areas in the south-eastern United States, than may the geographical cycle (Hack 1960, 1975).

In practice, the idea of dynamic equilibrium is difficult to apply to landscapes and other forms of equilibrium have been advanced (Howard 1988). Dynamic metastable equilibrium suggests that, once perturbed by environmental changes or random internal fluctuations, a landscape will respond in a complex manner (Schumm 1979). A stream, for instance, if it should be forced away from a steady state, will adjust to the change; but the nature of the adjustment may vary in different parts of the stream and at different times. Douglas Creek in western Colorado, for example, has been cutting into its channel bed since about 1882 (Womack and Schumm 1977). The manner of incision has been complex, with discontinuous episodes of downcutting interrupted by phases of deposition, the erosion–deposition sequence varying from one cross-section to another. Unpaired terraces have formed that are discontinuous downstream. This kind of study serves to dispel for ever the simplistic cause-and-effect view of landscape evolution in which change is seen as a simple response to an altered input. It shows that landscape dynamics may involve abrupt and discontinuous behaviour involving flips between quasi-stable states.

More generally, steady states in the landscape may be rare because landscapes are inherently unstable. Any balance obtaining in a steady state is readily disrupted by any process that triggers positive feedback mechanisms. This idea is formalized as an 'instability principle'. This principle recognizes that, in many landscapes, accidental deviations from a 'balanced' condition tend to be self-reinforcing (Scheidegger 1983). This explains why cirques tend to grow, sink holes increase in size, longitudinal mountain valley profiles become stepped. The intrinsic instability of landscape is borne out by mathematical analyses that point to the chaotic nature of much landscape change (e.g. J. D. Phillips 1992; Scheidegger 1994).

Observed landform changes

Some topographic changes are rapid enough to be witnessed, as when an earthquake produces a fault scarp. But, for the geomorphologists studying topographic change, most changes are frustratingly slow. Current rates can be measured and they allow inferences about long-term changes to be made. Mathematical models are helpful here: they predict long-term effects of particular process regimes. However, given the possibility of high-magnitude events that are not presently recorded and the problem of environmental change, long-term predictions must be treated gingerly.

Perhaps the best guides to long-term topographic changes are topographic chronosequences. Charles Darwin used the chronosequence method to test his ideas on coral-reef formation. He thought that barrier reefs, fringing reefs, and atolls occurring at different places represented different evolutionary stages of island development applicable to any subsiding volcanic peak in tropical waters. William Morris Davis applied this evolutionary schema to landforms in different places and derived what he deemed was a time sequence of landform development – the geographical cycle. This approach is also applied to soil development sequences (p. 189). It is open to misuse (Summerfield 1991a: 17). The temptation is to fit the landforms into some preconceived view of landscape change, even though other sequences might be constructed. The significance of this problem is highlighted by a study of south-west African landforms since Mesozoic times (Gilchrist *et al.* 1994). It was found that several styles of landscape evolution were consistent with the observed history of the region. There is also a danger of making the erroneous assumption that all spatial differences are temporal differences – factors other than time exert a strong influence on landform, and landforms of the same age might differ through historical accident. Given these not inconsiderable difficulties, it is best to treat chronosequences circumspectly.

Trustworthy topographic chronosequences are conspicuous by their rarity. The best examples are normally found in artificial landscapes, though there are some landscapes in which, by quirks of history, spatial differences can be translated into time sequences. Occasionally, field conditions lead to adjacent hillslopes being progressively removed from the action of a fluvial or marine process. This has happened along a segment of the South Wales coast, in the British Isles (Savigear 1952, 1956). Originally, the coast between Gilman Point and the Taf estuary was exposed to wave action. A sand spit started to grow. Wind-blown and marsh deposits accumulated between the spit and the original shoreline, causing the sea to abandon the cliff base progressively from west to east. The present cliffs are thus a topographic chronosequence: the cliffs furthest west have been subject to subaerial denudation without waves cutting their base the longest, while those to the east are progressively younger (Figure 7.8). All the cliffs are formed in Old Red Sandstone.

Slope profiles along Port Hudson bluff, on the Mississippi River in Louisiana, reveal a chronosequence (Brunsden and Kesel 1973). The Mississippi River was undercutting the entire bluff segment in 1722. Since then, the channel has shifted about 3 km downstream with a concomitant cessation of undercutting. The changing conditions at the slope bases have reduced the mean slope angle from 40° to 22°.

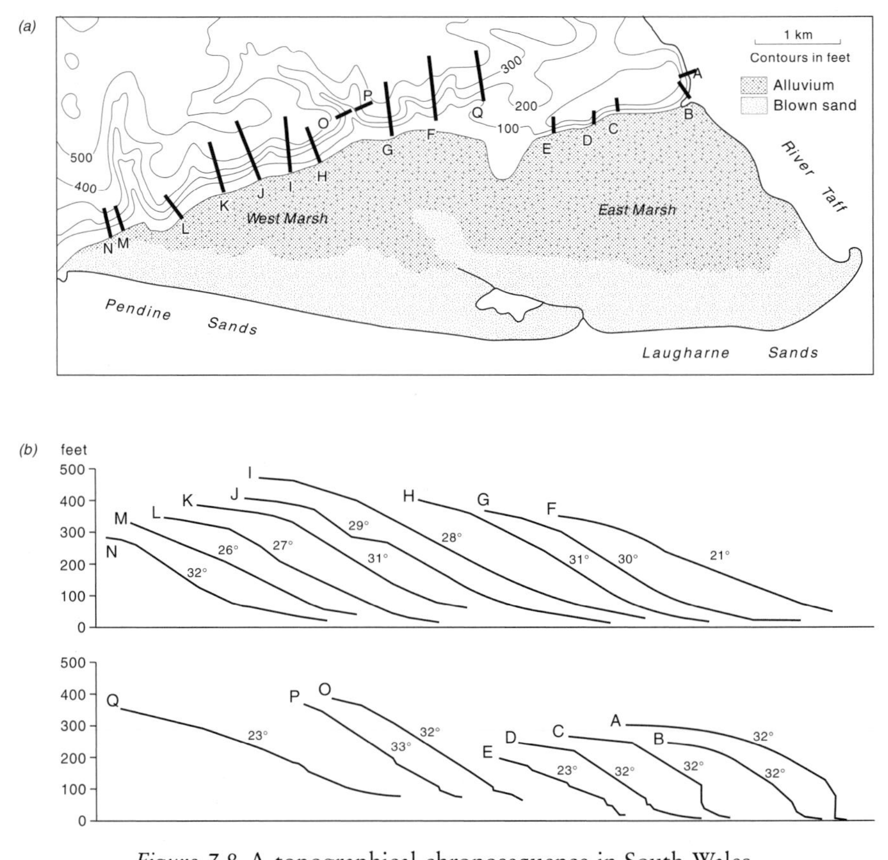

Figure 7.8 A topographical chronosequence in South Wales.
(a) The coast between Gilman Point and the Taff estuary. The sand spit has grown progressively from west to east so that the cliffs to the west have been longest protected from wave action.
(b) The general form of the slopes profiles located on Figure 7.8a. Cliff profiles become progressively older in alphabetical order, A–N
Source: After Savigear (1952, 1956)

RAIN AND RIVERS: CAUSES OF TOPOGRAPHIC CHANGE

Why do landscapes change? Are they moulded by surface processes under the control of environmental factors? Are they mainly a reflection of lithospheric processes? Or do they possess immanent change-inducing properties?

Geology and landform change

The ascension of internal energy originating in the Earth's core triggers a complicated set of geological processes. Deep-seated lithospheric, and ultimately baryspheric, processes and structures influence the shape and dynamics of the toposphere. The primary surface features of the globe are in very large measure the product of geological processes. This primary tectonic influence is manifest in the structure of mountain chains, volcanoes, island arcs, and other large-scale structures exposed at the Earth's surface. Passive margins, for example, have three main topographic and structural features: a marginal bulge falling directly into the sea, or bounded by a 'great escarpment' (Colour plate 7); a broad coastal plain and low plateau; or a complex of blocks and basins (Battiau-Queney 1991).

Many secondary topographic features are also an expression of geological processes and structures. A few, such as offset valleys, fault scarps, and volcanic landforms, are direct productions of endogenic processes. Several landscape features, patently of exogenic origin, have tectonic or endogenic predesign stamped on them (or, literally speaking, stamped under them). Tectonic predesign arises from the tendency of erosion and other exogenic processes to follow lithospheric stress patterns (Summerfield 1987, 1988; Scheidegger 1991: 245). The resulting landscape features are not fashioned directly by the stress fields. Rather, the exogenic processes act preferentially in conformity with the lithospheric stress. The conformity is either with the direction of a shear or, where there is a free surface, in the direction of a principal stress. Most textbooks on geomorphology abound with examples of structural landforms. The scarp and vale sequence of the Kentish Weald, England, is a classic case (Figure 7.9). Even in the Scottish Highlands, many present landscape features, which resulted from Tertiary etching, are closely adjusted to underlying rock types and structures (A. M. Hall 1991).

Mantle plumes seem to determine drainage patterns on rifted continental margins. Ancient lithospheric domes above plumes create radial drainage networks (Cox 1989; Kent 1991). A postulated Deccan plume beneath India caused the growth of a topographic dome, the eastern half of which is now gone (Figure 7.10a). Most of the rivers rise close to the west coast and drain eastwards into the Bay of Bengal, except those in the north that drain north-eastwards into the Ganga and a few that flow westwards or south-

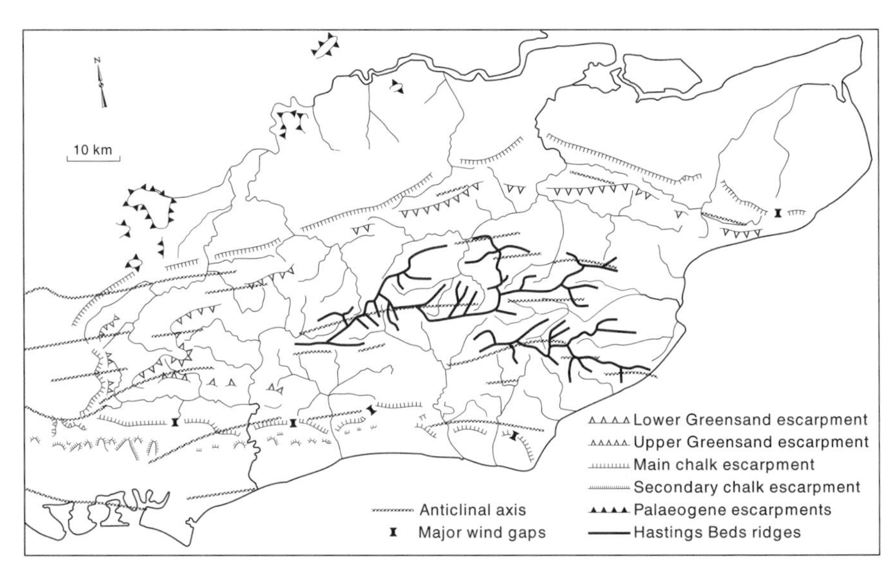

Figure 7.9 Some structurally influenced topographic features in south-east England
Source: After Jones (1981)

westwards (possibly along failed rift arms). Mantle plumes beneath southern Brazil and southern Africa would account for many features of the drainage patterns (Figure 7.10b–c).

Many smaller landforms, once ascribed to exogenic processes, are now being reinterpreted as essentially endogenically created forms that have been modified by weathering and erosion. Some valleys and river benches appear to form in this way (Kvet 1993). So, too, do river gorges (Scheidegger and Hantke 1994). The Aare Gorge in the Bernese Oberland, the Moutier–Klus Gorge in the Swiss Jura, the Samaria Gorge in Crete, hill-klamms in the Vienna Woods, Austria, and the Niagara Gorge in Ontario and New York State all follow pre-existing faults and clefts. Erosive processes may have deepened and widened them, but they are essentially endogenic features and not the product of antecedent rivers.

It is not uncommon for endogenic processes, and in some cases exogenic processes, to set in train a sequence of events that ultimately inverts the relief – valleys become hills and hills become valleys. Lava flows, for example, tend to flow down established valleys. Erosion then reduces the adjacent hillside leaving the more resistant volcanic rock as a ridge between two valleys. Such inverted relief is remarkably common (Pain and Ollier 1995). Inversion caused by lava flows occurs in California, France, Australia, and in all places where lava flows are common. Exogenic processes that create resistant material in the regolith may also promote

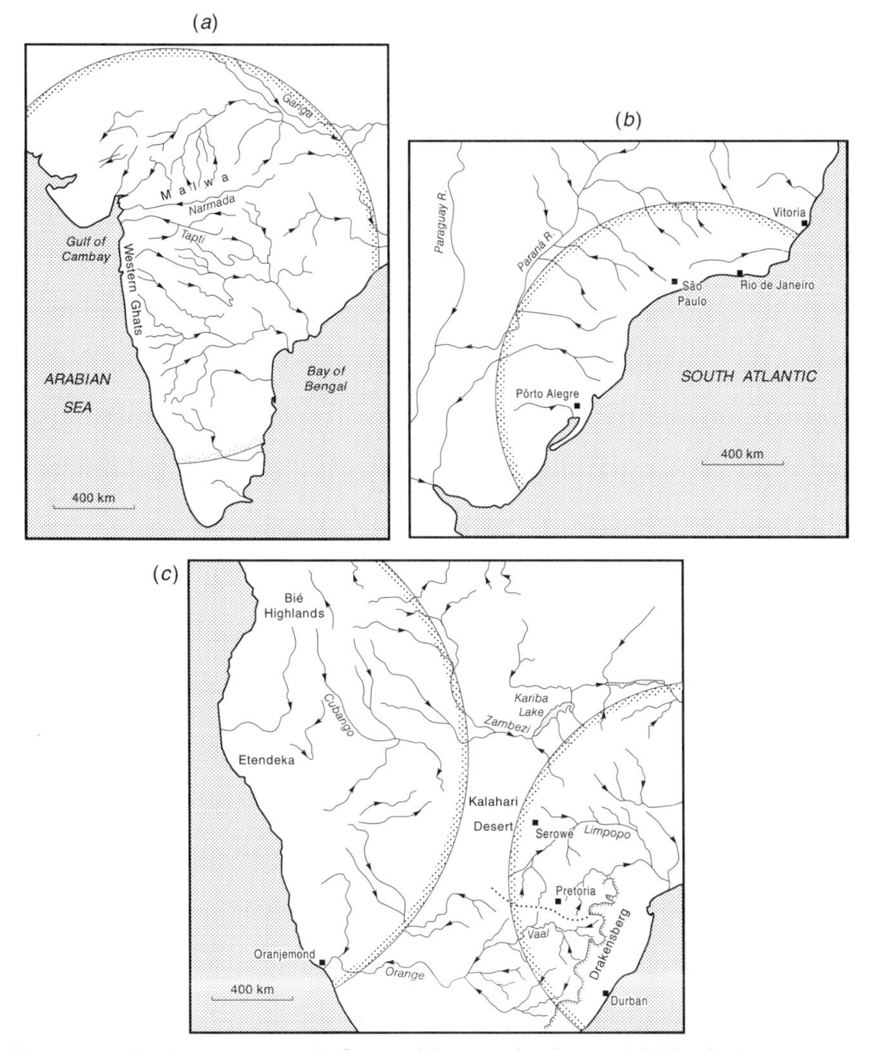

Figure 7.10 Drainage patterns influenced by mantle plumes. (a) The drainage pattern of peninsular India with the postulated Deccan plume superimposed. Most of the peninsula preserves dome-flank drainage. The Gulf of Cambay, Narmada, and Tapti systems exhibit rift-related drainage. (b) The drainage pattern of southern Brazil with superimposed plume dome-flank drainage is dominant except near Pôrto Alegre. (c) The drainage pattern in south-eastern and south-western Africa with the Paraná plume (left) and Karoo plume (right) superimposed. Rivers over the Paraná plume show an irregular dome-flank pattern drainage eastwards into the Kalahari. Notice that the Orange River gorge is formed where antecedent drainage has cut through younger uplift. Rivers over the Karoo plume display preserved dome-flank drainage west of the Drakensberg escarpment. The dotted line separates dome-flank drainage in the south from rift-related drainage in the north
Source: After Cox (1989)

relief inversion. Duricrusts are commonly responsible for inverting relief. Old ferricreted valley bottoms resist erosion and eventually come to occupy hilltops. Even humble alluvium may suffice to cause relief inversion (Mills 1990). Floors of valleys in the Appalachian Mountains, eastern United States, become filled with large quartzite boulders, more than 1 m in diameter. These boulders protect the valley floors from further erosion by running water. Erosion then switches to sideslopes of the depressions and, eventually, ridges capped with bouldery colluvium on deep saprolite form. Indeed, the saprolite is deeper than that under many uncapped ridges.

Exogenic processes and landform change

Exogenic processes (weathering, erosion, transport, and deposition) work on the raw relief created by lithospheric processes to produce the actual toposphere. Several factors influence the nature and rate of exogenic processes, climate and the action of the sea being especially important.

Climate influences

It is commonly suggested, though not universally accepted, that each climatic zone produces a distinctive suite of landforms (e.g. Tricart and Cailleux 1972; Büdel 1982; Bremer 1988). To be sure, exogenic processes are influenced by climate. Whether the set of exogenic processes within each climatic zone creates characteristic landforms is open to question. Much climatic geomorphology has been criticized for using temperature and rainfall data, which provide too gross a picture of the relationships between rainfall, soil moisture, and runoff, and for excluding the magnitude and frequency of storms and floods, which are important in landform development (Stoddart 1969). Some landforms are more climatically zonal in character than others. Arid, nival, periglacial, and glacial landforms are quite distinct. Relief types in the central Sahara are best explained by climatic factors (Hagedorn 1993). Other morphoclimatic zones have been distinguished but their constituent landforms are not clearly determined by climate. In all morphoclimatic regions, the effects of geological structure and etch processes are significant, even in those regions where climate exerts a strong influence on landform development (Twidale and Lageat 1994). So, climate is not of overarching importance for landforms development in over half the world's land surface. Indeed, some geomorphologists opine that landforms, and especially hillslopes, will be the same regardless of climate in all geographical and climatic zones (King 1957; Frye 1959; Ruhe 1975).

The conclusion is that, mainly because of ongoing climatic and tectonic change, the climatic factor in landform development is not so clear cut and simple as climatic geomorphologists have on occasions suggested. Responses to these difficulties go in two directions – towards complexity and towards simplicity. The complexities of climate–landform relations are explored in at least two ways. One way is to attempt a fuller characterization of climate. A recent study of climatic landscape regions of the world's mountains used several pertinent criteria: the height of timberline, the number and character of altitudinal vegetational zones, the amount and seasonality of moisture available to vegetation, physiographic processes, topographic effects of frost, and the relative levels of the timberline and permafrost limit (Thompson 1990). Another way of delving into the complexity of climatic influences is to bring modern views on fluvial system dynamics to bear on the question. One such study has taken a fresh look at the notion of morphogenetic regions and the response of geomorphic systems to climatic change (Bull 1992).

A simpler model of climatic influence on landforms is equally illuminating (Ollier 1988). It seems reasonable to reduce climate to three fundamental classes: humid where water dominates, arid where water is in short supply, and glacial where water is frozen (Table 7.4). Each of these 'climates' favours certain weathering and slope processes. Deep weathering occurs where water is unfrozen. Arid and glacial landscapes bear the full brunt of climatic influences because they lack the protection afforded by vegetation in humid landscapes. Characteristic landforms do occur in each of these climatic regions (see Table 7.1). Nevertheless, correlations in the field are confounded by relict features, some persisting 100 million years or much more, resulting from climatic swings. This point will be elaborated in the section of superslow landscape change.

Marine influences

Marine transgressions almost certainly refashion the land surface. During each transgression, marine waves and currents work over large parts of continents. In doing so, they cut cliffs, gouge out valleys, and so on. A transgressive sea cannot perform the catastrophic erosion that rapid floods can, but it slowly works over continents and can be expected to leave

Table 7.4 A simple scheme for relating climate and exogenic processes

Climate	*Weathering process*	*Weathering depth*	*Mass movement*
Glacial	Frost (chemical effects reduced by low temperatures)	Shallow	Rock glacier Solifluction (wet) Scree slopes
Humid	Chemical	Deep	Creep Landslides
Arid	Salt	Deep	Rockfalls

Source: After Ollier (1988)

traces in the landscape, even though the traces might be subsequently buried beneath sediments. It may even be capable of producing relatively flat areas of land, which are today expressed as planation surfaces, but that is a mootable point. A transgression creates a new sedimentary basin. The sheer weight of basin sediment may lead to tectonic changes that affect the adjacent land.

Much of the geomorphological research in the first half of the twentieth century in Britain focused on denudation chronology and the changing level of the sea (E. H. Brown and Waters 1974). Denudation chronologists hunted down marine planation surfaces. The theory of marine planation was unfashionable for many years. Its advocates were accused of letting their eyes deceive them: most purely morphological evidence, they were told, is 'so ambiguous that theory feeds readily on preconception' (Chorley 1965: 151). It is back in vogue, albeit in a new shape. Historical geomorphologists have recognized the importance of marine transgressions in landscape evolution.

Cretaceous, and particularly Early Cretaceous, marine transgressions played an important part in the evolution of the Australian landscape (Twidale 1991, 1994). It seems that the Early Cretaceous marine transgression flooded large depressed basins on the Australian land mass (Figure 7.11). The transgression covered about 45 per cent of the present continent. The new submarine basins subsided under the weight of water and sediment. Huge tracts of the Gondwanan landscape were preserved beneath the unconformity. Hinge lines (or fulcra) would have formed near shorelines. Adjacent land areas would have been uplifted, raising the Gondwanan palaeoplain, and basin margins warped and faulted. Parts of this plain were preserved on divides as palaeoplain remnants. Other parts were dissected and reduced to low relief by rivers graded to Cretaceous shorelines. Subsequent erosion of the Cretaceous marine sequence margins has exhumed parts of the Gondwanan surface, which is an integral part of the present Australian landscape.

Some other recent work on the effects of marine transgressions is less empirical and far more speculative. Old mountain belts are likely to have been flooded repeatedly by marine incursions. The old landscapes may denude at a rate proportional to their regional elevation (Pitman and Golovchenko 1991), though this assumption is highly questionable (e.g. Summerfield 1991b). If the denudational proportionality constant, k, is 1.0 per cent per million years, several conclusions may be reached. First, without large (>200 m) sea-level changes, it will take more than 300 million years, and probably more than 450 million years, to degrade a Himalayan-size mountain belt to a peneplain. Second, in a landscape where rivers are graded to base level, the rate of downcutting is decreased by the rate of uplift. This means that large-scale peneplanation can only occur during episodes of significant sea-level rise (>250 m) over 50 million years or more. Consequently, episodes of peneplanation indicate sea-level changes of that magnitude.

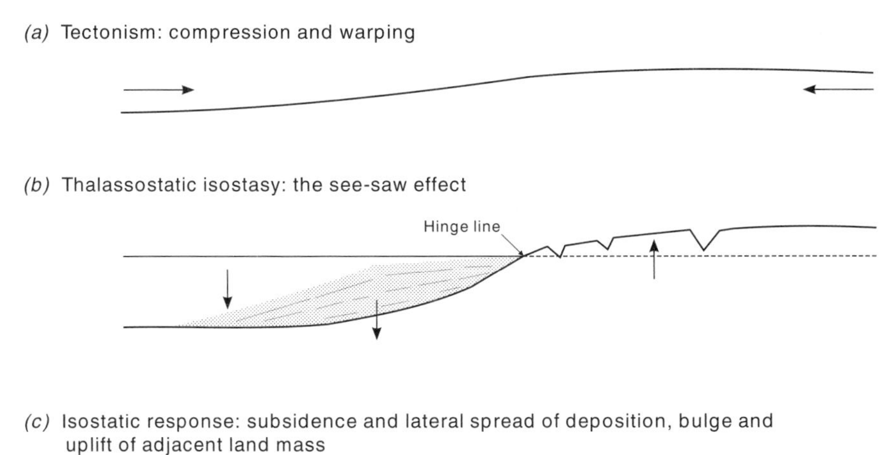

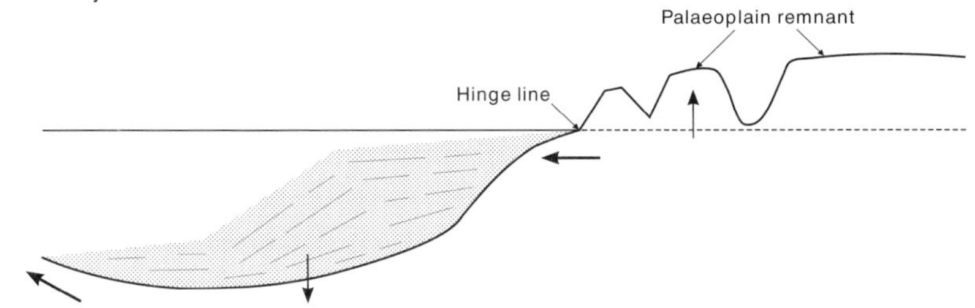

Figure 7.11 The sequence of events following a marine incursion into an Australian cratonic basin, and consequent uplift of adjacent land
Source: After Twidale (1994)

Internal mechanisms of geomorphic change

Endogenic and exogenic forces undoubtedly encourage change in geomorphic systems. Most landforms and landscapes possess a degree of inertia, and resist environmental changes. But resistance is possible only within certain bounds – once thresholds of environmental change are crossed, the drive towards geomorphic change is irresistible and a response in the landscape is inevitable. All landforms and landscapes have their own internal thresholds and environmental changes may certainly cause these to be crossed. However, internal geomorphic processes may also push geomorphic systems beyond local limits of stability and so induce landform and landscape changes (p. 236).

Discussions of self-organization have recently enlivened the question of internal changes in geomorphic systems (e.g. Huggett 1988; J. D. Phillips 1995a). Self-organizing geomorphic systems tend to produce orderly, repetitive patterns in the landscape that result from internal processes and not from any external control. Examples are patterned ground formed in periglacial environments (Hallet 1990), geochemical lateritic weathering (Nahon 1991), at-a-station hydraulic geometry (J. D. Phillips 1995a), riffle–pool sequences in rivers (Clifford 1993), and beach cusps (Werner and Fink 1994). Landscape evolution as a whole displays self-organizing and non-self-organizing (homogenizing) tendencies (J. D. Phillips 1995a). Down-wasting denudation is non-self-organizing (perhaps a better term would be self-destructing) – it leads to homogeneous landscapes in which relief decreases and elevations generally converge. Dissection is self-organizing – it may lead to heterogeneous landscapes in which relief increases and elevations diverge.

THE EVERLASTING HILLS: RATES OF TOPOGRAPHIC CHANGE

How fast does topography change? Is it a slow and steady process running over megayears? Or does it occur in brief episodes of ultra-fast, very-high-magnitude events?

Everyday, rare, and very rare events

The 'background' rate of chemical and mechanical denudation is slow. Overall denudation rates are about 0.05 mm/yr in non-mountainous regions and 0.5 mm/yr in mountainous regions (e.g. A. Young 1974; Saunders and Young 1983). The average rate of uplift is around 1.0 mm/yr (Ollier and Pain 1996: 215). The highest 'natural' denudation rates, as in the Himalayas, are considerably greater. Fluvial strata in the Cenozoic basin of northern Pakistan have eroded very fast – 1–15 mm/yr (Burbank and Beck 1991).

Currently, some landforms do change relatively swiftly compared with the background rate of land-surface denudation. The coastline of East Anglia is receding, recent cliff falls being obvious to the casual observer. Coastal spits change shape and position, sometimes within a single year. Volcanic islands emerge from the sea to form new land within human lifetimes. Soil erodes from many bare areas at an alarming rate, and a network of gullies is often created in the process. Impact events instantaneously blast out a topographic hole, witness Meteor Crater, Arizona. Most of these topographic changes may be observed or studied from old maps, and the rates of change established with confidence.

Normal denudation and deposition rates, on land and in the sea, vary greatly. Indeed, the variations are often of such a magnitude that a 'norm' is practically impossible to define. The problem is made worse by environmental change, which forces rate changes in geomorphic processes. For instance, many large meandering palaeochannels and meandering valleys appear to have resulted from wetter climates during the late

Pleistocene and early Holocene epochs (e.g. Dury 1953). The floods that fashioned the palaeochannels were considerably larger than the present floods. In the Ogeechee River, south-east Georgia, large meandering palaeochannels were formed by floods of at least twice the discharge of modern floods (Leigh and Feeney 1995).

The effects of rare, high-magnitude events in shaping landforms have received much attention over the last several decades. None the less, these rare events simply mark the very top of the 'normal' rate range. Moreover, their impact on the landscape is sometimes surprisingly modest. In 1993, a ~100-year flood occurred in the Mississippi River (Plate 7.4). This was a high-magnitude event of long duration. A study of a 70-km-long reach of the river, near Quincy, Illinois, showed that it had only minor geomorphic effects, with at most 4 mm of vertical floodplain accretion, even where there were no levees (Gomez *et al.* 1995).

Even rarer events of ultra high magnitude are now thought to occur, though none has been witnessed. These rarest of events, which at the tame end of the spectrum include large lake bursts, act suddenly with mighty

Plate 7.4a Flood of the Mississippi River, 4 August 1993. The oblique aerial photograph shows flood flow in the main channel of the Mississippi River and through the Miller City levee-break near river mile 35 (35 miles upstream from Cairo, Illinois). Flow through the levee-break was ~7,000 cms (cubic metres per second), while total flood discharge was ~26,000 cms. The 1993 flood peaked at this site on 7 August 1993 with a total flood discharge of ~27,600 cms
Photograph by Robert B. Jacobson, US Geological Survey

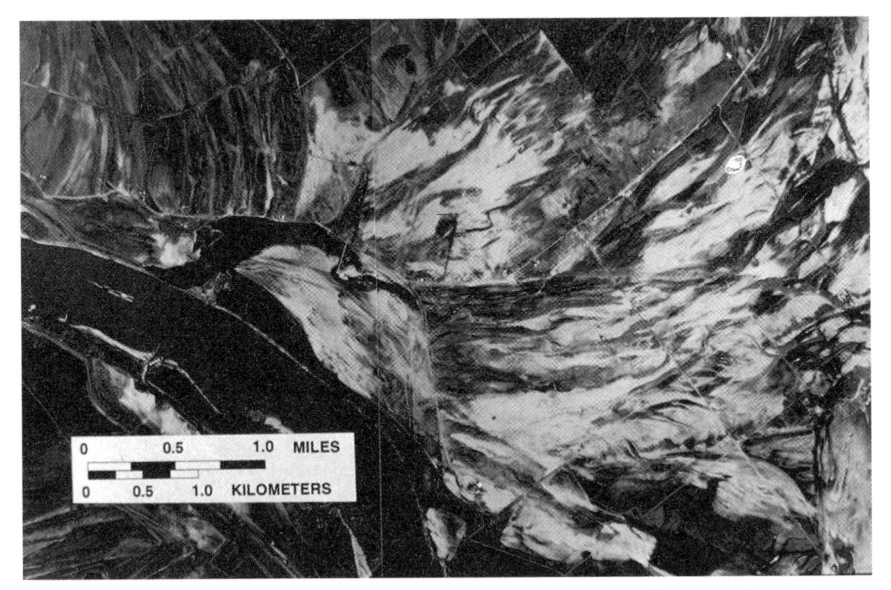

Plate 7.4b Montage of vertical aerial photographs of the Miller City levee-break complex taken 8 January 1994. The montage shows the 2,000-m-long scour and most of the 75 km^2 area affected by sand deposition. Total sand deposition in the levee-break complex has been estimated to be equivalent to 22–36 per cent of the total flood sediment load (Jacobson and Oberg in press)

Photograph available through the St Louis District Office, US Army Corps of Engineers

Plate 7.4c Low-altitude, oblique aerial photograph of typical levee-break complex formed on the Missouri River during the flood of 1993. This site is near Columbia, Missouri. The Missouri River is about 340 m wide in this photograph; flow was from left to right. The photograph was taken in November 1993

Photograph by Robert B. Jacobson, US Geological Survey

force. They are truly catastrophic events that cause swift and substantial changes of topography, refashioning the land surface within a matter of hours. These events may be studied by searching for their traces in the landscape and by theoretical modelling.

This section will consider two rates of landscape change: superfast and superslow.

Superfast landscape change

Votaries of neodiluvialism claim that spectacularly large superfloods play a major role in the development of some landscapes (Huggett 1989a, 1989b, 1994). There is a mounting body of evidence that superwaves, produced either by submarine landslides or by the impact of asteroids or comets in the ocean, have flooded continental lowlands. On continents, there is firm evidence that the catastrophic release of impounded water, possibly resulting from impact events, has led to grand deluges (Figure 7.12). Both lines of evidence add weight to the idea that some landscape features are caused by diluvial, rather than by fluvial, processes. Thus neodiluvialism

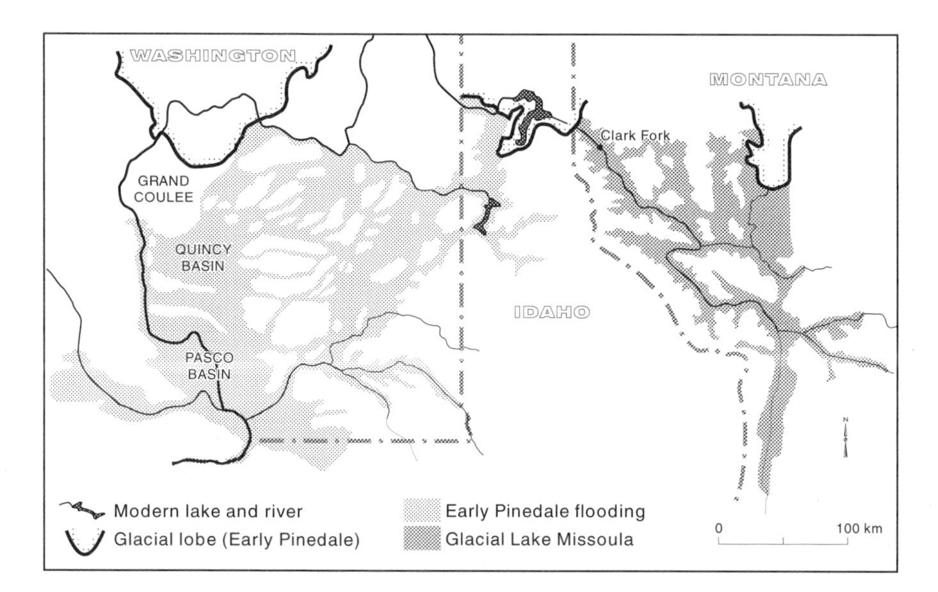

Figure 7.12 Glacial Lake Missoula and the Channeled Scabland. Two outbursts from glacial Lake Missoula, which took place between 18,000 and 13,000 years ago, produced massive floods in south-eastern Washington State. Evidence of these debacles includes abandoned waterways, cataract cliffs and plunge basins, potholes and deep rock basins, giant bars and giant ripples

Source: After V. R. Baker (1978a)

goes some way to vindicating the beliefs held by the old diluvialists before they were silenced by the dogma of uniformitarianism. It would perhaps be unwise to draw too close a parallel between the old diluvialism and the new diluvialism. Suffice it to remark that the diluvial metamorphosis of the landscape that might be inflicted by superwaves and superfloods, and the diluvial metamorphosis of the landscape wrought by the conjectural floods of the old diluvialists, are in most respects very similar. Before Charles Lyell, it was commonplace to interpret virtually all gravel and boulder deposits as a sure sign of the catastrophic action of flowing water. After 150 years of gradualistic explanations of flood deposits, it seems fair, with the general acceptance of the bombardment hypothesis, to look again at catastrophic explanations. Interestingly, evidence for a diluvial origin of landscapes and sediments comes not so much from the finding of previously unnoticed phenomena, as from the reinterpretation of well-known landscape and sedimentary features.

If these impact-induced superwaves should have occurred at the end of the last ice age, what would the likely effects have been when they reached lowland areas in western Europe? It is possible that they might have produced the very extensive spreads of late Pleistocene gravels found, for instance, in Cheshire and Lancashire. Equally, it is not beyond the bounds of possibility that vast quantities of water rapidly draining off lowland areas would produce gorge-like valleys such as those found in Cornwall, Devon, and Dorset. In Cornwall, several workers have noticed steep-sided valleys cutting into wide plateaux. Several such river-gorges in the Tintagel area, including the Rocky Valley and its picturesque St Nectan's Kieve, might have been formed by rejuvenation following uplift of an upland plain currently standing at about 130 m above sea level (H. Dewey 1916; see also Hendriks 1923). A catastrophic origin by receding superflood waters is another possibility. The draining of superflood waters might also account for the chines of Bournemouth — narrow gorges in which streams from the Bournemouth Plateau tumble into the sea. Like the river-gorges in Cornwall, the Bournemouth chines are generally thought to have been created by rejuvenation (e.g. Arkell 1947: 318), but superflood waters could have produced the same features.

The river-gorges and chines of southern England are reminiscent of the drowned valleys or *calas* found on the island of Mallorca. The *calas* are a prominent landscape feature around the eastern and southern coast of the island. The ephemeral streams that feed the *calas* originate in the Sierra de Levante. They cut into a plain formed in horizontal Miocene limestones. There is no evidence that water running off the Sierra de Levante has reached the sea within living memory, and analysis of fluvial gravels confirms minimal transport along the streams in the recent past. The *calas* were not produced by marine submergence during the Holocene epoch, but are fossil forms created at the end of the Tertiary period or the beginning of the Pleistocene epoch (Butzer 1962). It is possible that they are not purely fluvial forms, but were largely fashioned by the passage of superflood waters.

Superslow landscape change

'Little of the earth's topography is older than the Tertiary and most of it no older than Pleistocene' (Thornbury 1954: 26). This view was for many decades widely held by geomorphologists. Research over the last twenty years has revealed that a significant part of the land surface is surprisingly ancient, surviving in either relict or exhumed form.

Relict landscapes and landforms endure for millions of years. In tectonically stable regions, land surfaces, especially those capped by duricrusts, may persist a hundred million years or more, witness the Gondwanan and post-Gondwanan erosion surfaces in the Southern Hemisphere (King 1983). Remnants of a ferricrete-mantled land surface surviving from the early Mesozoic era are widespread in the Mount Lofty Ranges, Kangaroo Island, and the south Eyre Peninsula of South Australia (Twidale *et al.* 1974). Indeed, much of south-eastern Australia contains many very old topographical features (R. W. Young 1983; P. Bishop *et al.* 1985; Twidale and Campbell 1995). Some upland surfaces originated in the Mesozoic era and others in the Early Palaeogene period; and in some areas the last major uplift and onset of canyon cutting occurred before the Oligocene epoch. In southern Nevada, early to middle Pleistocene colluvial deposits, mainly darkly varnished boulders, are common features of hillslopes formed in volcanic tuff. Their long-term survival indicates that denudation rates on resistant volcanic hillslopes in the southern Great Basin have been exceedingly low throughout Quaternary times (Whitney and Harrington 1993).

In Europe, signs of ancient saprolites and duricrusts, bauxitic and lateritic sedimentation, and the formation and preservation of erosional landforms, including tors, inselbergs and pediments, have been detected (Summerfield and Thomas 1987). The palaeoclimatic significance of these finds has not passed unnoticed: for much of the Cenozoic era, the tropical climatic zone of the Earth extended much further polewards than it does today (e.g. Linton 1955; Godard 1965; Dury 1971; Thomas 1978). Indeed, evidence from deposits in the soil–landscape, like evidence in the palaeobotanical record, indicates that warm and moist conditions extended to high latitudes in the North Atlantic during the Late Cretaceous and Palaeogene periods. Julius Büdel (1982) was convinced that Europe has suffered extensive etchplanation during Tertiary times. Traces of a tropical weathering regime have been unearthed (e.g. Battiau-Queney 1996). In the British Isles, several Tertiary weathering products and associated land-forms and soils have been discovered (e.g. Battiau-Queney 1984, 1987;

Mitchell 1980; Isaac 1981, 1983; A. M. Hall 1985; Catt 1989). On Anglesey, which has been a tectonically stable area since at least the Triassic period, inselbergs, such as Mynydd Bodafon, have survived several large changes of climatic regime (Battiau-Queney 1987). In Europe, Asia, and North America many karst landscapes are now interpreted as fossil landforms originally produced under a tropical weathering regime during Tertiary times (Büdel 1982; Bosák *et al.* 1989).

Exhumed landscapes and landforms are also common, preserved for long periods beneath sediments then uncovered by erosion (Plates 7.5–7.7). They are common on all continents (e.g. Lidmar-Bergström 1989, 1993, 1995, 1996; Twidale 1994; Thomas 1995). The relief differentiation on the Baltic Shield, once thought to result primarily from glacial erosion, is considered now to depend on basement-surface exposure time during the Phanerozoic aeon (Figure 7.13). In northern England, a variety of active, exhumed, and buried limestone landforms are present (I. Douglas 1987). They were originally created by sedimentation early in the Carboniferous period (late Tournaisian and early Viséan ages). Subsequent tectonic changes associated with a tilt-block basement structure have effected a complex sequence of landform changes (Figure 7.14). The Waulsortian knolls are exhumed mounds of carbonate sediment formed about 350 million years ago. They were covered by shales and later by chalk, and then exhumed during the Tertiary period to produce reef knoll hills that are features of the present landscape. In the Clitheroe region, they form a series of isolated hills, up to 60 m high and 100–800 m in diameter at the base, standing above the floor of the Ribble Valley. The limestone fringing reefs formed in the Asbian and

Plate 7.5 A granitic landscape overwhelmed by, and then partly re-exposed from beneath, Early Cretaceous sediments, just east of Port Hedland in the north of Western Australia
Photograph by C. R. Twidale

Plate 7.6 An exhumed surface displayed by an unconformity in the Isa Highlands, north-west Queensland, Australia. Laterized Late Jurassic and Early Cretaceous sediments lie upon Proterozoic metasediments
Photograph by C. R. Twidale

Plate 7.7 Folded strata and ridge and valley topography in the Isa Highlands, north-west Queensland, Australia. The bevelled crests are remnants of the exhumed sub-Cretaceous surface shown in Plate 7.6
Photograph by C. R. Twidale

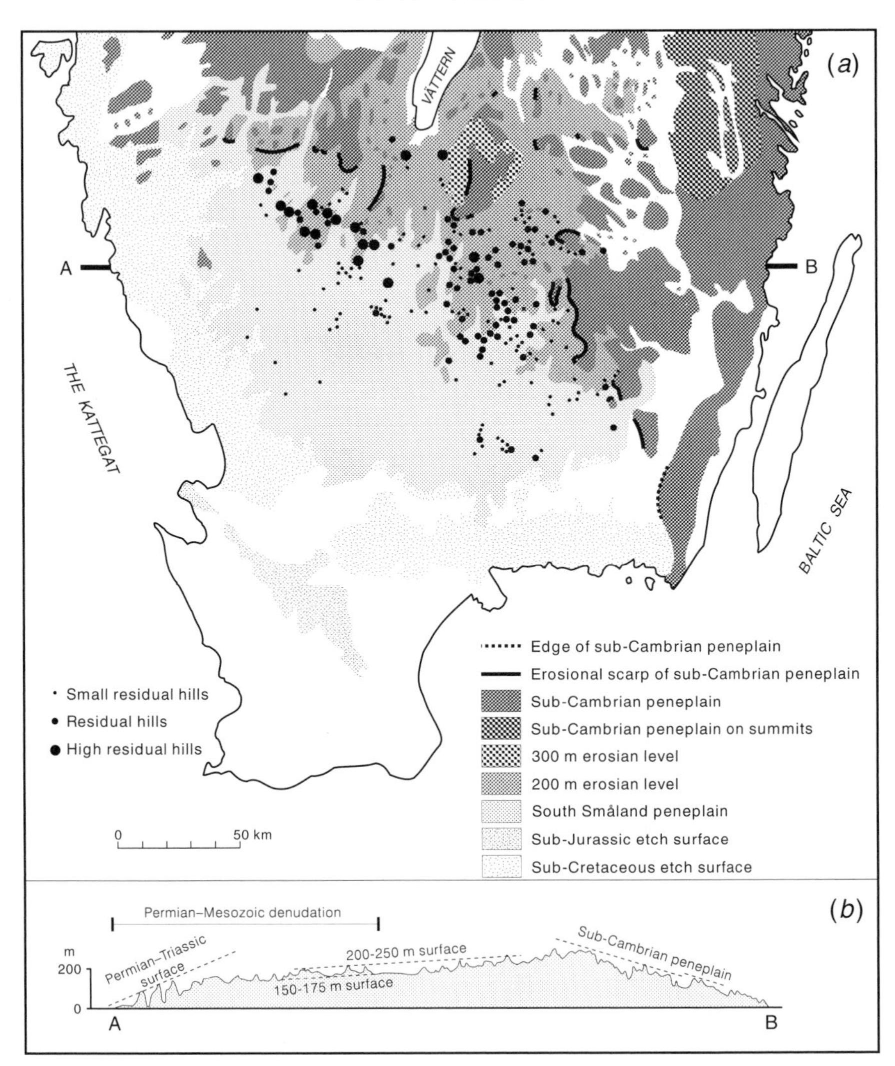

Figure 7.13 Denudation surfaces and tectonics of southern Sweden. (a) Mapped features. (b) West–east profile across the dome-like uplift of the southern Baltic Shield. Note the exhumed sub-Cretaceous hilly relief evolved from the Permo-Triassic surface

Source: After Lidmar-Bergström (1993, 1996)

Brigantian ages today form prominent reef knoll hills close to Cracoe, Malham, and Settle.

Just what proportion of the Earth's land surface pre-dates the Pleistogene period has yet to be ascertained, but it looks to be a not insignificant figure.

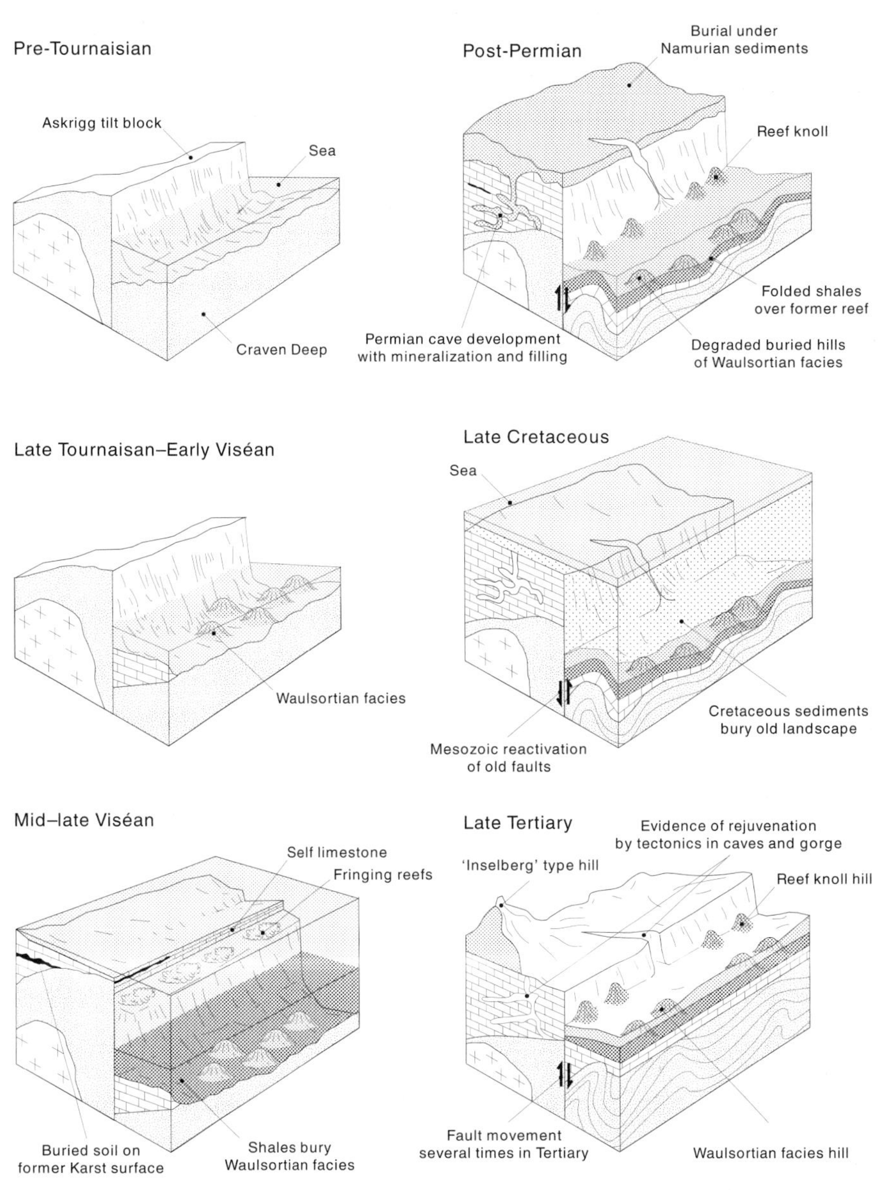

Figure 7.14 Schematic diagram showing the evolution of limestone landscapes in northern England. The Askrigg tilt block is separated from the Craven Deep by the Craven Fault, a major east–west fault across the northern Pennines
Source: After I. Douglas (1987)

In Australia, Gondwanan land surfaces constitute 10–20 per cent of the contemporary cratonic landscape (Twidale 1994). An important implication of all this work is that some landforms and their associated soils can survive through various climatic changes when tectonic conditions permit. A problem arises in accounting for the survival of these palaeoforms. Most modern geomorphological theory would dictate that denudational processes should have destroyed them long ago. It is possible that they have survived under the exceptional circumstance of a very-long-lasting arid climate, under which the erosional cycle takes a vast stretch of time to run its course (Twidale 1976). A controversial explanation is that much of the Earth's surface is geomorphologically rather inactive: the ancient landscape of south-eastern Australia, rather than being an exceptional case, may be typical of a very substantial part of the Earth's surface (R. W. Young 1983). If this contention should prove correct, then cherished views on rates of denudation and on the relation between denudation rates and tectonics would require a radical revision, and the connections between climate and landforms would be even more difficult to establish.

YOUTH, MATURITY, AND OLD AGE: DIRECTIONS OF TOPOGRAPHIC CHANGE

Landscape cycles and episodes

Several geomorphologists believe that landscape history has been cyclical or episodic. The Davisian system of landscape evolution combined periods dominated by the gradual and gentle action of exogenic processes interrupted by brief episodes of sudden and violent tectonic activity (p. 230). A land mass would suffer repeated 'cycles of erosion' involving an initial, rapid uplift followed by a slow wearing down. The Kingian model of repeated pediplanation envisaged long-term cycles, too. Remnants of erosion surfaces can be identified globally (King 1983). They correspond to pediplanation during the Jurassic, Early to middle Cretaceous, Miocene, Pliocene, and Quaternary times. However, King's views are not widely accepted, and have been challenged (e.g. Summerfield 1984).

A popular theme, with several variations, is that the landscape alternates between stages of relative stability and stages of relative instability. An early version of this idea, which still has considerable currency, is the theory of biostasy and rhexistasy (Erhart 1938, 1956). According to this model, landscape change involves long periods of 'biostasy' (biological equilibrium), associated with stability and soil development, broken in upon by short periods of 'rhexistasy' (disequilibrium), marked by instability and soil erosion. The theory of K-cycles is an elaboration of this view (B. E. Butler 1959, 1967). It proposed that landscapes evolve through a succession of cycles called K-cycles. Each cycle comprises a stable phase, during which soils develop forming a 'groundsurface', and an unstable phase, during which erosion and deposition occur.

More recent versions of the stability–instability model take account of regolith, tectonics, sedimentation, and sea-level change. A 'cratonic regime' model, based on studies carried out on the stable craton of Western Australia, envisaged alternating planation and transgression occurring without major disturbance for periods of up to a billion years (Fairbridge and Finkl 1980). During this long time, a 'thalassocratic regime' (corresponding to Erhart's biostasy and associated with high sea levels) is interrupted by short intervals dominated by a 'epeirocratic regime' (corresponding to Erhart's rhexistasy and associated with low sea levels). The alternations between thalassocratic and epeirocratic regimes may occur every 10–100 million years. However, more frequent alternations have been reported. A careful study of the Koidu etchsurface in Sierra Leone has shown that interruptions mirror environmental changes and occur approximately every 1,000–10,000 years (Thomas and Thorp 1985).

During biostasy, which is the 'normal' state, streams carry small loads of suspended sediments but large loads of dissolved materials: silica and calcium are removed to the oceans, where they form limestones and chert, leaving deep ferrallitic soils and weathering profiles on the continents. Rhexistatic conditions are triggered by bouts of tectonic uplift and lead to the stripping of the ferrallitic soil cover, the headward erosion of streams, and the flushing out of residual quartz during entrenchment. Intervening plateaux become desiccated owing to a falling water table and duricrusts form. In the oceans, red beds and quartz sands are deposited.

A variant of the cratonic regime model explains the evolution of many Australian landscape features (Twidale 1994; see p. 240). Indeed, the cratonic regime model is chiefly applicable to the southern continents. In the Northern Hemisphere, glaciation has disrupted the cratons and scoured them of much old regolith. None the less, recent work has shown that a deal of the palaeorelief and the saprolites on the Baltic Shield, for instance, is very ancient (Lidmar-Bergström 1995; see p. 252).

Evolutionary landscapes

Few theories of Earth surface development refer to very-long-term directional changes. The idea of directional change is implicit in Davis's cycle of erosion, in which the landscape system gradually runs down as potential energy, imparted during a short burst of diastrophism, is used up. It is also implicit in the notion of dynamic equilibrium, though there appears to be no preferred direction of change in models that employ this idea.

Directional change in landscape development is made explicit in the non-actualistic system of Earth surface history known as 'evolutionary geomorphology' (Ollier 1981, 1992). The argument runs that the land surface has changed in a definite direction through time, and has not suffered an 'endless' progression of erosion cycles. In other words, the Earth's landscapes have evolved as a whole. In doing so, they have been through several geomorphological 'revolutions' that have led to distinct and essentially irreversible changes of process regimes. These revolutions probably occurred during the Archaean aeon, when the atmosphere was reducing rather than oxidizing, during the Devonian period, when a cover of terrestrial vegetation appeared, and during the Cretaceous period, when grassland appeared and spread. The breakup and coalescence of continents would also alter landscapes. The geomorphology of Pangaea was, in several respects, unlike present geomorphology (Ollier 1991: 212). Vast inland areas lay at great distances from the oceans, many rivers were longer by far than any present river, terrestrial sedimentation was more widespread. When Pangaea broke up, rivers became shorter, new continental edges were rejuvenated and eroded, continental margins warped tectonically. Once split from the supercontinent, each Pangaean fragment followed its own history. Each experienced its own unique events. These included the creation of new plate edges and changes of latitude and climate. The landscape evolution of each continental fragment must be viewed in this very long-term perspective. In this evolutionary context, the current fads and fashions of geomorphology – process studies, dynamic equilibrium, and cyclical theories – have limited application (Ollier 1991: 212).

A good example of evolutionary geomorphology is afforded by tectonics and landscape evolution in south-east Australia (Ollier and Pain 1994; Ollier 1995). Morphotectonic evolution in this area appears to represent response to unique, non-cyclical events. Today, three major basins (the Great Artesian Basin, the Murray Basin, and the Gibbsland–Otway Basins) are separated by the Canobolas and Victoria Divides, which are intersected by the Great Divide and putative Tasman Divide to the east (Figure 7.15). These Divides are major watersheds. They evolved in several stages from an initial Triassic palaeoplain sloping down westwards from the Tasman Divide (Figure 7.16). First, the palaeoplain was downwarped towards the present coast, forming an initial divide. Then the Great Escarpment formed and retreated westwards, facing the coast. Much of the Great Divide is at this stage. Retreat of slopes from the coast and from inland reduced the palaeoplain to isolated High Plains, common on the Victoria Divide. Continued retreat of the escarpment consumed the High Plains and produced a sharp ridge divide, as is seen along much of the Victoria Divide. The sequence from low relief palaeoplain to knife-edge ridge is the reverse of peneplanation. With no further tectonic complications, the present topography would presumably end up as a new, lower level plain. However, the first palaeoplain is Triassic in age and the 'erosion cycle' is unlikely to end given continuing tectonic changes to interrupt the erosive processes.

EROMANGA BASIN
SOUTH AUSTRALIA
Canobolas Divide
Great Divide
MURRAY BASIN
NEW SOUTH WALES
Sydney
Adelaide
Tasman Divide
VICTORIA
Victoria Divide
N
Melbourne
200 km
GIPPSLAND BASIN
OTWAY BASIN

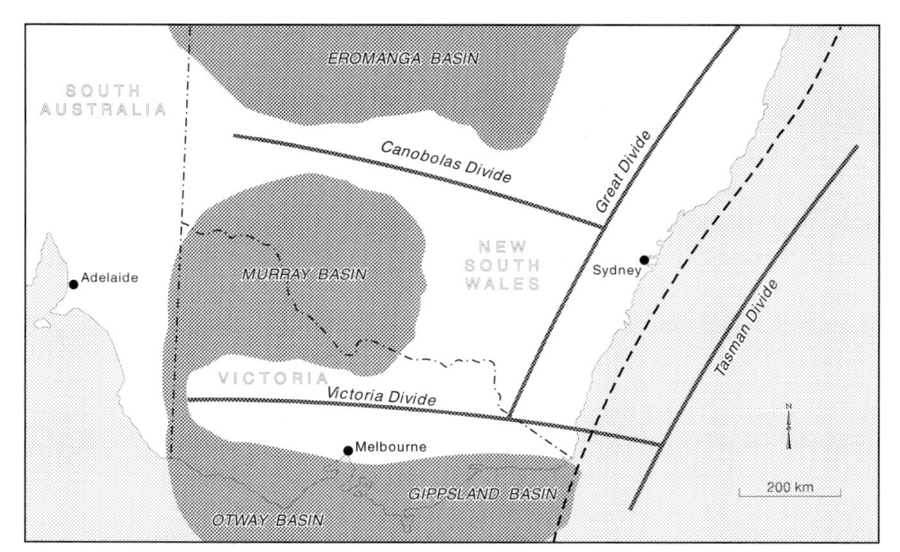

Figure 7.15 The major basins and divides of south-east Australia. The Eromanga Basin is part of the Great Artesian Basin
Source: After Ollier (1995)

(a) Initial palaeoplain sloping down from the Tasman Divide
(b) Downwarp of palaeoplain to coast, forming the initial divide
(c) Formation and retreat of the Great Escarpment facing the coast
(d) Retreat of slopes from the coast and inland reduces the palaeoplain to isolated High Plains
(e) Continued retreat of the escarpment of the inland slopes consumes the High Plains and produces a sharp divide

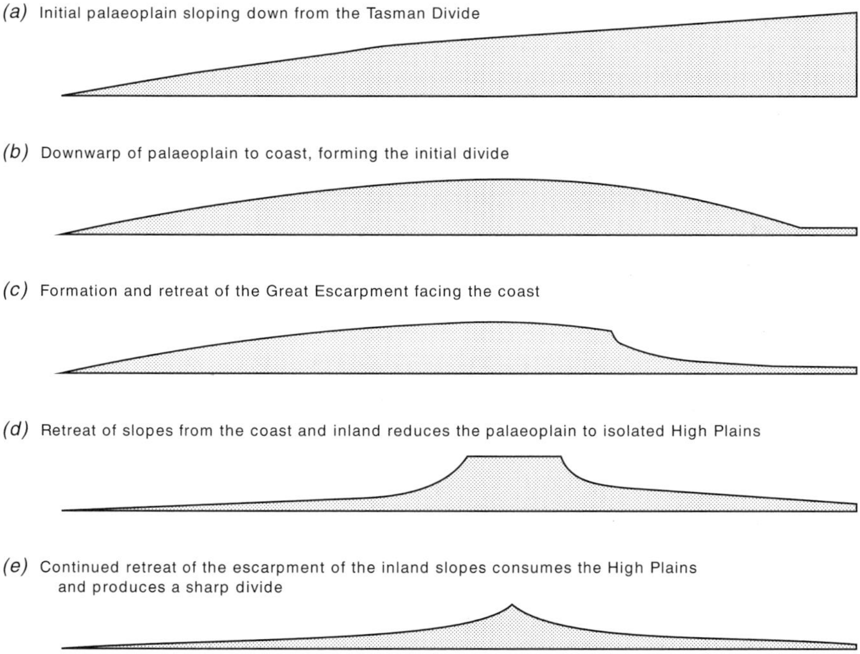

Figure 7.16 Evolution of the south-east Australian drainage divides
Source: After Ollier (1995)

The morphotectonic history of the area is associated with unique events. These include the sagging of the Murray Basin, the opening of the Tasman Sea and creation of a new continental margin, the eruption of the huge Monaro volcano, and the faulting of huge blocks in Miocene times. The geomorphology is evolving, and there are no signs of erosional cycles or steady states.

SUMMARY

The toposphere is the Earth's relief features. Continents, extended to the edge of continental shelves, and oceans form a major topographic division. Terrestrial, shallow marine, and deep marine environments bear characteristic landforms. Topography changes with the action of endogenic processes (tectonic and volcanic) and exogenic processes (weathering, erosion, transport, and deposition). The continuing interaction of endogenic and exogenic forces is called the 'principle of antagonism'. Hillslopes decline (the Davisian model) or retreat (the Penckian and Kingian models). Chemical etching is important in explaining some landscape features. Topographic chronosequences, though rare, provide valuable information on hillslope change. Geology, climate, and the sea influence landscape evolution. Endogenic landforms are common. Radial drainage on mantle-plume domes is an example. Exogenic processes fashion landforms, although different climates do not produce distinctive landforms, except in the case of glacial landscapes and arid landscapes. Marine transgressions affect landscape development. They produce erosional features, such as marine beaches, and, by creating new sedimentary basins, tectonic changes. Thresholds within landscape systems lead to topographic change. Most topographic change is imperceptibly slow. Catastrophic lake bursts and impact-induced superfloods produce ultra-fast topographic changes. In places, topographic change may be superslow – some relict landscapes and landforms have lasted for hundreds of millions of years with little modification. Landscape history involves cycles and episodes of change. It may also involve evolutionary trends caused partly by evolving exogenic process regimes (for instance, following the colonization of land and, later, the expansion of grasslands) and partly by unique events (accidents of geography and history).

FURTHER READING

A splendid introduction to the toposphere is *Global Geomorphology: An Introduction to the Study of Landforms* (Summerfield 1991a). Current reviews of specific landforms include *Cold Climate Landforms* (Evans 1994), *Geomorphology of Desert Dunes* (Lancaster 1995), and *Geomorphology in the Tropics: A Study of Weathering and Denudation in Low Latitudes*

(Thomas 1994). *An Introduction to Theoretical Geomorphology* (Thorn 1988) is a thoughtful account of theories of landscape evolution, ancient and modern. *Tectonics and Landforms* (Ollier 1981), *Ancient Landforms* (Ollier 1991), *Regolith, Soils and Landforms* (Ollier and Pain 1996) are highly recommended for thought-provoking reading. Readers interested in neodiluvialism might like to begin with *Cataclysms and Earth History: The Development of Diluvialism* (Huggett 1989b). Rates of landscape change are discussed in *The Changing Earth: Rates of Geomorphological Processes* (Goudie 1995).

8

BIOSPHERE AND ECOSPHERE

THE LIVING WORLD: INDIVIDUALS, COMMUNITIES, AND ECOSYSTEMS

Living together: communities and ecosystems

Individual organisms interact with their biotic environment (other individuals) and with their abiotic environment (the non-living parts of the ecosphere). A group of interacting individuals belonging to the same species is a population; a group of interacting individuals belonging to different species is a community. In both populations and communities, the 'groups' are tightly linked associations of organisms, rather than loose associations of individuals that happen to live in the same neighbourhood.

Communities are sometimes called biocoenoses. Individuals and communities live in particular physical surroundings known as the biotope. Ecosystems are individuals or communities interacting with their physical environment – biocoenoses plus biotopes. Biotopes will normally contain several distinct habitats that occur as patches in the landscape. The Dorset heathlands, England, are a biotope (Plate 8.1). They form a patchwork of different vegetation types. The patch distribution was originally determined largely by topography and soils, but human activities have produced 'holes' in the heathland biotope so that many patches are now isolated. Each species living in the heathland has particular habitat requirements (N. R. Webb and Thomas 1994). The silver-studded blue butterfly (*Plebejus argus*) has a broad niche (Colour plate 6). It requires dry or humid heath, of any aspect, that is in the pioneer or building phase of a succession (within about ten years of a disturbance by fire or cut). The red ant, *Myrmica sabuleti*, has narrow requirements; it needs dry heath with a warm south-facing aspect, containing more than 50 per cent grass species, and which is in the pioneer phase (within the first five years after a disturbance).

Plate 8.1 A heathland biotope: Hartland Heath, Dorset, England
Photograph by N. R. Webb

Community units

Communities and ecosystems range from microscopic to global. The tiers in the community hierarchy are local communities, communities, biomes, zonobiomes, and the biosphere. There are no specific names for equivalent units in the ecosystem hierarchy, but the terms local ecosystems (which occupy landscape patches), ecosystems, geozonal ecosystems (equivalent to zonobiomes), and ecosphere could be used.

The biosphere is made of living things. It includes lower parts of the air, the oceans, seas, lakes, and rivers, the land surface, and the soil. Some organisms live deep in the lithosphere and in sediments beneath the deep ocean floor (p. 174). All life interacts with its surroundings. This pervasive interaction of living and non-living kingdoms creates and conserves an ecosphere, a zone fit for terrestrial-type life forms.

A biome is a regional community of animals and plants (Clements and Shelford 1939). The humid temperate zone of western Europe is an example. It supports a deciduous forest biome, with areas of heath and moorland. The equivalent term for a community of plants is a plant formation. A community of animals at the biome scale has no special designation; it is simply an animal community. Smaller communities within biomes are normally based on plant distribution and are called plant associations. In England, associations within the deciduous forest biome include beech forest, lowland oak forest, and ash forest. A zonobiome (Walter 1985) consists of like biomes. Its phytogeographical equivalent is a formation-type. It has no zoogeographical counterpart. The broad-leaved temperate forests of western Europe, North America, eastern Asia, southern Chile, south-east Australia and Tasmania, and most of New Zealand comprise a humid temperate zonobiome. Zonobiomes are also called ecozones (Schultz 1995) and ecoregions (R. G. Bailey 1995, 1996).

The modern terrestrial biosphere consists nine ecozones or zonobiomes (Figure 8.1):

1 Polar and subpolar zone. This zone includes the Arctic and Antarctic regions. It is associated with tundra vegetation.
2 Boreal zone. This is the cold-temperate belt that supports coniferous forest (taiga).
3 Humid mid-latitude zone. This zone is a maritime warm-temperate (maritime) belt and supports temperate evergreen forests and a typical temperate zone (nemoral) with a short period of frost supporting broad-leaved deciduous forests.
4 Arid mid-latitude zone. This is the zone of cold-temperate (continental) belt supporting temperate grasslands and cold deserts and semi-deserts.
5 Tropical and subtropical arid zone. This is a hot desert climate that supports thorn and scrub savannahs and hot deserts and semi-deserts.
6 Mediterranean subtropical zone. This is a belt with winter rains and summer drought that supports sclerophyllous vegetation.
7 Seasonal tropical zone. Savannah or tropical grassland.
8 Humid subtropical zone. This supports tropical deciduous forest.
9 Humid tropical zone. Rain all year. Evergreen tropical rain forest.

Freshwater communities (lakes, rivers, marshes, and swamps) are part of continental zonobiomes. They may be subdivided in various ways. Lakes, for instance, may be well mixed (polymictic or oligomictic) or permanently layered (meromictic). They may be wanting in nutrients and biota (oligotrophic) or rich in nutrients and algae (eutrophic). A thermocline (where the temperature profile changes most rapidly) separates a surface-water layer mixed by wind (epilimnon), from a more sluggish, deep-water layer (hypolimnon). And, as depositional environments, lakes are divided into a littoral (near-shore) zone and a profundal (basinal) zone.

The marine biosphere also consists of ecozones (or domains). The main surface-water marine ecozones are the polar zone, the temperate zone, and the tropical zone (R. G. Bailey 1996: 161):

1 Polar zone. Ice covers the polar seas in winter. Polar seas are greenish, cold, and have a low salinity.

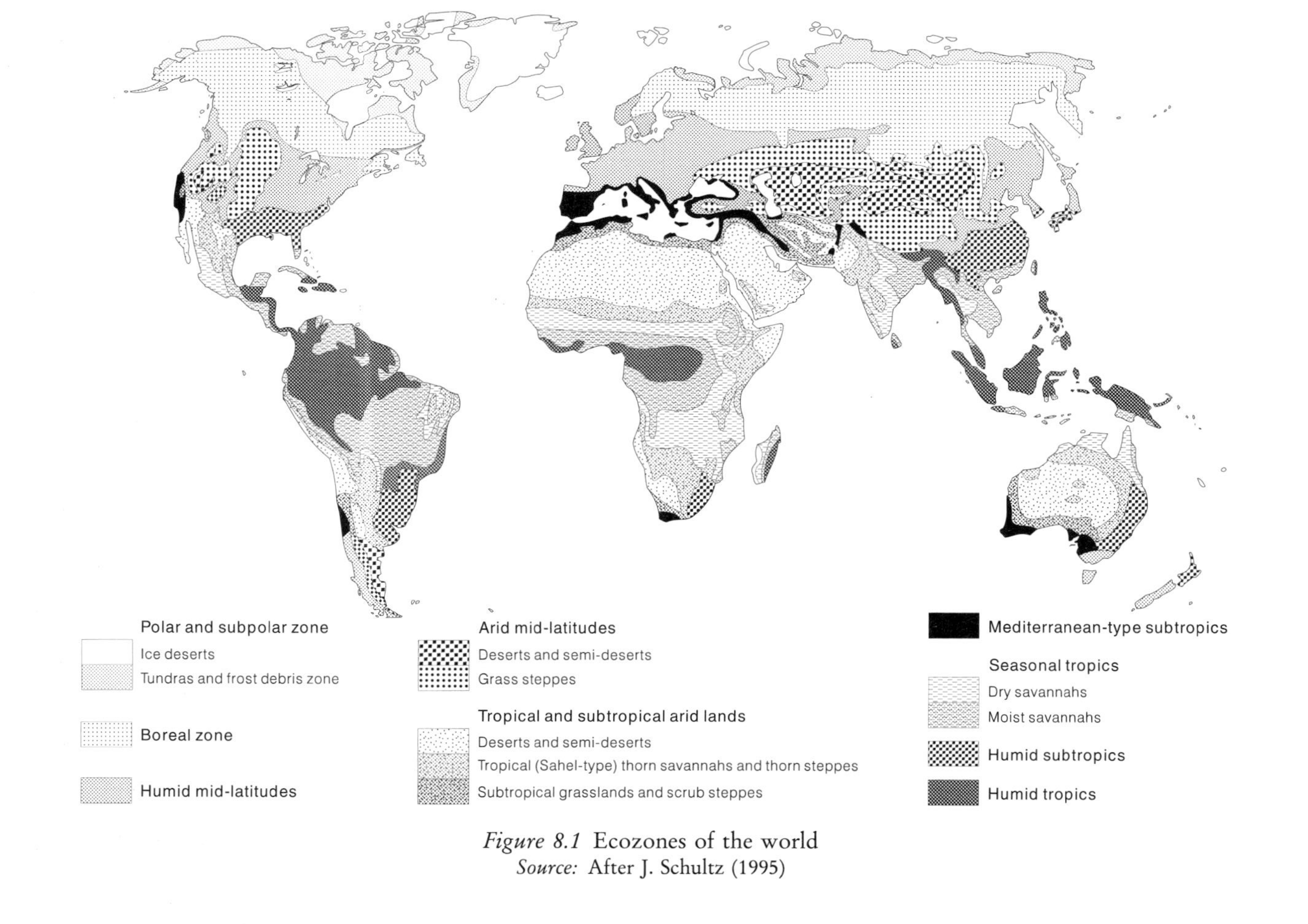

Figure 8.1 Ecozones of the world
Source: After J. Schultz (1995)

2 Temperate zone. Temperate seas are very mixed in character. They include regions of high salinity in the subtropics.
3 Tropical zone. Tropical seas are generally blue, warm, and have a high salinity.

These ecozones, and the deep-water regions, consist of biomes (equivalent to terrestrial zonobiomes). The chief marine biomes are the intertidal (estuarine, littoral marine, algal bed, coral reef) biome, the open sea (pelagic) biome, the upwelling zone biome, the benthic biome, and the hydrothermal vent biome.

Community properties

Communities and ecosystems possess emergent properties that evolve from the multeity of individual interactions. The emergent properties include biodiversity, production and consumption, nutrient cycles, and food webs.

Biodiversity

A community has a stock of species. The size of the stock is variously styled species diversity, species richness, and species numbers. Genetic diversity and habitat diversity are usually included as parts of biodiversity.

Production and consumption

Green plants use solar energy, in conjunction with carbon dioxide, minerals, and water, to build organic matter. The organic matter so manufactured contains chemical energy. Photosynthesis is the process by which radiant energy is converted into chemical energy. Production is the total biomass produced by photosynthesis within a community. Part of the photosynthetic process requires light, so it occurs only during daylight hours. At night, the stored energy is consumed by slow oxidation, a process called respiration in individuals but consumption for a community.

Sunlight comes from above, so ecosystems tend to have a vertical structure (Figure 8.2). The upper production zone is rich in oxygen. The lower consumption zone, especially that part in the soil, is rich in carbon dioxide. Oxygen is deficient in the consumption zone, and may be absent (Table 8.1). Such gases as hydrogen sulphide, ammonia, methane, and hydrogen are liberated where reduced chemical states prevail. The boundary between the production zone and the consumption zone, which is known as the compensation level, lies at the point where there is just enough light for plants to balance organic matter production against organic matter utilization.

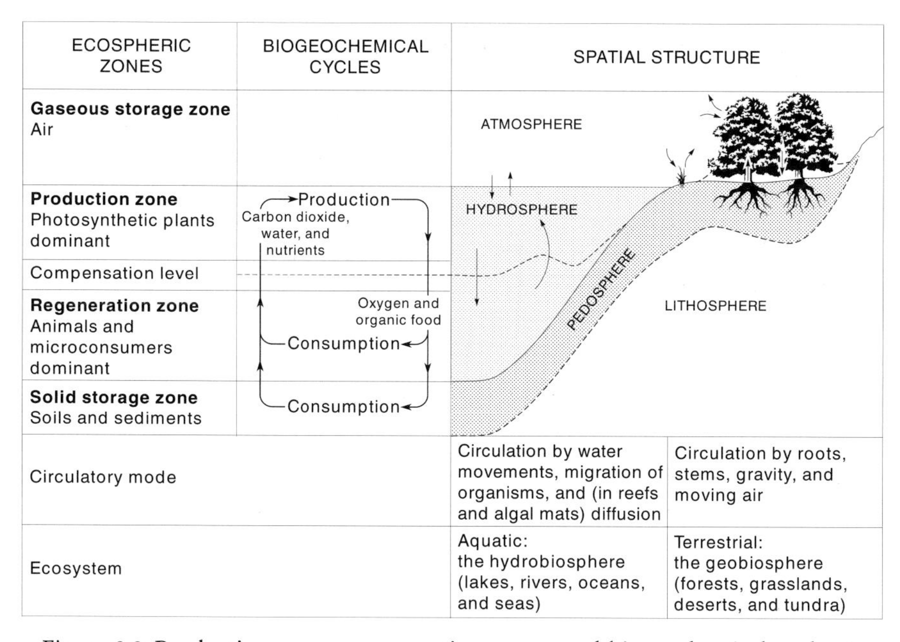

Figure 8.2 Production zones, consumption zones, and biogeochemical cycles in ecosystems
Source: Partly after Odum (1971)

Table 8.1 Oxygen in the environment

Oxygen status	*Environmental conditions*	*Oxygen concentration in water (ml/l)*
Anoxic	Anaerobic	<0.1
Dysoxic	Dysaerobic	0.1–1.0
Oxic	Aerobic	>1.0

Nutrient cycles

The biosphere is made of three main elements – hydrogen (49.8 per cent by weight), oxygen (24.9 per cent), and carbon (24.9 per cent). Several other elements are found in the biosphere, and some of them are essential to biological processes – nitrogen (0.27 per cent), calcium (0.073 per cent), potassium (0.046 per cent), silicon (0.033 per cent), magnesium (0.031 per cent), phosphorus (0.03 per cent), sulphur (0.017 per cent), and aluminium (0.016 per cent). These elements are the basic ingredients for organic compounds, around which biochemistry revolves. Carbon, hydrogen, nitrogen, oxygen, sulphur, and phosphorus are needed to build nucleic acids (RNA and DNA), amino acids (proteins), carbohydrates (sugars, starches, and cellulose), and lipids (fats and fat-like materials). Calcium, magnesium, and potassium are required in moderate amounts. More than a dozen elements are required in trace amounts, including chlorine, chromium, copper, cobalt, iodine, iron, manganese, molybdenum, nickel, selenium, sodium, vanadium, and zinc. The biosphere has to obtain these chemical elements from its surroundings.

Material exchanges between life and life-support systems are a part of biogeochemical cycles (Figure 8.2). At their grandest scale, biogeochemical cycles involve the entire Earth. An exogenic cycle, involving the transport and transformation of materials near the Earth's surface, is normally distinguished from a slower and less-well-understood endogenic cycle involving the lower crust and mantle. Cycles of carbon, hydrogen, oxygen, and nitrogen are gaseous cycles – their component chemical species are gaseous for a leg of the cycle. Other chemical species follow a sedimentary cycle because they do not readily volatilize and are exchanged between the biosphere and its environment in solution.

Food webs

All organisms have a community and ecological role. The chief roles are (Figure 8.2):

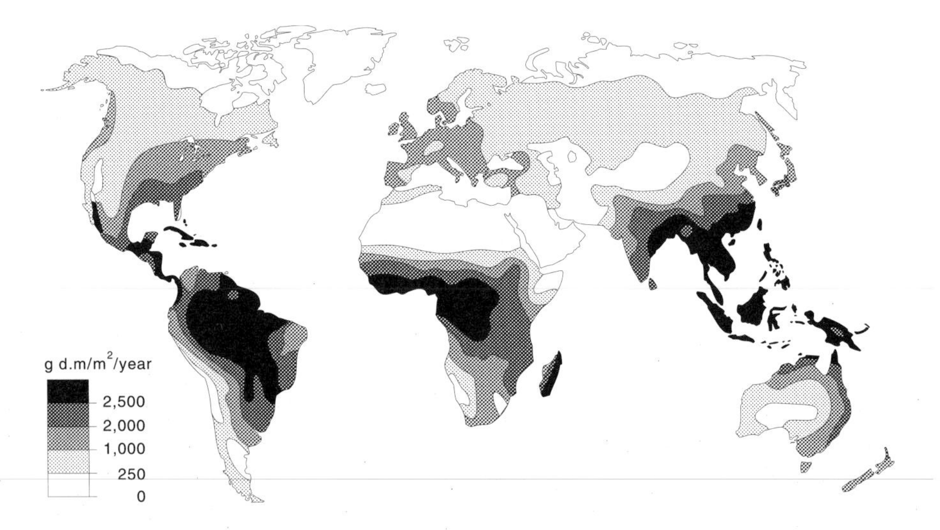

Figure 8.3 Net primary production on land
Source: After Box and Meentemeyer (1991)

1 Producers (autotrophs) – algae, coccoliths, diatoms, dinoflagellates, plants, and photosynthetic bacteria.
2 Consumers (heterotrophs) – herbivores, carnivores, and top carnivores.
3 Decomposers (microconsumers or saprophages).

Gross primary production is the amount of material synthesized by autotrophs per unit time. Some of the primary production is transported through the phloem to other parts of plants, especially to the roots. It powers the day-to-day running. Net primary production is the gross primary production less the energy used in all the processes that constitute plant respiration. The world pattern of terrestrial net primary production is shown in Figure 8.3.

Other organisms feed on the net primary production. They are collectively called heterotrophs or consumers. The energy stored in consumers is called secondary production. Decomposers or saprophages feed on waste products and dead tissues. An ecosystem contains two chief types of food web: a grazing food web (plants, herbivores, carnivores, top carnivores) and a decomposer or detritus food web (Figure 8.4).

These community properties enable the biosphere to perform three important tasks (Stoltz *et al.* 1989). First, the biosphere is able to harness energy to power itself. Second, it is able to win elements essential to life from the atmosphere, hydrosphere, and pedosphere. And third, it is able to respond to cosmic, geological, and biological perturbations by adjusting or reconstructing food webs.

A PERPETUAL MOTION MACHINE: BIOTIC CHANGE

Change in the biosphere occurs at the level of individuals, at the level of communities, and at the level of ecosystems. Biological evolution is a change in individuals, or more properly in an interbreeding population of individuals sharing a common gene pool. It is not discussed in this book. Community evolution is a change in community form and function. Ecosystem evolution is a change in ecosystem form and function.

Changing communities

Do communities succeed one another along a predetermined path? Do open lakes, through hydrosere succession, end up as forest? The traditional view, which held sway until the early 1970s, was that they do. But ecologists have now come round to an individualistic interpretation of community change along the lines suggested by Henry A. Gleason early this century (e.g. Drury and Nisbet 1973). They also recognize the vital role played by disturbance in community change, and the patchy nature of the landscapes in which communities live. Communities are now thought to consist of

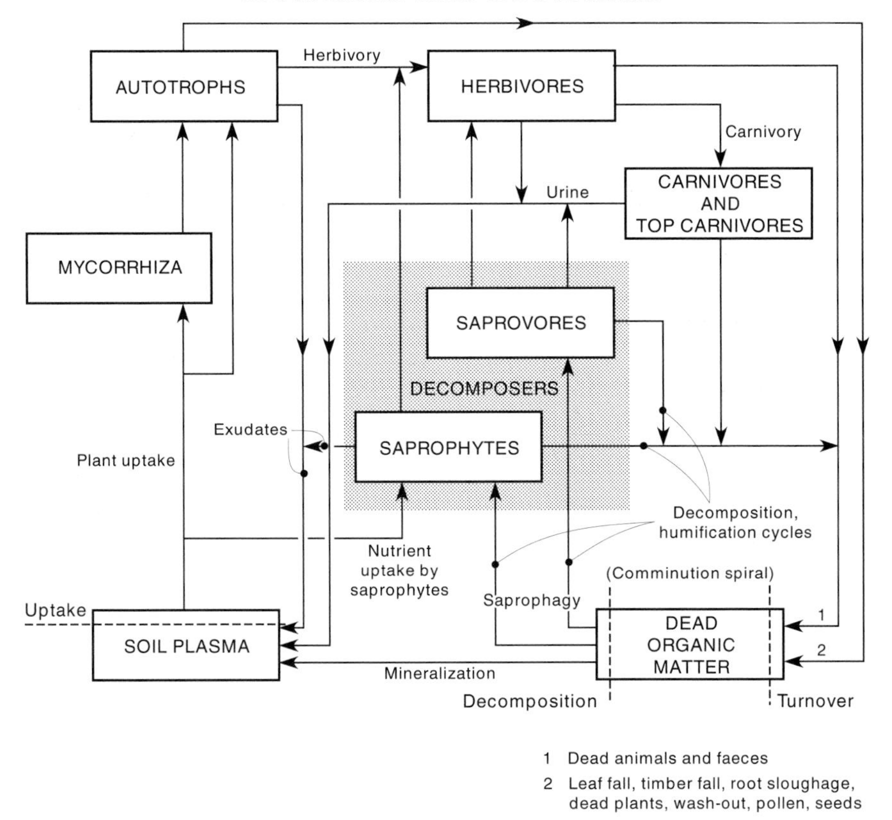

Figure 8.4 Feeding relations in an ecosystem
Source: After Huggett (1980)

species, each of which 'does its own thing'. They occupy a mosaic of landscape patches, patches of different age supporting a different successional 'stage' of community evolution. And they are disturbed by various agencies – fire, storms, humans, and so forth – that reset the successional trajectories in individual patches. It is also likely that there are several stable communities for a given set of environmental constraints. In short, modern research shows that there is nothing inevitable about successional sequences, and that climatic climax communities do not move bodily in the wake of shifting climatic zones.

Community comings and goings

It is helpful to view a community as a stock of species. The stock changes because new species arrive and old species are lost. New species appear in speciation events and through immigration. Old species vanish through extinction and through emigration. Community change is an unceasing process of community assembly, in which species invade, persist, or go extinct (Hang and Pimm 1993). Each species has its own propensity for dispersal, invasion, and population expansion. Thus community assembly is an inherently individualistic enterprise. But that does not mean to say that communities are unorganized assemblages of individuals. Recent mathematical models have confirmed what ecologists have felt for a long time but been unable to prove. Out of the host of individual interactions in an assembling community, such emergent community properties as food webs and a resistance to invasion by alien species arise (e.g. Drake 1990; Pimm 1991). These emergent properties are holistic but not vitalistic.

Communities, like individuals, are fleeting. Species abundances and distributions constantly change, each according to its own life-history characteristics, largely in response to a capriciously vicissitudinous environment. This is seen in the changing distributions of three small North American mammals since the late Quaternary (Figure 8.5). The northern plains pocket gopher (*Thomomys talpoides*) lived in south-western Wisconsin around 17,000 years ago and persisted in western Iowa until at least 14,800 years ago. Climatic change associated with deglaciation then caused it to move west. The same climatic change prompted the least shrew

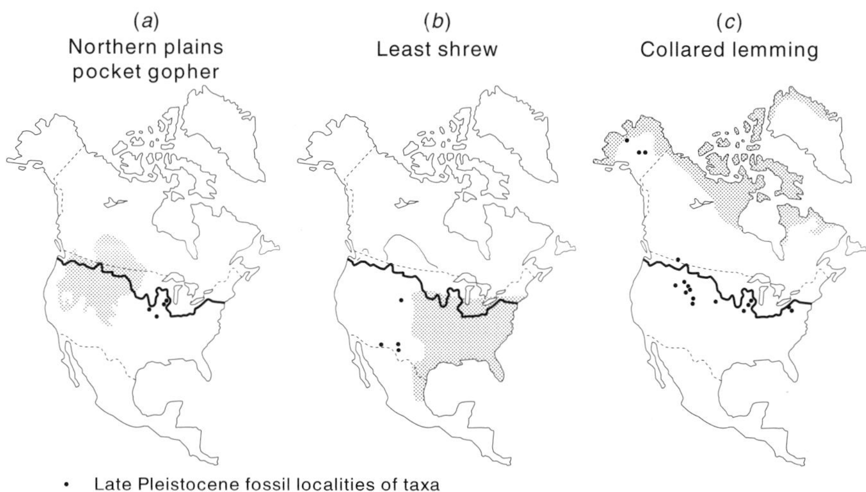

Figure 8.5 Individualistic responses of some small North American mammals since the late Quaternary. (a) Northern plains pocket gopher (*Thomomys talpoides*). (b) Least shrew (*Cryptotis parva*). (c) Collared lemming (*Dicrostonyx* spp.)
Source: After R. W. Graham (1992)

(*Cryptotis parva*) to shift eastwards. The collared lemming (*Dicrostonyx* spp.), which lived in a broad band south of the Laurentide ice sheet, went 1,600 km to the north. With each species acting individually, it follows that communities, both local ones and biomes, will come and go in answer to environmental changes (e.g. R. W. Graham and Grimm 1990). This argument leads to a momentous conclusion: there is nothing special about present-day communities and biomes. However, this bold assertion should be tempered with a cautionary note – insect species and communities have shown remarkable constancy in the face of Quaternary climatic fluctuations (Coope 1994).

Some modern communities and biomes are similar to past ones, but most have no exact fossil counterparts. Contrariwise, many fossil communities and biomes have no precise modern analogues. The list of past communities that lack modern analogues is growing fast. In the Missouri–Arkansas border region between 13,000 and 8,000 years ago, the eastern hornbeam (*Ostyra virginiana*) and the American hornbeam (*Carpinus caroliniana*) were significant components of plant communities (Delcourt and Delcourt 1994). These communities, which were found between the Appalachian Mountains and the Ozark Highlands, bore little resemblance to any modern communities in eastern North America. They appear to have evolved in a climate characterized by heightened seasonality and springtime peaks in solar radiation. Further north, in the north-central United States, a community rich in spruce and sedges existed from about 18,000 to 12,000 years ago. This community was a boreal grassland biome (Rhodes 1984). It occupied a broad swath of land south of the ice sheet and has no modern counterpart, though it bore some resemblance to the vegetation found in the southern part of the Ungava Peninsula, in northern Quebec, today.

Older communities are even less likely to possess modern counterparts. Late Ordovician and Devonian terrestrial plant formations cause a particular problem. The closest modern communities are probably polar and desert vegetation on well-drained soils. These are called polsterland where lichens or plants lacking true roots dominate, and brakeland where herbaceous plants with rhizomes and roots that fail to form a dense turf (as in grasslands) dominate (Retallack 1992b).

Disharmonious communities

The fauna and flora of communities with no modern analogues are commonly described as 'disharmonious communities'. This inapt name inadvertently conjures an image of animal and plants struggling for survival in an alien environment. It is meant to convey the idea that these communities had evolved in, and flourished under, climatic types that no longer exist anywhere in the world (R. W. Graham and Mead 1987). In the boreal grassland, for example, species that now inhabit grassland or deciduous woodland – including prairie voles, sagebrush voles, and the eastern chipmunk – lived cheek-by-jowl with species that now occur in boreal forests and Arctic tundra – including Arctic shrews, lemmings, voles, and ground squirrels (Lundelius *et al.* 1983; e.g. Figure 8.5). During the late Pleistocene epoch, disharmonious animal communities were found over the whole of the United States, except for the far west where vertebrate faunas bore a strong resemblance to their modern-day equivalents, and date from at least 400,000 years ago to the Holocene epoch. These disharmonious communities evolved from species responding individually to changing environmental conditions during late Pleistocene times (R. W. Graham 1979). At the end of the Pleistocene, new environmental changes led to the disassembly of the communities: the climate became more seasonal and individual species had to readjust their distributions and communities of a distinctly modern mark emerged during the Holocene epoch.

North America does not have a monopoly in disharmonious communities. In Australia, an Early Pliocene fauna from Victoria – the Hamilton local fauna – contains several extant genera whose living species live almost exclusively in rainforest or rainforest fringes (Flannery *et al.* 1992). The indication is, therefore, that the Pliocene fauna lived in rainforest environment, but a more complex rain forest than exists today. Modern representatives of four genera (*Hypsiprymnodon*, *Dorcopsis*, *Dendrolagus*, and *Strigocuscus*) are almost entirely rainforest dwellers, but they live in different kinds of rain forest. Living species of *Dorcopsis* live in high mountain forests, lowland rain forests, mossy montane forests, and mid-montane forests. Living species of *Dendrolagus* live mainly in montane rain forests. *Hypsiprymnodon* is restricted to rain forest where it prefers wetter areas. The modern New Guinea species *Strigocuscus gymnotus* is chiefly a rainforest dweller, though it also lives in areas of regrowth, mangrove swamps, and woodland savannah. Living species of *Thylogale* live in an array of environments, including rain forests, wet sclerophyll forests, and high montane forests. Other modern relatives of the fossils in the Hamilton fauna, which includes pseudocheirids, petaurids, and kangaroos and wallabies, live in a wide range of habitats. It contains two species, *Trichosurus* and *Strigocuscus*, whose ranges do not overlap at present. It is thus a disharmonious assemblage. Taken as a whole, the Hamilton mammalian assemblage suggests a diversity of habitats in the Early Pliocene. The environmental mosaic consisted of patches of rain forest, patches of other wet forests, and open area patches. Nothing like this environment is known today.

Observed community changes

Changes in present-day communities, ecosystems, and the biosphere are observed in two ways: remotely sensed information and ground-based surveys. Longer-term community and ecosystem changes are more difficult to establish. As a rule, the older the community or ecosystem, the more difficult it is to reconstruct: it is easier to piece together Holocene environments than Precambrian environments. None the less, some finely resolved geological changes in communities and ecosystems are available. Examples come from Carboniferous coal deposits (DiMichele and Nelson 1989; DiMichele and Phillips 1994) and the Paleogene of Wyoming–Montana and the Neogene of Pakistan (Badgley and Behrensmeyer 1995a, 1995b).

Organic and inorganic sedimentary sequences record change in communities and ecosystems, but, unless several cores or sections from the same area should be available, they say little about past geographies. For ecosystems that no longer exist, such sequences are all that is left from which to reconstruct community change. Ground and aerial surveys yield intricate detail on the geography of communities and ecosystems, but are uninformative on medium- and long-term changes. The present distribution of organisms and soils may be used to reassemble community and ecosystem change. Two methods are used to do this. Both involve constructing a community chronosequence by substituting space for time. Both are beset with problems.

The first method assumes that communities evolve through a succession of changes. When the chance arises, usually as a result of disturbance, the succession restarts and passes through the same stages as before. At any time, therefore, different communities living in similar habitats should represent different stages in the successional sequence. For example, present communities in the fenlands of England suggest a succession of community change known as a hydrosere (e.g. Tansley 1939). Open water is colonized by aquatic macrophytes, and then by reeds and bulrushes. Decaying organic matter from these plants accumulates to create a reed swamp in which water level is shallower. Marsh and fen plants become established in the shallower water. Further accumulation of soil leads to even shallower water. Such shrubs as alder then invade to produce 'carr' (a scrub or woodland vegetation), and ultimately, it was claimed, carr would change into mesic oak woodland.

The second method is applicable in situations where land surfaces of different but known age carry communities and ecosystems. Suitable land surfaces are sand dunes, river terraces, and volcanic rocks. The assumption is that the community on the oldest land surface is the most evolved, and that the community on the youngest land surface is the least evolved, with other communities filling in some of the intermediate stages of community evolution. As with the first method, geographical differences in current communities are arranged as a time sequence. However, it is securer than the first method insofar as it works with communities whose age is determined by independent means, rather than with communities whose absolute, and even relative, ages are uncertain. It thus builds a chronosequence from dates rather than by inference. A good example is the hydrosere succession on the sand-dune complex next to south-eastern Lake Michigan, Indiana, United States (e.g. Cowles 1899; Shelford 1911, 1913; Clements 1916). The dune–pond complex is assumed to form a time sequence, with the youngest dunes being nearest to the lake. The communities supported by the dunes and ponds are, therefore, a chronosequence, the youngest communities being associated with the youngest dunes and ponds. The hydrosere was originally interpreted as starting with an open beach pool surrounded by bare dune sand (Shelford 1911, 1913). This was invaded by a mixture of aquatic vegetation that included submersed, floating-leaved, and emergent life-forms. After a while, the emergent vegetation came to dominate and, eventually, a switch to vegetation dominated by the shrub *Cephalanthus occidentalis* occurred.

New work on this chronosequence highlighted the pitfalls of substituting space for time without some knowledge of environmental changes. A sediment core extracted from one of the oldest ponds (Pond 51) led to a radical revision of the classic hydrosere succession (Jackson *et al.* 1988). Pond 51 formed 2,700 years ago. Plant macrofossils in the core showed a community comprising an admixture of submersed, floating-leaved, and emergent macrophytes that remained virtually unchanged until about AD 1800 when Euro-American pioneers began clearing land in the region (Figure 8.6). The human disturbance is marked by an increase in ragweed (*Ambrosia*) pollen, a decline in jack pine (*Pinus banksiana*) microsporangia, and, significantly, a noticeable increase in cattail (*Typha angustifolia*) and buttonbush (*Cephalanthus occidentalis*) pollen, and an equally noticeable decrease in macrofossil remains of the numerous submerged and floating-leaved macrophytes. Thus, the classical hydrosere chronosequence in the ponds of the Indiana Dunes may have resulted from differences in the type and magnitude of human disturbance, rather than from autogenic vegetation change.

NATURE AND NURTURE: CAUSES OF BIOTIC CHANGE

What is the motor of change in communities and ecosystems? Does it lie within the biosphere? Or does it rest in the environment? Or is it a combination of the two?

Autogenic community change

Communities and ecosystems are capable of changing themselves. This appears to happen in the process of autogenic succession. In its classical conception, autogenic succession is a unidirectional sequence of commu-

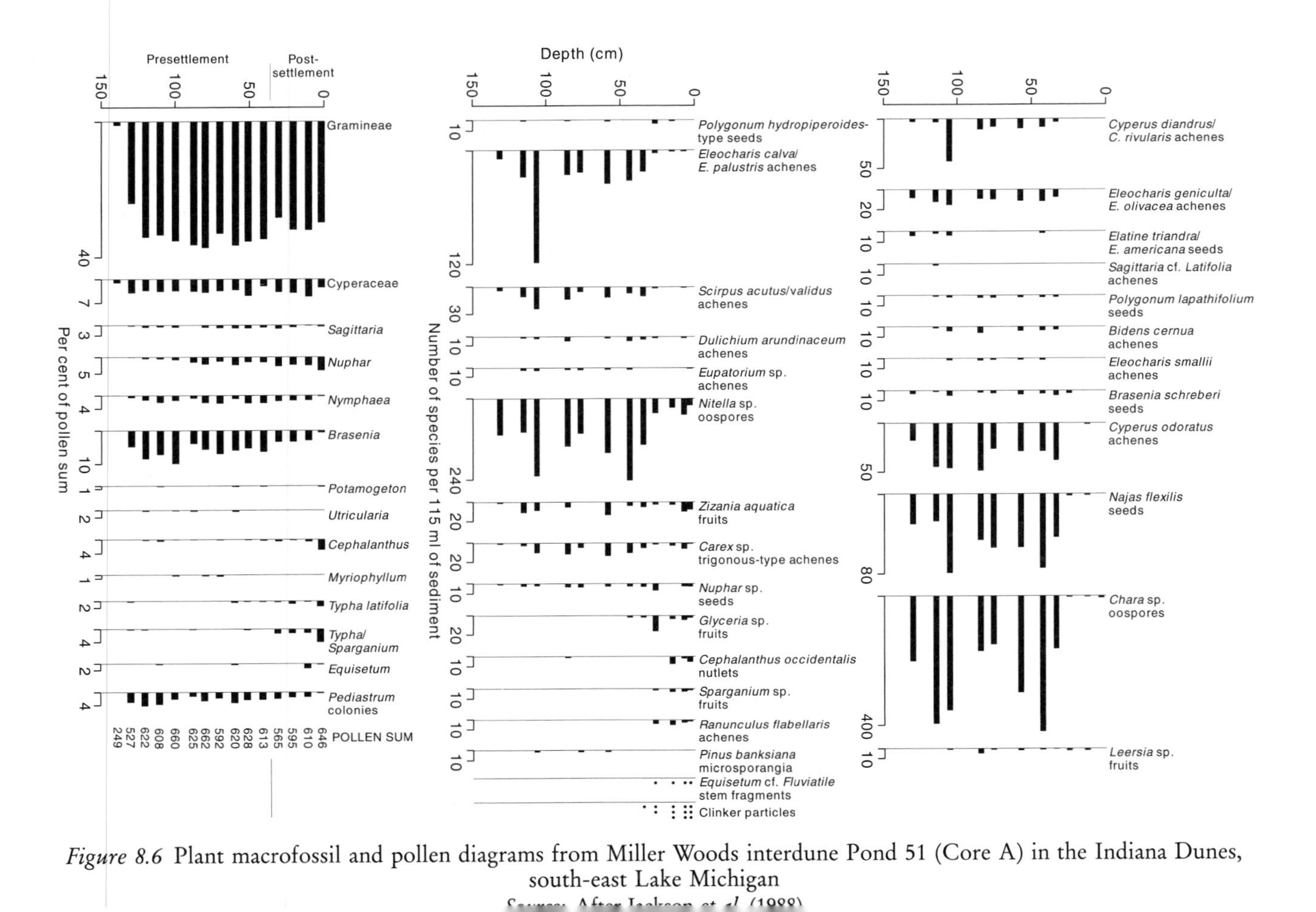

Figure 8.6 Plant macrofossil and pollen diagrams from Miller Woods interdune Pond 51 (Core A) in the Indiana Dunes, south-east Lake Michigan

nity, and related ecosystem, changes that follow the opening up of a new habitat. The sequence of events takes place even where the physical environment is unchanging. The classical hydrosere mentioned in the previous section is an example. However, several modern studies tell a cautionary tale. They confirm the individualistic nature of species change, and show that succession seldom sticks unfailingly to a predestined sequence of events. The inherent unpredictability of successional events is largely due to a constantly changing physical environment and the action of various disturbing agents. But even where autogenic factors are the primary motive force of community change, succession may lead in several different directions. Moreover, autogenic factors may play an overruling role in community change, even where the physical environment changes considerably.

A study of wetland evolution in twenty sites from Great Britain revealed that hydrosere succession did not willy-nilly culminate in mesic woodland, even where autogenic factors were the chief determinants of change (Walker 1970). Throughout the Holocene epoch, changes in plant-macrofossil assemblages at the sites were largely unconnected with episodes of pronounced climatic change. Nor was there evidence to suggest that open lakes ended up as mesic forest. The main successional sequence, which occurred in 46 per cent of the observed cases, was from micro-organisms or submerged macrophytes, to floating-leaved macrophytes, to reed swamp, to fen, to fen carr, to bog. In the other 56 per cent of the cases, the successional sequences were different. In many instances, the determinative factor directing succession was the timing of bog moss (*Sphagnum*) establishment, which plant can alter the pH of the site and sustain its own growth independently of the lake water-table. If there should be a 'climax' community produced by hydroseres, then it is bog dominated by *Sphagnum*, rather than by mesic forest.

Autogenic forces sometimes lead to cycles of community change. A case in point is the peatland hummock–hollow cycle (Figure 8.7). As classically interpreted (e.g. Tansley 1939), the cycle starts with a peat hummock, supporting a community of white-beaked sedge (*Rhynchospora alba*), heather (*Calluna* spp.), common cotton-grass (*Eriophorum vaginatum*), narrow cotton-grass (*E. angustifolium*), cross-leaved heath (*Erica tetralix*), and bog asphodel (*Narthecium ossifragum*). The hummock is flanked by two wet hollows, supporting the bright-green semi-aquatic species of bog moss, *Sphagnum cuspidatum*. Peat accumulates in the hollow. As the site becomes less wet, *Sphagnum papillosum* and *S. magellanicum* begin to replace *S. cuspidatum*. Eventually, as *S. papillosum* and *S. magellanicum* build low hummocks, the peaty hollow rises above the level of the original hummock and is colonized by common cotton-grass (*Eriophorum vaginatum*) and deer sedge (*Trichophorum cespitosum*). The upper part of the hummock is colonized by bog mosses that do not require much water, very often the crimson *Sphagnum rubellum*. *Calluna* and the lichens *Cladonia arbuscula* (*C. sylvatica*) and *C. impex* are also important on the dry hummock top. The summit of the original hummock eventually dries out, the bog moss dies, and the heather becomes old and leggy, and in time dies. The hollows flanking the senescent hummock grow, turning it into a wet hollow containing *S. cuspidatum*, and the cycle starts afresh.

The cause of the hummock–hollow cycle, originally thought to be autogenic, is disputable. Raised bogs, which contain hummocks and hollows, obtain their moisture directly from the atmosphere and follow allogenic growth cycles. Draved raised bog, Denmark, exhibits a 260-year growth cycle that matches climatic changes, but individual hummocks have stayed in the same place for over 2,500 years, a finding which raises doubts about the effectiveness and prevalence of an autogenic hummock–hollow cycle (Aaby 1976). Similar doubts were cast by a study of Bolton Fell Moss, England (Barber 1981). Over the last 2,000 years, climate, rather than autogenic processes, has driven the evolution in this bog by determining the occurrence of wet and dry phases. In Clearwater Lake peatland, northern Quebec, Canada, layers of *Sphagnum* peat have alternated with layers of spruce (*Picea*) detritus (Payette 1988). The alternations were fairy orderly though not strictly periodic, the switch taking place roughly every 100 years between 4,010 and 1,710 years ago. They are ascribable to an autogenic peat-construction cycle that occurred without fire disturbance. The community

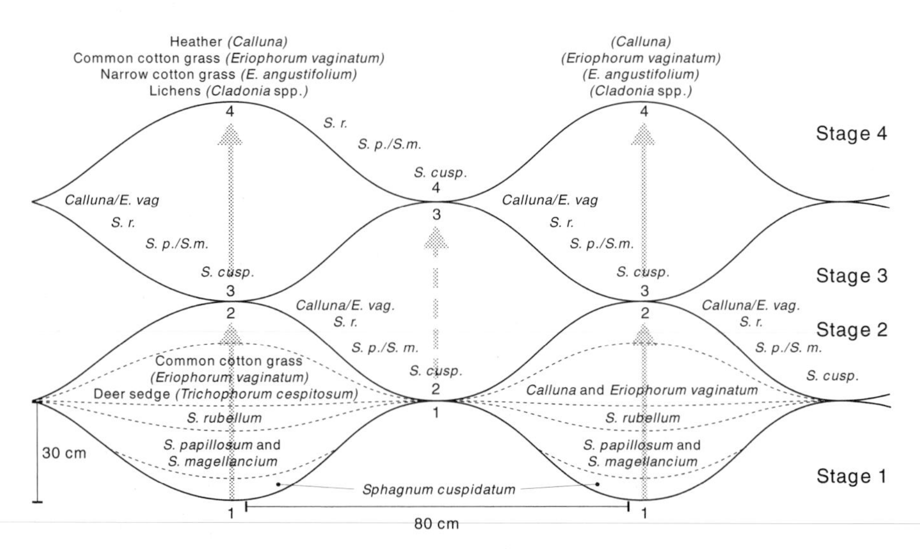

Figure 8.7 The hummock–hollow cycle in an English peat bog. Heather and common cotton-grass hummocks alternate with bog moss (*Sphagnum cuspidatum*) pools
Source: After Barber (1981)

switch is self-reinforcing: bog moss supplies a medium for spruce to grow; stunted spruce supplies a favourable microclimate for increased snow accumulation and promotes bog-moss growth. However, owing to differences of climate, topography, and hydrology, longer-term successional trends vary at other sites within the region, and longer-term bog evolution is influenced by allogenic factors, particularly climatic change. Studies in the Pennines, England, confirmed the importance of environmental changes in interpreting peatland evolution (e.g. Tallis 1994).

Allogenic community change

Many community and ecosystem changes are driven by fluctuations in the physical environment. A host of environmental factors may disturb communities and ecosystems by disrupting the interactions between individuals and species. Climatic change (in air, water, and ice) is a key disruptor of marine and terrestrial communities. In turn, climatic change is a disruptor of the climate system by cosmic and geological processes, and, possibly to a lesser extent, internal mechanisms of the world weather machine. It is to be expected, therefore, that cycles of cosmic and geological activity, which are mirrored in climatic cycles (Chapters 4 and 5), should express themselves in the biosphere. To be sure, the dynamics of individuals, communities, and ecosystems are pulled along by cyclical and directional changes in the physical environment, and especially changes of climate and geography, which themselves are locked into changes in the cosmic and geological environments.

Climatic change and life

The Moon, the Sun, and the planets all influence life on Earth. The 28-day lunar rhythm is fixed in many body functions, and in humans is demonstrable as a hormone cycle. It affects many life-history and behavioural traits of marine species. For example, it tends to orchestrate the spawning of hermatypic corals (Wyers *et al.* 1991), echinoids (Lessios 1991), cichlid fishes (Nakai *et al.* 1990); the settlement patterns of newly arrived Caribbean reef fishes on patches of reef (Robertson 1992), and the release of larvae in the decapod, *Carcinus maenus* (Queiroga *et al.* 1994). Birds and mammals are also attuned to monthly lunar cycles. During winter, lapwings (*Vanellus vanellus*) feed by day and roost at night during most of the lunar month, except for a few days around full moon, when the pattern is reversed (Milsom *et al.* 1990). In a population of Merriam's kangaroo rat (*Dipodomys merriami*), losses to diurnal versus noctural predators vary through the lunar cycle (Daly *et al.* 1992). The 18.6-year lunar-nodal pulse resonates throughout the biosphere. It is found in European fish catches and wine harvests (Currie *et al.* 1993).

The Sun is the modern-day 'primum mobile' of the terraqueous globe. The entire biosphere (with the possible exception of deep-sea and deep-rock ecosystems) throbs with daily and annual rhythms, and, less strongly, to the longer cycles of solar activity. The occurrence of warm-water fish, such as hake and red mullet, and cold-water fish, such as cod and haddock, in the English Channel follows the 11-year sunspot cycle, which causes periodic change of sea temperature (Southward *et al.* 1975). The annual blue crab (*Callinectes sapidus*) population in Chesapeake Bay varies with the lunar nodal cycle, the sunspot cycle, and an 8.8-year period of the Earth–Moon–Sun tidal force (Hurt *et al.* 1979). The tidal signals are also present in annual rainfall figures, suggesting that phases of minimum rainfall allow high tides to wash nutrients into the surface waters of the Bay, and allow the waters to become more saline, so promoting growth of the blue crab population. Similarly, the abundance of four barnacle species (*Chthamalus montagui*, *Chthamalus stellatus*, *Semibalanus balanoides*, and *Elminius modestus*) monitored near Cellar Beach, River Yealm, south Devon, since 1951, varied with annual mean inshore sea-surface temperature, which in turn followed the solar cycle, to 1975 (Southward 1991). After 1975, the biological data diverged from the solar cycle and now show a less good match with annual mean sea temperature, possibly owing to changing weather patterns and other effects of global climatic shift.

Orbital changes appear to have harmonized the broad shifts of climate during the Holocene epoch. These shifts varied from region to region. Climatic change in western Europe, for instance, was (and still is) sensitive to changes in the North Atlantic, and especially to changes that influenced winter conditions on the west of the continent (Huntley 1992). Holocene climatic changes forced animal and plant communities to disassemble and reassemble. In Europe, pollen data for the last 13,000 years indicate individualistic changes in species distributions and evanescent community composition (Huntley 1990; Huntley and Prentice 1993). These patterns are illustrated in Figure 8.8. Superimposed upon the broad sweep of Holocene climatic swings were singular climatic events produced by volcanoes. These caused temporary disruption of the climate system and left their mark in the growth rings of trees. Tree rings in European oaks (*Quercus*), for example, responded to a volcanic eruption (or possibly two) in AD 536 (Baillie 1994).

During the ice age, orbital variations generated a sequence of glacial–interglacial switches that marine and terrestrial communities were obliged to track. The result is that orbital signals are present in records of Quaternary vegetation change. Pollen in three cores from the Grand Pile bog, Vosges Mountains, north-eastern France, register the 23,000-year precessional cycle and its harmonics (Molfino *et al.* 1984). The abundances of oak (*Quercus*), grass (Poaceae), wormwood (*Artemisia*), and fir (*Abies*) pollen in a core from the Tyrrhenian Sea spanning 55,000–9,000 years ago are locked in phase with global ice volume, as measured in oxygen isotope ratios in

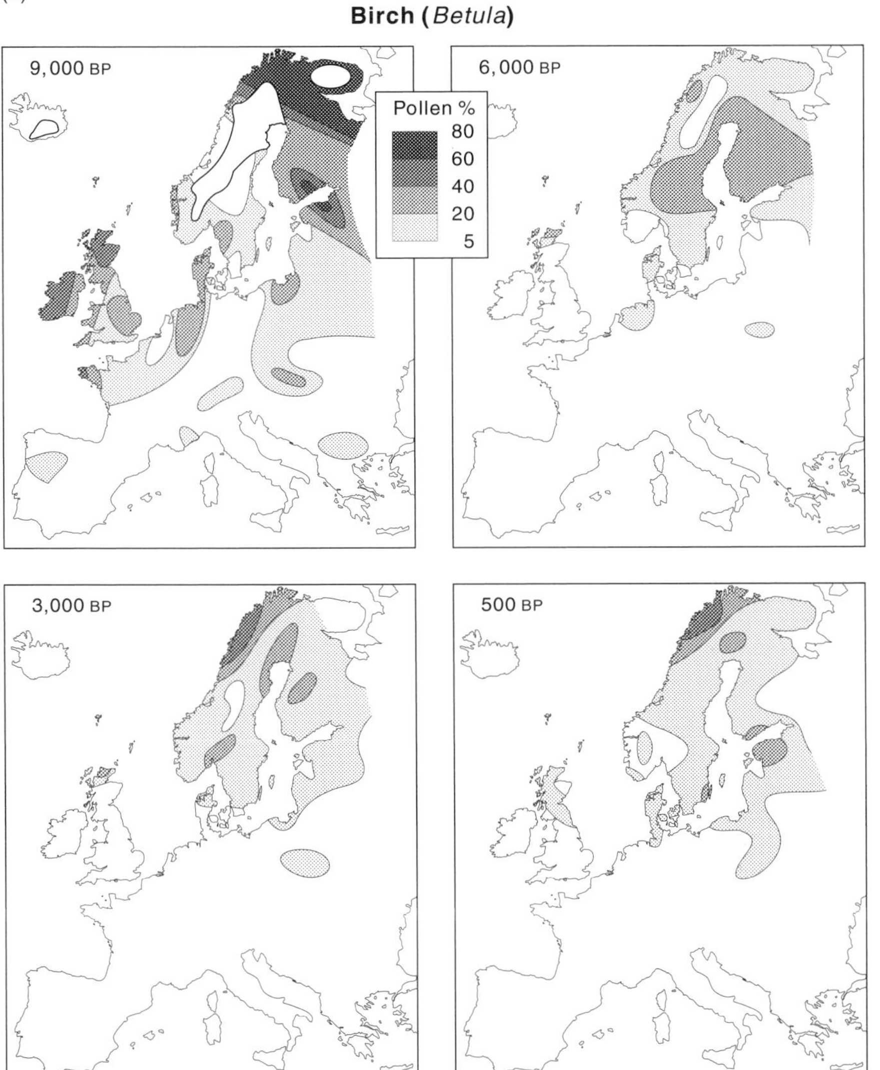

Figure 8.8 Isopoll maps for two tree species in Europe, 9,000 years ago to the present. (a) Birch (*Betula*) was abundant over much of unglaciated northern Europe 9,000 years ago. Its southern limit shifted northwards between 9,000 and 6,000 years ago. It readvanced slightly towards the present. This change is consistent with a general increase in growing-season warmth over central and northern Europe from 9,000 to 6,000 years ago, followed by a decline. (b) on page 282

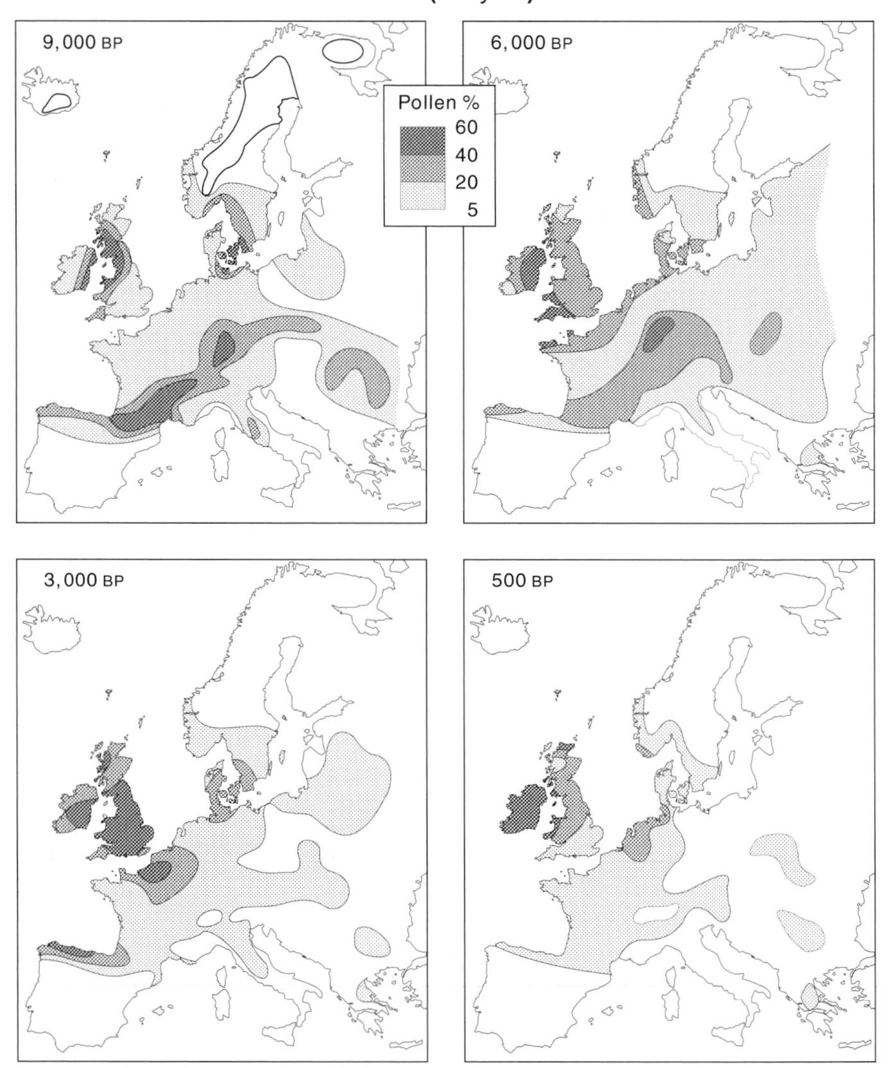

Figure 8.8 (continued) (b) Hazel (*Corylus*) was abundant in a broad band across central Europe 9,000–6,000 years ago, but then, as steppe vegetation expanded into eastern Europe, it retreated westwards
Source: After Huntley and Prentice (1993)

marine cores (Rossignol-Strick and Planchais 1989). Not all climatic changes during the ice age were forced by orbital variations. The Heinrich events, which were massive periodic advances of ice-streams from the eastern margin of the Laurentide ice sheet (p. 125), were not. They were significant enough climatic fluctuations to leave traces in the pollen record. In Lake Tulane, south-central Florida, sediment cores show alternating peaks of vegetation dominated by pine (*Pinus*) and vegetation dominated by oak and ragweed (*Quercus* and *Ambrosia*) (Grimm *et al.* 1993; Watts and Hansen 1994). The pine peaks correlate neatly with the first five Heinrich events.

The long-term hothouse–icehouse and warm-mode–cool-mode climatic shifts drove past communities and ecosystems through protracted successional sequences. The Cenozoic succession from the Cretaceous hothouse to the Quaternary icehouse is particularly well documented. Floral and faunal changes in many parts of the world point to a cooling and, in some regions, drying of climate during the Miocene epoch. In the broad expanses of the North American mid-continent, the shift to cooler and drier climatic conditions in Miocene times directly drove a change from forest to savannah and from savannah to steppe, and indirectly drove parallel changes in the diversity of browsing and grazing ungulate taxa (S. D. Webb 1983).

An excellent record of the long descent into the current icehouse Earth comes from the Siwalik Group sediments, in the Potwar Plateau region, northern Pakistan (Quade and Cerling 1995). These floodplain sediments span the last 18 million years. Carbon isotope ratios in palaeosols suggest the following changes in the floodplain ecosystem (Figure 8.9):

1 From 17 to 7.3 million years ago. A pure, or almost pure, C_3 biomass, probably a mosaic of closed canopy forest and grassland, dominated the floodplains.
2 From 7.3 to 6 million years ago (Late Miocene). A gradual expansion of C_4 plants (essentially grassland) on the floodplain occurred, probably in response to monsoon intensification. This interpretation is supported by concurrent changes in soils (depth of leaching decreased, the colour-leached zones changed from dominantly strong reds and oranges to yellower hues, and humic soil horizons, possibly owing to the appearance of grassland, became commoner), and in a major faunal turnover that occurred at about the same time, in which larger browsing animals (tragulids, suids, okapi-like giraffes, low-crowned bovids, and others) were replaced by grazing animals, such as a new tragulid with high-crowned dentition, and in which rodents underwent a rapid change (J. C. Barry *et al.* 1985; Flynn and Jacobs 1982; J. C. Barry and Flynn 1990).
3 From 6 to 0.4 million years ago. C_4 grasslands remained dominant throughout that time.

In East Africa, as in Asia, aridification led to grassland expanding at the expense of forests (Figure 8.10). The marked floral changes at around

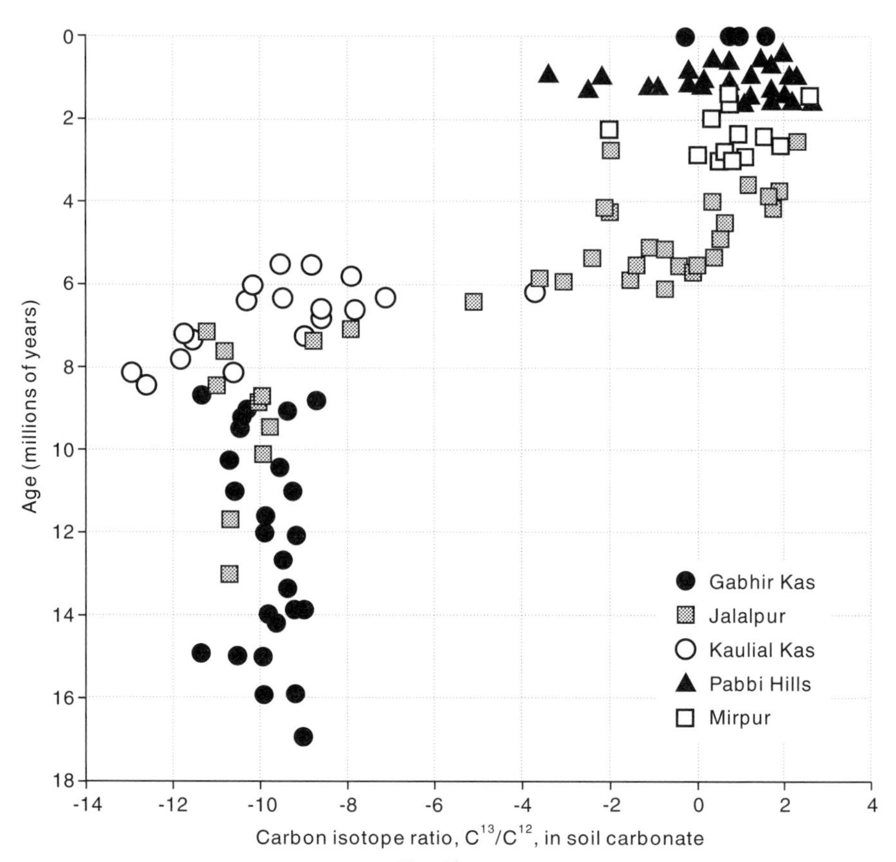

Figure 8.9 Carbon isotope ratios, C^{13}/C^{12}, in palaeosol carbonate nodules plotted against age in the Siwalik sequence, northern Pakistan. The negative values before 7.3 million years ago indicate that a C_3 biomass dominated the floodplain (perhaps trees and shrubs), whereas the more positive values in the Plio-Pleistocene palaeosols are suggestive of a C_4 grassland being dominant

Source: After Quade and Cerling (1995)

2.5 million years ago had a drastic effect upon mammals (e.g. Vrba *et al.* 1989). Many forest antelope species became extinct and new savannah-dwelling species evolved, most of which survive as elements of the modern African fauna. The climatic and floral change might also have nudged gracile australopithecines (*Australopithecus afarensis* and *A. africanus*) in an evolutionary direction that produced the first members of the human genus – *Homo* (Stanley 1992, 1995).

The trend towards drier climates (aridification) over the last 25 million years led to several adaptive radiations (Figure 8.11). The radiations displayed a relay effect. The relay started at the food-web base with the

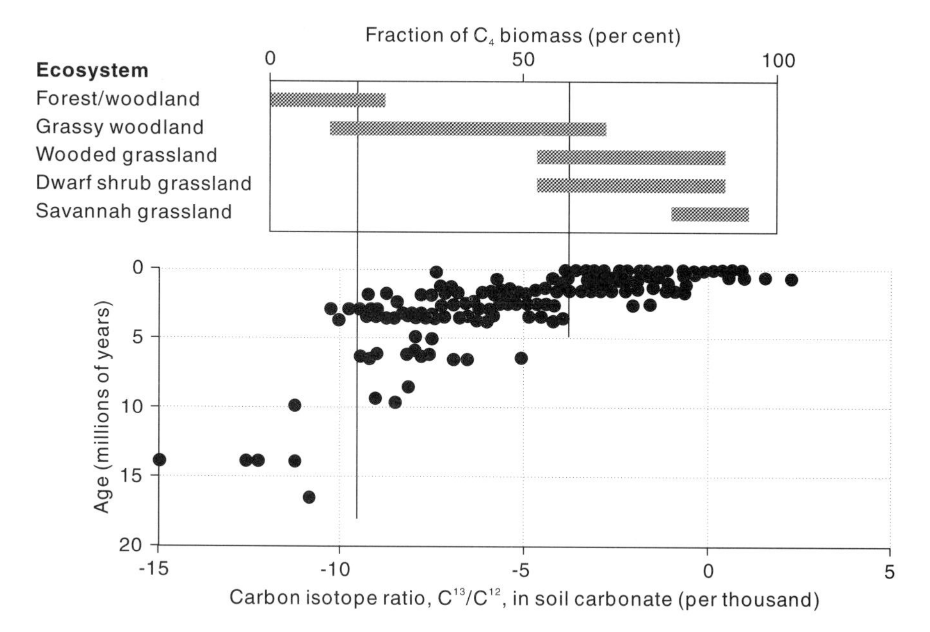

Figure 8.10 Carbon isotope ratios, C^{13}/C^{12}, in palaeosol carbonates and inferred floral changes in East African fossil localities. The top part of the diagram shows isotopic compositions of modern biomes
Source: After Cerling (1992)

diversification of grasses. Herbs and weeds diversified during the spread of grasses. Many songbirds and Old World rats and mice, which fed on the seeds of the expanding plant groups, then underwent a radiation. Lastly, modern snakes (many of which belong to the family Colubridae) expanded as the rats and mice, and eggs and chicks of songbirds, appeared on the menu.

A switch from an icehouse to a hothouse climate leads to many environmental changes. In the seas, warmer waters develop in polar regions. Deep ocean waters become warm, sluggish, and contain little or no oxygen (dysaerobic or anoxic). Such a change in the hydrosphere would cause widespread extinctions in the deep sea. During the mid-Cretaceous high-stand of sea level, when seas stood about 300 m above present levels, anoxic conditions in the ocean deeps touched the deeper portions of epicontinental seas. At this time, environmental perturbations created pulses of biotic destruction that comprise the Cenomanian–Turonian mass extinction (Kauffman 1995). Marine diversity was at a peak for the Cretaceous period immediately before the extinction events. Many lineages had evolved narrow tolerance limits in the hothouse conditions. This meant that marine life would be prone to extinction if environmental conditions should

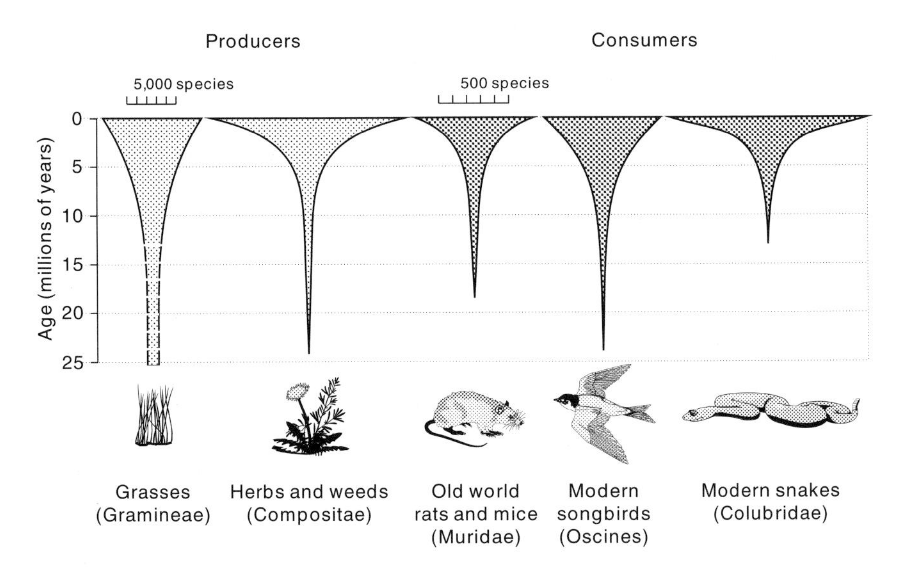

Figure 8.11 A relay of adaptive radiations in terrestrial ecosystems that were favoured by global aridification during the last 25 million years
Source: Stanley (1990a)

change. Tropical reef communities suffered heavy losses near the beginning of the early and middle Cenomanian boundary. Within 520,000 years of the Cenomanian–Turonian boundary, non-tropical biotas suffered a loss of species in the range 45–75 per cent, depending on the group. These losses were stepped, most of the species disappearing in short-term events. They started in the subtropical and warm temperate biotas and progressed to the cool temperate biotas. The extinction events were closely connected with abrupt, large-scale changes in the ocean and atmosphere that are recorded in wild fluctuations of trace element, stable isotope, and organic carbon levels. The expanded anoxic zone of the deep-ocean floor spread to deeper continental shelf and epicontinental sea habitats, initiating trace element advection and chemical stirring of the ocean. These changes might have been the result of ocean impacts of asteroids and comets during a Cenomanian impact storm.

Geological change and life

Several geological events and processes produce change in the biosphere. Many of these act indirectly through atmospheric and hydrospheric change, but continental drift and mountain building have direct effects. Continental drift and the growth of large mountain ranges influence the spread of organisms by creating and destroying dispersal routes. A number of effects can be identified: the effects of continental separation, of continental rafting, and of continental collisions. Overall, continental drift, and processes associated with it (mountain building), provides a good explanation of such features as evolutionary divergence and evolutionary convergence (cf. Hallam 1972).

Divergence is the sundering of a land mass into two or more blocks. The tearing asunder of previously contiguous land masses causes the separation of ancestral populations of animals and plants. Once parted, the populations evolve independently and diverge. Eventually, they may become quite distinct from the ancestral population that existed before the land masses broke apart. A good example of evolutionary divergence resulting from continental drift is afforded by the Cenozoic land animals. Fossil Cenozoic faunas, and particularly mammalian faunas, are relatively distinct on each continent. The most distinct faunas are those of South America, Africa, and Australia – the southern continents. There are some thirty orders of mammals, almost two-thirds of which are alive today. It is generally believed that all the mammalian orders have a common origin – an ancestor that lived in the Mesozoic era. The Cenozoic divergence of mammals may be due to the late Mesozoic breakup of Pangaea (Kurtén 1969; Hedges *et al.* 1996). Interestingly, the land-dwelling animals (mostly reptiles) that lived in the Cenozoic era displayed far less divergence than the mammals. By the end of the Cretaceous period, after about 75 million years of evolution, some seven to thirteen orders of reptiles had appeared, far fewer than the thirty or so produced by mammalian evolution over 65 million years. A possible explanation for this difference lies in the palaeogeography of the Mesozoic continents. For much of the age of reptiles, the continents were not greatly fragmented. There were two supercontinents: Laurasia lay to the north and Gondwana to the south (see Figure 3.4). Rifts between the continents existed as early as the Triassic period, but they were not large enough to act as barriers to dispersal until well into the Cretaceous period. In Late Cretaceous and early Tertiary times, when the mammals began to diversify, the rapid breakup of the former Pangaea, coupled with high sea levels, led to the separation of several land masses and genetic isolation of animal populations. The result was divergence of early mammalian populations.

Convergence is the coming together of two previously separate land masses. A land bridge between two formerly separated land masses may be forged by continental collision or other geological processes such as uplift and the fall of sea level. A moving plate may carry species (indeed, entire faunas and floras) over vast distances, acting like a gigantic raft or Noah's Ark. (It may also carry a consignment of fossil forms, and has been compared to a Viking funeral ship.) Biologically, the result of a convergence is the mixing of two faunas and floras, which come to resemble one another and lose some of their individuality. Major intercontinental immigration episodes may result from the broadening and opening of land bridges during low-stands of sea level, or phases of strong climatic change (S. D. Webb and Opdyke 1995). Immigration episodes occurred in North America (and most other continents) around 20 million, 5 million, and 2.5 million years ago. These episodes correlate with times of global climatic cooling and lower sea levels. On the other hand, two periods of global cooling – 35–30 million years ago and 16–6 million years ago – show no signs of unusually high immigration rates in North America. This is perhaps because no immigration routes were open, or more likely because the faunas at those times were immensely stable and resistant to invasion.

FAST AND SLOW: RATES OF BIOTIC CHANGE

How fast can communities change? Is there a slow and steady turnover of species, or is the history of life punctuated by biotic crises?

Species on the move

Providing the environment changes slowly enough, populations can simply ride with the tide of change. That does not mean that communities will stay intact, but there is evidence to suggest that communities stay stable for lengthy periods, say a few million years, when environmental changes are small and slow. During the Cenozoic era, for example, land mammals in North America consisted of stable sets of taxa (chronofaunas) that evolved slowly and persisted for some 10 million years (S. D. Webb and Opdyke 1995).

Fast environmental changes mean hectic times in the biosphere: communities may be destroyed in a flash by some cosmic and geological forces, and altered radically within years by such biotic agents as pests and pathogens. Community assembly and disassembly rates are constrained by the migration rates of component species. Migration rates vary considerably between species. Three basic migration strategies seem to exist and are analogous to fugitive, opportunist, and equilibrium population strategies:

1 Fugitive species. These are the 'weeds' of the animal and plant kingdoms. They colonize temporary, disturbed habitats, reproduce rapidly, and soon depart before the habitat disappears or before competition with other organisms overwhelms them. The common dandelion (*Taraxacum officinale*) is a fine example. A fugitive-migration strategist occupies a patchwork of habitat islands. The tamarack (*Larix laricina*) is a fugitive-strategist (P. A. Delcourt and Delcourt 1987: 319–22). Its changing distribution mirrors the availability of wetland habitats, and any snapshot of its distribution shows low dominance in most areas, with dominance hot-spots in local wetland sites (e.g. Payette 1993).

2 Opportunist species. Fast migrators spread rapidly, pushing forward along a steep migration front, but failing to maintain large populations in the wake. Spruce (*Picea*) is an *r*-migration strategist and displays these characteristics (P. A. Delcourt and Delcourt 1987: 306–12). With the retreat of the Laurentide ice sheet, spruce spread rapidly onto the newly exposed ground. The rate of migration reached 165 m/yr 12,000–10,000 years ago. The spruce's northwards advance was halted only by the ice. Behind its migration front, populations diminished as climates warmed. Black spruce (*Picea mariana*) was the first conifer species to invade northern Quebec immediately after the ice had melted 6,000 years ago (Desponts and Payette 1993).
3 Equilibrium species. Slow migrators rise more slowly to dominance after an initial invasion, the highest values of dominance lying well behind the 'front lines' and reflecting an unhurried build-up of populations. Oak (*Quercus*) is a *K*-migration strategist. It was a minor constituent of late glacial forests in eastern North America. It did reach the Great Lakes in late glacial times, but its rise to dominance was a slow process that reached a ceiling in the Holocene epoch in the region now occupied by the eastern deciduous forest (P. A. Delcourt and Delcourt 1987: 313–16).

The variations in the speed of movements of species lead to the disassembly of communities. As a rule, this process will alter community composition and structure. In some cases, communities retain their essential character and shift bodily in the wake of shifting climatic zones. In Africa, for instance, vegetation belts swung swiftly northwards 9,000 years ago, after an increase in the Atlantic monsoon led to greater rainfall (p. 145).

Biotic crises

A biotic crisis, or mass extinction, occurs when species biodiversity falls to low levels. It may arise from a higher than normal extinction rate, from a lower than usual speciation rate, from species loss through net outward migration, or from a combination of all these.

There are varying levels of crisis. Mild crises involve an elevated turnover of species. Severe crises involve a loss of 20 per cent or more of all species. When such severe crises act globally, they are called mass extinctions. But crises can occur at all geographical scales. Landscape patches commonly suffer species loss after a disturbance, and explosions of volcanic islands cause local mass extinctions. The biospheric stresses that cause extinction events and the post-crisis recovery processes have parallels at local, regional, and global scales.

Extinction is the shared fate of all species. Like the death of an individual, it is an inevitability. The fossil record points to a continuum of extinction events that range from everyday background levels to mass extinctions (cf. Raup 1993). However, it also suggests that mass extinctions are not the chance coincidence of independent extinction events, but regular episodes of mass killings. Additionally, extinction is to some extent selective, but in many cases it appears to act randomly in that the survivors are apparently not better adapted to life in the post-crisis environment than the victims – they are simply luckier.

There are a few key questions that need addressing here. First, why should background extinction levels suddenly 'go critical'? In other words, what causes mass extinctions? Second, are mass extinction events grand global dyings occurring within days, weeks, months, or years? Or are they clusters of independent extinction episodes occurring over hundreds of millennia? And third, what happens to the biosphere in the aftermath of a mass extinction? The first two questions are too closely allied to separate and will be considered together.

Times of crisis

Data on marine genera covering the last 270 million years show peaks where the extinction rate has risen above background levels twelve times (Figure 8.12). Eight of these peaks match well-known extinction events, and nine are roughly periodic, occurring once every 26 million years (Sepkoski 1989). But are these 'mass extinctions'? When does a surge in extinction rate (or sharp drop in speciation rate) become a mass extinction? Three mass extinctions seem indisputable, for they involved an estimated loss of more than 63 per cent of all species. They are the Late Permian extinction event, the upper Norian (Triassic–Jurassic) extinction event, and the Maastrichtian (Cretaceous–Tertiary) extinction event. The Late Permian event is the mother of all mass extinction events: it entailed an estimated loss of at least 80 per cent (Stanley and Yang 1994), and perhaps 93 per cent (Sepkoski 1989) of all species. The other six events run at extinction rates about twice or thrice the background level.

It is difficult to read rates of extinction directly from the fossil record (Benton 1994). The existing evidence gives a mixed message about the rate at which mass extinctions occurred: some evidence points to protracted extinction episodes; other evidence suggests sudden extinction. That statement begs an obvious but tricky question: how sudden is sudden? Some researchers think that sudden means sudden – a year or so (McLaren 1988). This view of suddenness was given strong support by the discovery of a marker horizon at the Cretaceous–Tertiary boundary (Alvarez *et al.* 1980). The marker horizon was taken by some geologists as concrete evidence that the terminal Cretaceous extinction event was indeed geologically instantaneous, and was caused by an asteroid colliding with the Earth. The association of impact-event signatures within boundary layer sediments

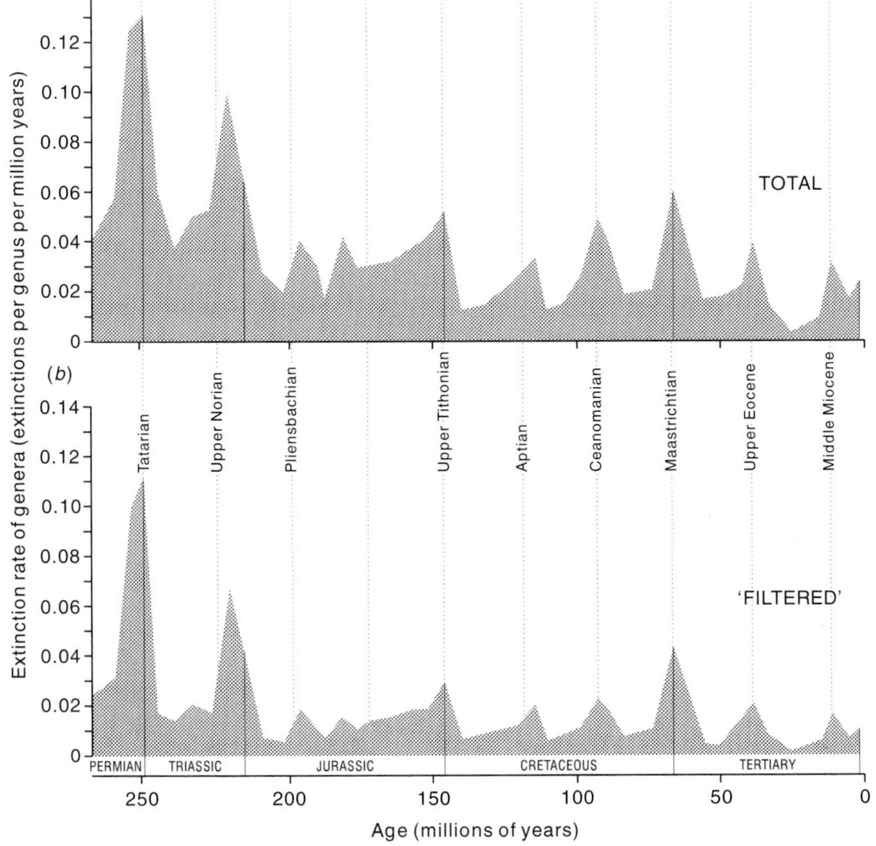

Figure 8.12 Extinction rate per genus for forty-nine sampling intervals from the mid-Permian (Leonardian) to Recent. A periodicity of 26 million years is indicated by the vertical lines. (a) Time series for the entire data set of 17,500 genera. (b) 'Filtered' time series for a subset of 11,000 genera, from which genera confined to single stratigraphical intervals are excluded
Source: After Sepkoski (1989)

lends much weight to the view that impacts did occur contemporaneously with boundary-layer clay formation. Some boundary clays contain organic chemicals with a composition highly suggestive of a cosmic origin (e.g. Zhao and Bada 1989), and some contain glass spherules of probable impact origin (e.g. Claeys and Casier 1994). To be sure, some geochemical signatures do change suddenly at boundary events. An example is the carbon isotope ratio at Permo-Triassic boundary in British Columbia, Canada (K. Wang *et al.*

1994). That does not mean that the impacts were the primary cause of the extinctions; they might simply have been the knockout blow.

The fossil record almost invariably does record long-lasting extinction episodes. Investigations of many boundary sites show that mass extinctions occurred in a series of discrete steps spread over a few million years (stepwise extinction), and not in an instant. Mass extinctions during Late Ordovician, Late Devonian, and Late Permian times were long affairs in which tropical marine biotas, including stenothermal calcareous algae, declined greatly, and reef communities were decimated (Stanley 1988a, 1988b). The detailed pattern of Late Cretaceous extinctions suggests a relatively gradual extinction-rate increase for many groups of organisms, followed by a catastrophe lasting a few tens of thousands of years. In the marine realm, the extinction of planktonic foraminiferal species spanned 300,000 years below, and some 200,000 to 300,000 years above, the Cretaceous–Tertiary boundary (Keller 1989). The dinosaurs might have suffered a gradual extinction (M. E. Williams 1994), but a sudden extinction is suggested by the Hell Creek Formation in eastern Montana and western North Dakota (Sheenan *et al.* 1991; but see Hurlbert and Archibald 1995).

Causes of crises

Climatic change was for a long time deemed the principal cause of extinctions (e.g. Simpson 1953). While the link between climate and extinction has been watered down in recent decades, there is still much evidence that points to climatic change as a potential disruptor of ecological stability. But it is probably wrong to seek a single cause for extinction events. Admittedly, mass extinctions in the marine realm usually occur at times of reduced ecosystem productivity (e.g. Paul and Mitchell 1994), and presumably this would be the case in terrestrial mass extinctions, too. Reduced productivity suggests a general deterioration of the environment during mass extinctions. But this simply defers the question of what causes environmental deterioration. The list of possibilities is long (Table 8.2).

Disaster in the biosphere wrought by an impact event is a popular explanation of mass extinction in some quarters. The ramifications of a large-body impact are many and various – there are many paths to impact-induced mass extinction (Figure 8.13; see also Toon *et al.* 1994). Over the last 540 million years, marine genera have experienced about twenty-four bouts of extinction (when rates are computed per geological stage or substage). Five of these were major and some nineteen were minor, judging by the percentage of genera that went extinct (Figure 8.14). A coincidence between many of these extinction peaks and known impact events strongly implies a causal relationship between them (Figure 8.14).

Table 8.2 Possible causes of mass extinctions

Ultimate cause	*Proximate cause*	*Effects*	*Examples of mass extinctions*
Cosmic causes			
Single large impacts	Shock-waves, heat-waves, wildfires, impact winters (shutting down photosynthesis), super-acid rain, toxic oceans, superwaves and superfloods (from an oceanic impact)	A grand global dying	Cretaceous–Tertiary event (but fossil record suggests a stepwise extinction) (Alvarez *et al.* 1980; Smit 1994)
Comet storms	Shock-waves, heat-waves, wildfires, impact winters (shutting down photosynthesis), super-acid rain, toxic oceans, superwaves and superfloods (from an oceanic impact)	Stepwise extinction events	Cenomanian–Turonian, Cretaceous–Tertiary, Eocene–Oligocene (Donovan 1987)
Radiation from supernovae	Direct exposure to cosmic rays and X-rays Ozone destruction and exposure to excessive amounts of ultraviolet solar radiation	Sterilizes and kills organisms, causes mutations. Selective mass extinctions (exposed animals, including shallow-water aquatic forms, but not plant life)	Possibly any event (Schindewolf 1963; Terry and Tucker 1968)
Large solar flares	Exposure to large doses of ultraviolet radiation, X-rays, and photons Ozone depletion	Mass extinctions	Events during magnetic reversals (Reid *et al.* 1978) Sporadic faunal extinctions (Hauglustaine and Gérard 1990)
Geological causes			
Geomagnetic reversals (with spin rate changes)	Increased flux of cosmic rays	Mass extinctions	Late Ordovician, Late Permian, Late Devonian, and Late Cretaceous events (Whyte 1977)
Continental drift	Climatic change: glaciations when continents encroach upon the poles	Global cooling	Late Ordovician, Late Devonian, and Late Permian marine events associated with encroachment of land masses on poles (Stanley 1988a, b)

Table 8.2 (continued)

Ultimate cause	*Proximate cause*	*Effects*	*Examples of mass extinctions*
Continental drift (continued)	Aridity increase on moving into low latitudes	Extinctions because species find themselves in inhospitable climatic zones	Many land plants died as India drifted northwards (Knoll 1984)
Volcanism	Cold conditions, acid rain, and reduced alkalinity of oceans resulting from release of sulphate volatiles. Toxic trace elements. Climatic change from release of ash and carbon dioxide	Stepwise mass extinctions	Terminal Cretaceous flood basalt eruptions (McLean 1981, 1985; Officer *et al.* 1987)
Sea-level change	Loss of habitat	Mass extinctions of susceptible species (e.g. marine reptiles)	Cretaceous–Tertiary event (Bardet 1994)
Arctic spill-over (release of cold fresh or brackish water from an isolated Arctic Ocean)	Ocean temperature falls by about 10°C	Mass extinctions in marine ecosystems	Late Cretaceous event (Thierstein and Berger 1978)
	Atmospheric cooling and drought	Change of vegetation with drastic effect on large reptiles	Devonian marine events (Copper 1986)
Salinity changes	Reduced salinity	Mass extinction in marine realm	Permian extinctions (Fischer 1964; Stevens 1977) Triassic extinctions (Holser 1977, 1984)
Anoxia	Shortage of oxygen	Mass extinctions in oceans	Frasnian–Famennian event (Geldsetzer *et al.* 1987; Buggisch 1991) Late Palaeocene (Kennett and Stott 1995)
Biospheric causes			
Spread of disease and predators	Direct effects (made possible by changes in geography)	Mass extinctions	Late Cretaceous mass extinction (Bakker 1986)
Evolution of new plant types	Changed biogeochemical cycles reducing the ocean nutrient supply	Gradual extinctions of marine biota	Late Permian mass extinction (Tappan 1982, 1986)

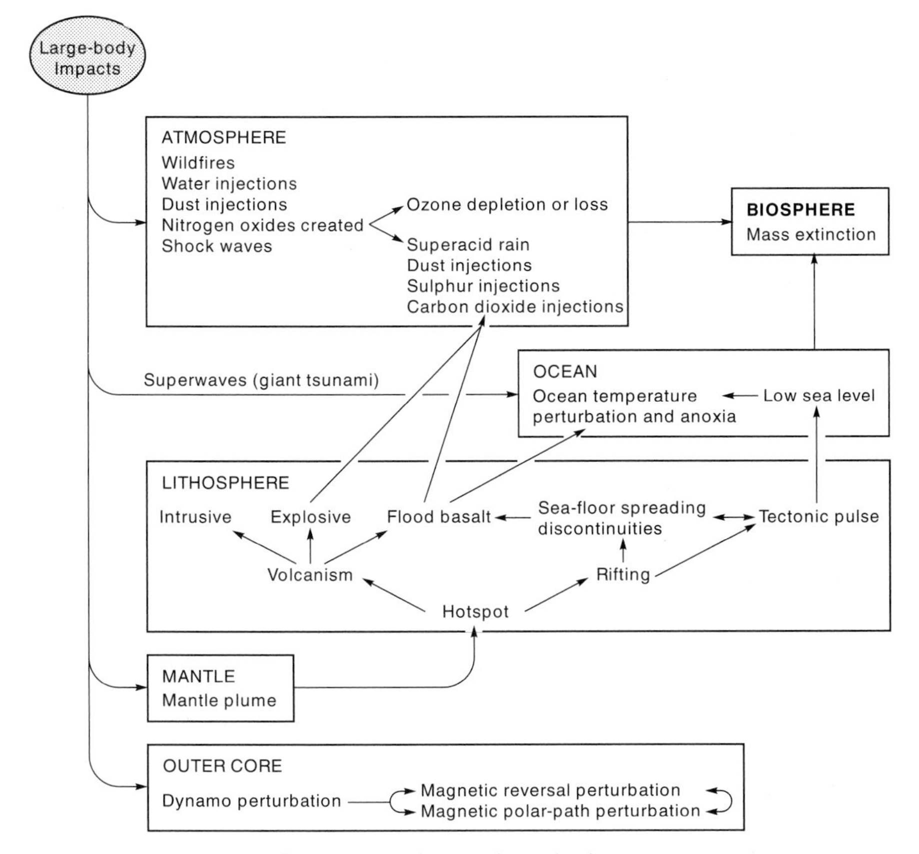

Figure 8.13 Speculative flow diagram linking large-body impacts to mass extinctions
Source: Inspired by Stothers and Rampino (1990)

In the aftermath of crises

Even the severest of mass extinctions were no match for the biosphere. If they had been, you would not be reading this book. The biosphere has managed to recover from all biotic crises in its long history. Recovery after the heavy crises is commonly delayed, but it does take place. Mass extinctions have regulated reef diversity during the Phanerozoic aeon: reef development after extinction events is negligible and diversity stays low, sometimes almost zero, for long periods (Kauffman and Fagerstrom 1993). Recovery may be delayed by the persistence of post-crisis inimical conditions, or else the slow restoration of vulnerable taxa (Stanley 1990b). Adverse conditions in the post-crisis environment may reduce ecosystem

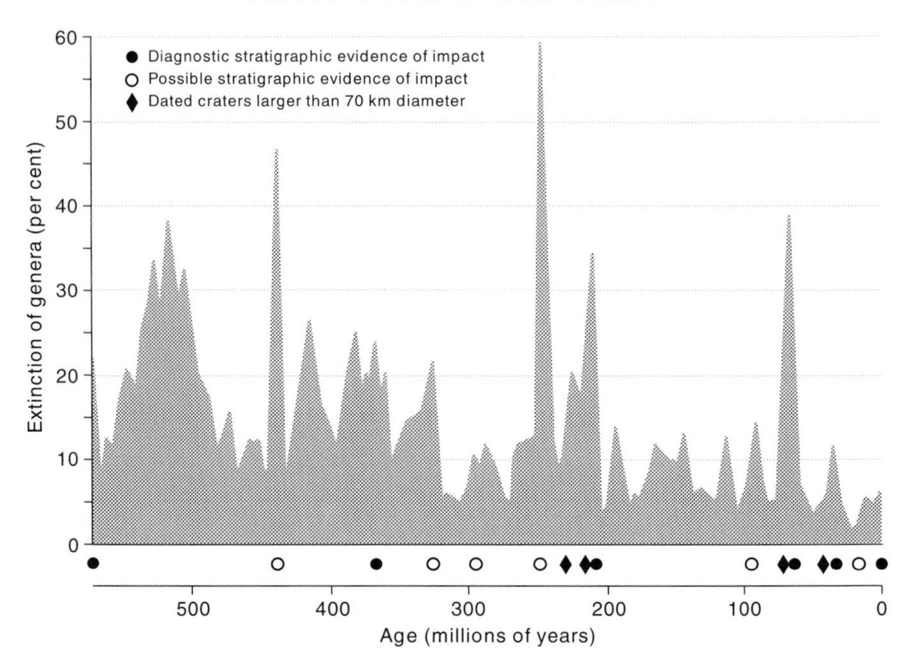

Figure 8.14 Per cent extinction of marine genera per geological stage (or substage) during the Phanerozoic aeon, and possible connections with impact events
Source: After Rampino and Haggerty (1994)

primary productivity. In British Columbia, Canada, a marine sequence through the Permo-Triassic mass-extinction event shows an abrupt drop in the carbon isotope ratio after the event, indicating a reduction in surface-water productivity (K. Wang *et al.* 1994). A similar drop occurred at Botomian time (Lower Cambrian) in Siberia, where it coincided with the mass extinction of archaeocyathan reef biota (Brasier *et al.* 1994).

The species surviving a biotic crisis have one thing in common – they are lucky. If they can grasp the opportunity for expansion and diversification, they may be spectacularly successful. The Norian (Upper Triassic) mass extinction saw the demise of the mammal-like reptiles and rhynchosaurs. This event was quickly followed, still within the Norian age, by the adaptive radiation of the dinosaurs. It is unlikely, therefore, that the dinosaurs out-competed the mammal-like reptiles, as was traditionally claimed (Tucker and Benton 1982). No, the original success of the dinosaurs was simply due to opportunism – they happened to be in the right place at the right time (cf. Emiliani 1982). The same argument applies to the rise of the mammals after the dinosaur extinctions (e.g. Sheenan and Russell 1994).

Extinction cycles?

One of the more contentious issues about mass extinctions is the assertion that they recur at regular intervals through geological time. This claim can at least be tested against the fossil record. Most of the evidence for periodicity in the fossil record has come from biostratigraphic data on marine organisms. Marine genera and families display a very dominant period of 26.0 million years (Sepkoski 1989; Rampino and Caldeira 1993). Extinctions in the Upper Permian, upper Norian, and Maastrichtian were truly massive and bouts of widespread volcanism or episodes of heavy bombardment seem likely causes. However, many of the periodic extinction events occurred over several stratigraphical stages or substages. Generic extinctions follow two patterns, both of which suggest a gradualistic mechanism behind the periodicity (Sepkoski 1989). First, most of the extinction peaks, especially for filtered data, have almost identical amplitudes, well below the amplitudes of the three major events. Second, the widths of most of the peaks span several stages. Taken with the three massive extinction events, two distinct causes of mass extinctions are indicated. The first cause is a 26-million-year oscillation of the Earth's oceans, or climates, or both that leads to higher extinction rates over long periods, either continuously or in high-frequency, stepwise episodes. The second cause involves independent agents or constraints upon extinction, including bombardment episodes, volcanic episodes, and sea-level changes, that boost the periodic oscillation of extinction rate when they happen to occur at times of increasing extinction. In other words, large impacts and massive outpourings of lava may trigger mass extinctions under some circumstances (cf. Yabushita 1994). And it would be wrong to surmise that all extinctions run to a rigid 26-million-year timetable. Cycles of extinction in mid-Palaeozoic reef communities fail to 'run on time', and appear to relate to climatic change (and possibly ocean anoxic events) associated with plate tectonic processes (Copper 1994).

THE WAY OF THE WORLD: DIRECTIONS OF BIOTIC CHANGE

Cycles and steady states in the living world

Whenever the environment oscillates strongly, be it through glacial–interglacial cycles or the longer solar cycles, then so do communities and ecosystems.

On daily, annual, and decadal scales, the oscillations appear to occur about a steady state. But oscillations there are – all life appears firmly locked into lunar, solar, and planetary beats. In addition, many communities display cycles of change over tens to hundreds of years. Forest patches, for instance, pass through building, maturing, and declining phases, the last being associated with the toppling of old trees. The falling trees tear gaps in the forest and the cycle begins anew. These cycles involve directional changes where the freshly created patches are captured by a different new plant community. Such forest-gap succession is a mass extinction writ small. Climatic changes in the Croll–Milankovitch frequency band also have a profound influence upon life (e.g. K. D. Bennett 1996).

Communities and ecosystems may display directional changes over millions of years. Such megatrends are sometimes a response to secular changes in environmental factors, and sometimes a response to the crossing of internal community and ecosystem thresholds. For example, the pulse of Cenozoic mammal communities displays a 'syncopated equilibrium', with long-lasting and stable chronofaunas separated by rapid turnover episodes involving radical reorganizations of terrestrial ecosystems (S. D. Webb and Opdyke 1995). The result is a large-scale and long-term succession of terrestrial ecosystems. The same process has operated throughout the Phanerozoic (Sheenan 1991; Sheenan and Russell 1994). Biotic crises abruptly ended long periods of stability in major varieties of dominant organisms. After each crisis, rapid diversification and ecological reorganization ushered in a new period of stability.

Protracted cycles of 150 million and 300 million years appear to be locked into climatic warm-mode–cool-mode and hothouse–icehouse cycles. For instance, the sites of greatest carbon sequestering switch from low latitudes during icehouse times to higher latitudes (poleward of 40°) during hothouse times (Spicer 1993). Marine palaeontological data hint at a 30-million-year cycle (Fischer 1984a). This cycle is expressed in the global diversity of planktonic and nektonic taxa, including globigerinacean foraminifers and ammonites, and in the episodic development of superpredators with body lengths of 10–18 m (Fischer 1981). Long-lasting phases of increasing diversity ('oligotaxic' phases), when oceans were cool, were punctuated roughly every 30 million years by high-diversity crises ('polytaxic' phases), when the oceans were warm. Each polytaxic pulse brought a new group of superpredators – the superpredator niche opened up after each biotic crisis was successively filled by ichthyosaurs, pliosaurian plesiosaurs, mosasaurs, whales, and sharks.

Secular trends in the living world

Life on Earth first appeared at least 3.8 billion years ago. Astonishingly, it now seems likely that it was present on Mars around 3 billion years ago. On Earth, life has followed secular trajectories of increasing complexity and diversity. Some major 'first appearances' are shown in Figure 1.4. The first living things were prokaryotes; they were anaerobic, fermenting heterotrophic bacteria. The first autotrophic bacteria evolved by about 3.5 billion years ago, and the first aerobic photoautotrophic bacteria and the nitrate-reducing and sulphate-reducing bacteria by 2.5 billion years ago. Eukaryotes had evolved by around 1.5 billion years ago, and metazoans by 600 million years ago. During the Phanerozoic aeon, key events in the biosphere were the appearance of the following: calcareous and siliceous skeletons (570 million years ago), the origin of vertebrates (510 million years ago), land plants (440 million years ago), wingless insects (420 million years ago), winged insects (310 million years ago), mammals (225 million years ago), birds (150 million years ago), flowering plants (140 million years ago), and humans (3 million years ago).

Communities and ecosystems, as well as individuals, have become more complex, insofar as they contain more, and a greater variety of, species. And they have become more diverse, mainly because the abiotic and biotic environments have become increasingly patchy – geodiversity and biodiversity have both increased. The increase in geodiversity is partly caused by the action of organisms, especially by those that, after their death, form the material of such sedimentary rocks as limestone.

Life has always interacted with its environment, but occasionally unusual events have caused spurts of evolution and biodiversity. An example of such an event is the late Palaeozoic oxygen pulse (mid-Devonian, Carboniferous, and Permian). This involved a marked rise (possibly to a hyperoxic 35 per cent) and then fall (possibly to 15 per cent) in atmospheric oxygen and associated changes in atmospheric carbon dioxide. It was probably caused by bottlenecks in lignin cycling, and in the cycling of other refractory compounds synthesized by the newly evolved land plants (J. M. Robinson 1990, 1991). Its effect was to quicken the terrigenous organic-carbon cycle and to enable terrestrial production to increase with a concomitant rise in atmospheric oxygen levels. The oxygen pulse influenced diffusion-dependent features of organisms (including respiration and lignin biosynthesis), and may have fuelled diversification and ecological radiation, permitting greater exploitation of aquatic habitats and the newly evolving terrestrial biosphere (J. B. Graham *et al.* 1995).

An increase in biodiversity has paralleled the evolutionary burgeoning of life forms. This is seen in rising number of fossil taxa recorded in successive stages during the Phanerozoic aeon (Figure 8.15). An initial upsurge marks the Cambro-Ordovician explosion. A decline sets in towards the close of the Palaeozoic era and ends in the rapid drop in the Late Permian period. From the Triassic period the trend is consistently upwards. This curve is thought to record real changes in the diversity of life, and is not an artefact of sampling (Valentine *et al.* 1978).

The causes of these long-term diversity changes are not known for sure (see Benton 1990). One suggestion is that after the Cambro-Ordovician explosion, which might have filled the available niches, diversity tracked changes in the disposition of the continents. Moderately high diversity was associated with moderately separated continents during the Palaeozoic era;

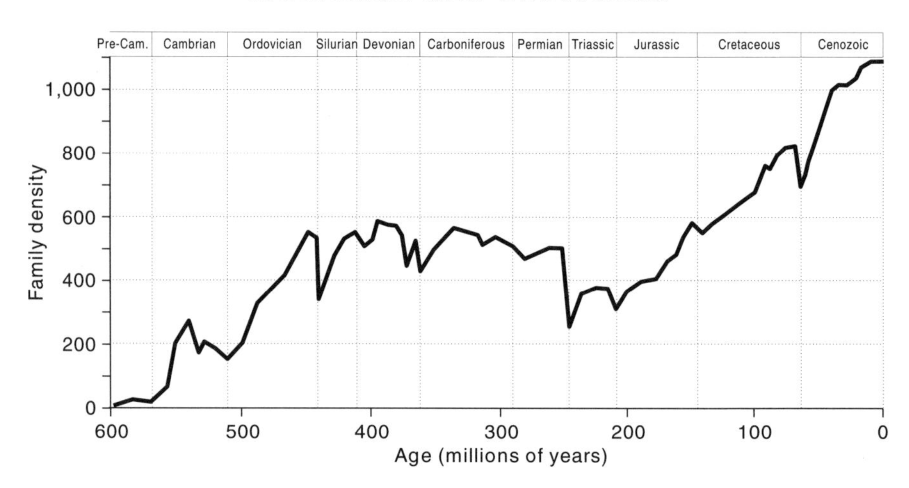

Figure 8.15 Diversity curve for marine faunal families
Source: Sepkoski (1993)

diversity dropped as the continents came together to form Pangaea; and diversity rose after the Permian period as the continents broke up (but see Raup 1972). A more recent study of Phanerozoic diversity used five major and essentially independent estimates of lower taxa (trace fossil diversity, species per million years, species richness, generic diversity, and familial diversity) in the marine fossil record (Sepkoski *et al.* 1981). Strong correlations between the independent data sets indicated that there is a single underlying pattern of taxonomic diversity during the Phanerozoic. The pattern is low diversity during the Cambrian period; a higher but not steadily increasing diversity through the Ordovician, Silurian, Devonian, Carboniferous, and Permian periods; low diversity during the early Mesozoic era, notably in the Triassic period; and increasing diversity through the Mesozoic culminating in a maximum diversity during the Cenozoic era (see Signor 1994). Congruent patterns have been discerned in the record of marine vertebrates (Raup and Sepkoski 1982), non-marine tetrapods (Figure 8.16a), vascular land plants (Figure 8.16b), and insects (Labandeira and Sepkoski 1993).

It is tempting to suppose that there is an upper limit to biodiversity, or a global carrying capacity. However, this diversity ceiling is cranked up by evolutionary innovations and by climatic and geological changes. Evolutionary innovations occasionally lead to the raising of this global carrying capacity and biodiversity can be expected to have risen. Climatic and geological processes incessantly increase the complexity of the physical environment – they drive geodiversity to ever-greater levels. The biosphere has always striven to reach this ever-rising biodiversity ceiling, but it has

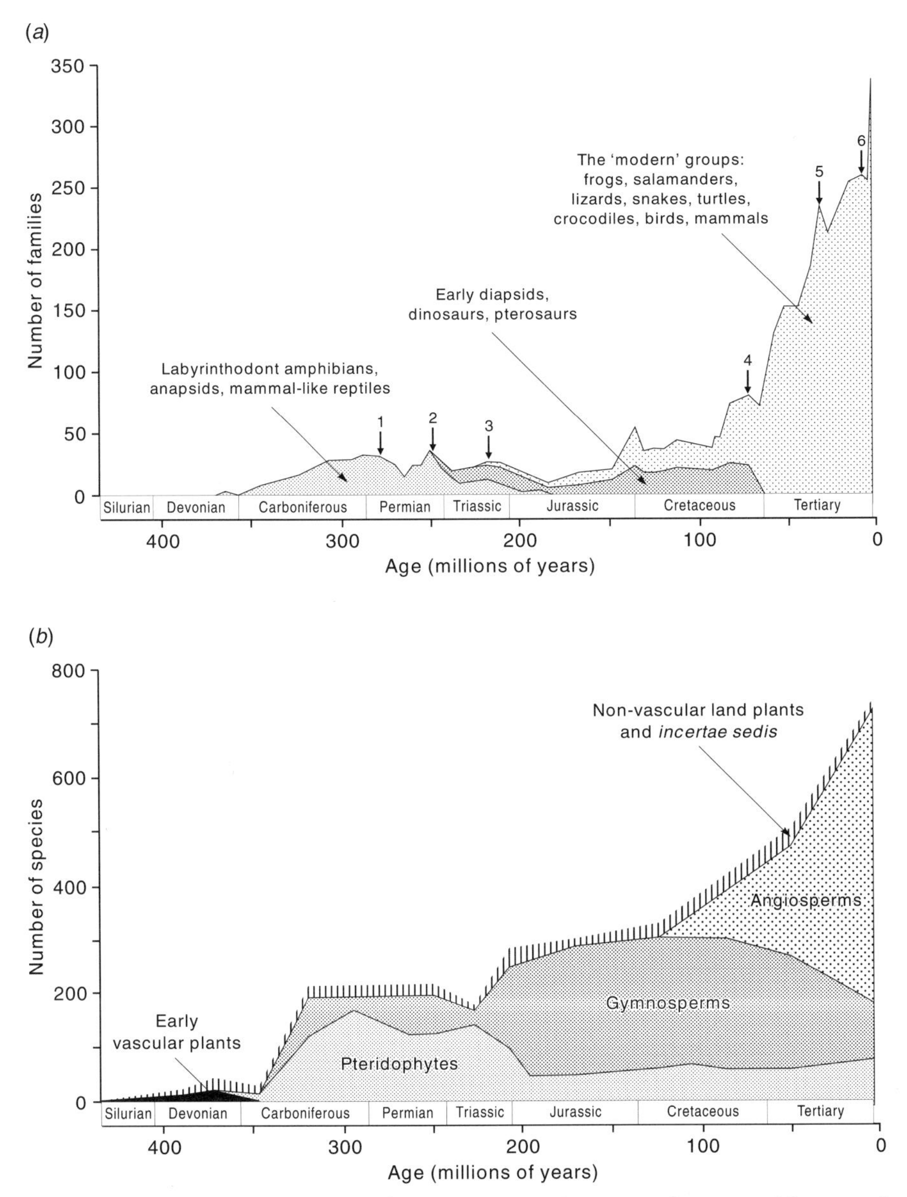

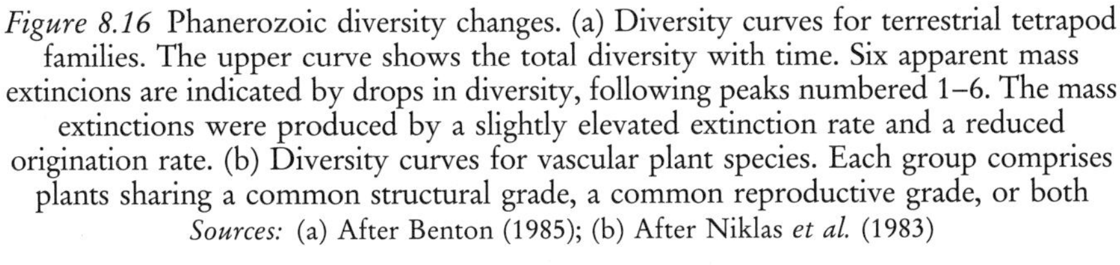

Figure 8.16 Phanerozoic diversity changes. (a) Diversity curves for terrestrial tetrapod families. The upper curve shows the total diversity with time. Six apparent mass extincions are indicated by drops in diversity, following peaks numbered 1–6. The mass extinctions were produced by a slightly elevated extinction rate and a reduced origination rate. (b) Diversity curves for vascular plant species. Each group comprises plants sharing a common structural grade, a common reproductive grade, or both
Sources: (a) After Benton (1985); (b) After Niklas *et al.* (1983)

been hindered by the major disturbances that have led to mass extinctions. Thus, the biodiversity ceiling is seldom reached by the biota: it is a theoretical maximum towards which the biosphere strives between perturbations (cf. Kitchell and Carr 1985). For these reasons, biodiversity is unlikely ever to attain a true steady state; rather, it will increase through time, tracking the increasing diversity of the physical environment, with occasional setbacks caused by mass extinctions (Cracraft 1985). There might be a limit to physical complexity that is possible on the Earth (Valentine 1989) – geodiversity will then limit biodiversity.

Prolonged phases of steady-state biodiversity are found within the overall biodiversity increase through geological time (Rosenzweig 1995: 52). Over the last 5 million years of the Ordovician period, the species diversity of muddy benthos organisms from Nicolet River Valley, Quebec, was stable – it fluctuated around a mean of about 32 species. This was a steady state, and represented an approximate balance between speciation events and extinctions. Cenozoic mammal faunas display similar steady-state diversities for long periods. The diversity of large herbivores and carnivores in North America fluctuated around a steady value for 45 million years (Van Valkenburgh and Janis 1993).

Gaia or Hades?

To what extent has life governed environmental conditions? This is one of those basic questions to which a sound answer cannot yet be given. There are two diametrically opposed possibilities, and all gradations in between them. At one extreme, life is viewed as an inconsequential film that, historically, has had little impact on surface processes, at least until humans arrived (e.g. Holland 1984). It adapts as best it may through biological evolution, ducking and weaving to avoid the brunt of geological and cosmic forces, bowing to the whim of every volcanic eruption, sea-level change, and asteroid impact. This is a somewhat gloomy view of life's role on the Earth. It has been dubbed the Ereban hypothesis, after Erebus, the Greek personification of darkness and the underworld (see Retallack 1990, 1992a). Instead, it might be dubbed the Hadean hypothesis (after Hades, the Greek god of the underworld), in which life is seen to struggle painfully for existence amidst a geological and cosmic Hell (Figure 8.17). The Hadean view is far more geological than biological, and sees the biosphere as a dynamic system capable of responding to changes in its environment, but only in a limited and essentially passive way. The biosphere is thus painted as a somewhat fragile system that is susceptible of permanent disruption.

At the other extreme, the Gaia hypothesis asserts that, shortly after it first appeared, life has been at the helm, exercising near total homeostatic control of the terrestrial environment (Figure 8.17). The brighter Gaian view of life's connection with the planet is named after Gaia, the Greek goddess of

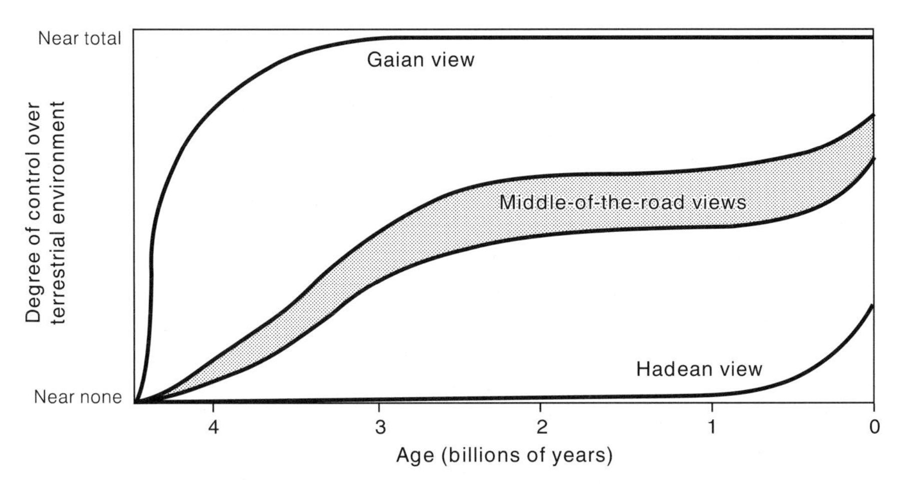

Figure 8.17 Views on the control wielded by life over the ecosphere
Source: Partly after Retallack (1992a)

the Earth, daughter of Chaos, mother and lover of the sky (Uranus), the mountains (Ourea), and the sea (Pontus). Between the Hadean and Gaian extremes, lie a gamut of middle-of-the-road views expressing varying degrees of control over the terrestrial environment. Some commentators feel these are more reasonable than the extreme views, pointing out that life did not rapidly assume a hegemony over Earth resources after its first appearance, nor has it been wholly at the mercy of cosmic and geological forces (e.g. Retallack 1992a). There is some evidence that life did take control of biogeochemical cycles very early in Earth history (Figure 8.18), which tends to favour the Gaian end of the life-control spectrum. However, the issue is extraordinarily complicated (see Boston and Thompson 1991). A report of a recent seminar at Green College, Oxford (Tickell 1996), suggests that middle-of-the-roadism is gaining supporters, with many geoscientists subscribing to the view that the entire Earth does have lifelike properties, though it is not alive.

There are at least two versions of the Gaia hypothesis – weak Gaia and strong Gaia (Kirchner 1991). Weak Gaia is the assertion that life wields a substantial influence over some features of the abiotic world. It does so mainly through playing a pivotal role in biogeochemical cycles. Its influence is sufficient to have produced highly anomalous environmental conditions in comparison with the flanking terrestrial planets, Venus and Mars. Notable anomalies include the presence of highly reactive gases (including oxygen, hydrogen, and methane) coexisting for long periods in the atmosphere, the stability of the Earth's temperature in the face of increasing solar luminosity, and the relative alkalinity of the oceans. These unusual

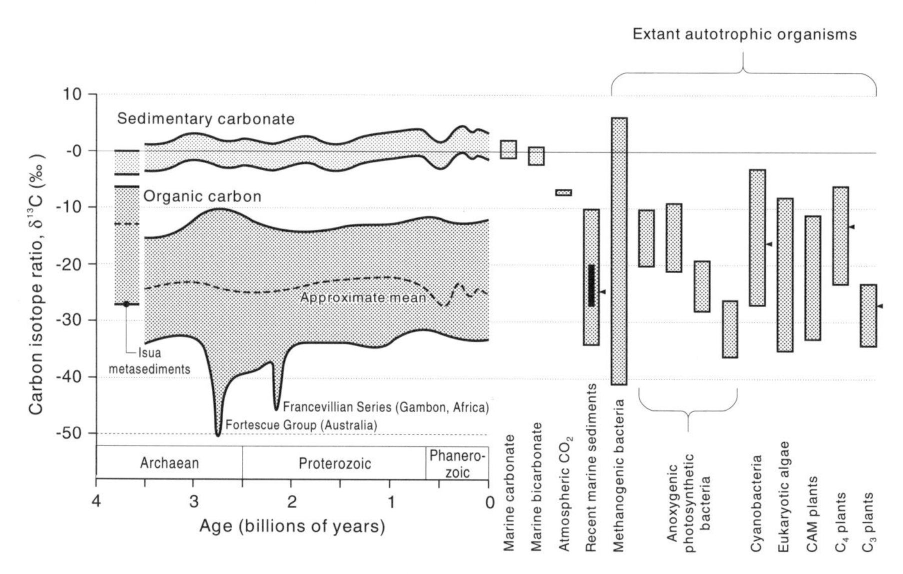

Figure 8.18 Isotopic composition of sedimentary carbonates and organic carbon over 3.8 billion years of Earth history. Note the constancy of the isotope ratios after about 3.5 billion years ago, suggesting that partial biological control of the carbon cycle was established very early and was in full operation when the oldest sediments were formed

Source: After Schidlowski (1988)

conditions of temperature, chemical composition, and alkalinity have been maintained over much of geological time by life's interacting with the surface materials of the planet. For this reason, to 'understand the Earth's surface we must understand the biota and its properties; we can no longer rely on physical sciences for a description of the planet' (Margulis and Hinkle 1991: 11). The weak Gaia hypothesis does not call upon anything other than mechanistic processes to explain terrestrial evolution, but does argue that the abiotic portion of the ecosphere has been built, and is largely maintained, by the biosphere.

Strong Gaia is the unashamedly teleological idea that the Earth is a superorganism that controls the terrestrial environment to suit its own ends, whatever they might be. James Lovelock, who with Lynn Margulis first suggested the Gaia hypothesis, seems to favour strong Gaia. He believes that it is useful to regard the planet Earth, not as an inanimate globe of rock, liquid, and gas, driven by geological processes, but as a sort of biological superorganism, a single life-form, a living planetary body that adjusts and regulates the conditions in its surroundings to suit its needs (e.g. Lovelock 1991). For Lovelock, Gaia includes the biosphere and the rest of the Earth. He explains that just 'as the shell is part of a snail, so the rocks, the air, and the oceans are part of Gaia' (Lovelock 1988: 10). It is not known whether life is able to exert an influence deep inside the Earth, whether the biosphere is merely the epidermis of a living global creature. However, the evolving biosphere has maintained an intercourse with the Earth's interior through plate-tectonic processes, the magmatic system being the medium and the mechanism of communication (Shaw 1994: 246).

Gaian scientists claim that traditional biology and geology offer ineffective methods with which to study the planetary organism. The right tool for the job, they contend, is geophysiology – the science of bodily process writ large and applied to the entire planet, or at least that outer shell encompassing the biosphere. The differences of approach and emphasis are fundamental – if the strong Gaia hypothesis should be correct, and the Earth really is an integrated superorganism, then the biosphere will regulate and maintain itself through a complex system of homeostatic mechanisms, just as your body adjusts to the vicissitudes of its surroundings. Consequently, the biosphere may be a far more robust and resilient beast than has often been suggested. For instance, homeostatic mechanisms may exist for healing the hole in the ozone layer or preventing the global thermometer from blowing its top.

The Gaia hypothesis, in all forms, suggests three important things (Lovelock 1991). First, life is a global, not local, phenomenon. It is not possible for a planet to be inhabited by sparse life – there must be a global film of living things because organisms must regulate the conditions on their planet to overcome the ineluctable forces of physical and chemical evolution that would render it uninhabitable. Second, the Gaia hypothesis adds to Darwin's vision by negating the need to separate species evolution from environmental evolution. The evolution of the living and non-living worlds are so tightly knit as to be a single indivisible process. Evolutionary success is measured as a coherent coupling between organisms and the material environment, not just in survival of the fittest. Third, the Gaia hypothesis may provide a way to view the planet in mathematical terms that 'joyfully accepts the nonlinearity of nature without being overwhelmed by the limitations imposed by the chaos of complex dynamics' (Lovelock 1991: 10). These ideas are currently being investigated by geophysiologists, a new and interdisciplinary breed of Earth and life scientists who probe the complex interdependent cycles that run through the geosphere and ecosphere.

SUMMARY

Individual organisms belong to a community and to an ecosystem. Community units are local communities, communities, biomes, and zonobiomes, and the biosphere. Ecosystem units are local ecosystems, ecosystems, geozonal ecosystems (equivalent to zonobiomes), and the ecosphere. Communities bear emergent properties (biodiversity, production and consumption, nutrient cycles, and food webs). Communities change through succession. It is now thought that successional sequences, such as hydroseres, fail to follow preordained paths. Rather, succession is an individualistic process that leads in many directions, with each species 'doing its own thing'. Communities are thus evanescent and evolve. This is borne out by communities reconstructed from soft sediments and sedimentary rocks, many of which have no modern analogues. Communities are subject to autogenic (internal) and allogenic (external) change. Hydrosere succession and the peatland hummock–hollow cycle are largely autogenic, but allogenic factors play a role. Individuals and communities commonly move with solar, lunar, and planetary cycles. Longer-term climatic swings, such as the trend to aridification during the Cenozoic era, drive changes in communities and ecosystems, and may spur adaptive radiations. Geological changes, especially continental drift, promote community and evolutionary change. The rate of community change is limited by species mobility. Fugitive species, such as the common dandelion, are the quickest dispersers. The fossil record is punctuated by several biotic crises (mass extinctions), the causes of which are hotly debated. Impact events almost certainly aided and abetted some crises. Biotic crises have a 26-million-year periodicity. During Earth history, communities and ecosystems have become more complex and more diverse, mainly because the abiotic and biotic environments have become increasingly patchy. Over shorter time-scales, they display steady states and long-running cycles, sometimes in response to secular environmental change, sometimes in response to internal community and ecosystem thresholds. Life exercises some control over life-support systems. The gloomy Hadean hypothesis claims that, until the recent proliferation of humans, life had minimal control over the environment. The edifying Gaia hypothesis claims that life has been at the helm since soon after it first appeared.

FURTHER READING

Anybody requiring a foundation in ecological thinking could do no better than *Environmental Science: Earth as a Living Planet* (Botkin and Keller 1995). The 'new ecology' is nicely explained in *Discordant Harmonies: A New Ecology for the Twenty-First Century* (Botkin 1990), and in the enjoyable and enlightening *Nature's Economy: A History of Ecological Ideas* (Worster 1994). Quaternary ecology is the subject of many books. I particularly like *The Holocene: An Environmental History* (Roberts 1989), *Quaternary Ecology: A Paleoecological Perpsective* (Delcourt and Delcourt 1991), *Late Quaternary Environmental Change: Physical and Human Perspectives* (Bell and Walker 1992), and *Environmental Change* (Goudie 1992). *Historical Ecology of the British Flora* (Ingrouille 1995) is a first-rate 'case study' of geological and historical floral changes. Likewise, *Palaeoecological Events During the Last 15,000 Years: Regional Syntheses of Palaeoecological Studies in Lakes and Mires in Europe* (Berglund *et al.* 1996) is a valuable tome. *Climate Change and Its Biological Consequences* (Gates 1993) provides a clear account of climatic change and its effect on the biosphere. Life's role in Earth history, as considered in the Gaia hypothesis, is discussed in several books. James Lovelock's books – *Gaia: A New Look at Life on Earth* (1979) and *The Ages of Gaia: A Biography of Our Living Earth* (1988) – are probably the best starting-point for readers interested in the Gaia hypothesis. However, this should be complemented by *Life as a Geological Force: Dynamics of the Earth* (Westbroek 1991), which is a well-balanced and highly readable introduction to ideas in the Gaia hypothesis, and *Scientists on Gaia* (Schneider and Boston 1991), which is a technical but most illuminating discussion of life's influence on the physical and chemical properties of the planet. A stunning fictional exploration of the Gaia hypothesis is found in Brian Aldiss's wonderful Helliconia trilogy – *Helliconia Spring* (1982), *Helliconia Summer* (1983), and *Helliconia Winter* (1985) – which I recommend most strongly.

9

ENVIRONMENTAL EVOLUTION

This book has progressed sphere by sphere. To conclude this broad survey of the evolving ecosphere, I shall briefly explore patterns shared by all terrestrial spheres. In doing so, I shall raise polemical issues, make a few suggestions for future work, and explain the value of an evolutionary, as opposed to developmental, approach. The structure adopted in the previous chapters will be followed – structure and function, causes of change, rates of change, and trends will be examined.

STRUCTURE AND FUNCTION OF THE ENVIRONMENT

The terrestrial spheres are normally conceived as separate entities. This seems logical, for that is how they appear to the observer. Humans live upon solid ground, beneath an ocean of air, surrounded by an ocean of water. Land, sea, and sky seem like permanent and autarkic fixtures of the planetary environment. They result from the predominant states of matter – solid, liquid, and gas – being discrete. In the Earth, gravity has sorted these states according to density to produce the lithosphere-cum-barysphere, hydrosphere, and atmosphere. The boundaries between these spheres correspond to state transitions. Admittedly, gases, liquids, and solids combine to form colloids. Examples are liquid aerosols (clouds, fog, mist), solid aerosols (smoke, dust, fumes), colloidal solutions (clay suspensions), gels (many soil materials, opal), and solid solutions (ice cream, glass). Colloids are important locally in the environment, but they do not form 'spheres'. Topsoil almost forms a colloidal sphere, but it has many other ingredients.

Although the terrestrial spheres are separate physically, they do change and interact. Many writers, starting with classical Greek natural historians, have argued that the terrestrial spheres are neither immutable nor independent. The water cycle links the seas, rivers, and skies. The rock cycle links the dark and light sides of the Earth, the bowels and the skin. Plume tectonics links crust and core. This pervasive interaction of the terrestrial spheres is the subject of Earth-system science. Its study is interdisciplinary. It has benefited considerably from developments in hard-rock and soft-rock stratigraphy (cf. Miall 1995).

Work in the twentieth century has revealed that the main spheres established by the Greeks (earth, water, and air) are themselves layered. 'X-rays' of the Earth's interior using seismic waves, probes (weather balloons) sent into the upper air, and instruments lowered from survey ships have uncovered a relatively consistent vertical structuring in the solid Earth, the atmosphere, and the hydrosphere. The pedosphere also has a vertical structure. The edaphosphere consists of soil horizons; the debrisphere consists of sedimentary layers (strata). Even ecosystems and communities are structured vertically. This is seen as upper production zones and lower consumption zones in ecosystems, and as the vertical structure of community life-forms – trees, shrubs, herbs.

The ubiquitous layering in the terrestrial spheres has evolved despite processes that act to destroy it. In the pedosphere, the processes of soil horizon construction (horizonation) that lead to anisotropy, under most circumstances outweigh the process of soil horizon destruction (haploidization) that lead to isotropy. The same rule appears to apply to processes in the other terrestrial spheres. If it did not, there would be no predominant vertical structure. Gravity is ultimately the factor responsible for this tendency to form horizontal layers. Heavy materials (solids) tend to sink towards the Earth's centre, lighter materials tend to rise to the outer portions of the planet. So much has been understood at least since John Woodward wrote *An Essay toward a Natural History of the Earth* (1695), in which he proposed that the Earth had dissolved in the Flood waters, and the materials sank from the turbid liquid so formed, each according to its specific gravity. In modern times, an analogy is drawn with a blast furnace. During early Earth history, the denser materials sank to form the core and mantle, the less dense material rose to form the lithosphere, the slag floated on the surface to form continental crust, and the gases escaped to form the atmosphere and hydrosphere.

The Earth is undoubtedly layered. But is it misleading to see the layers as separate entities? Would it be more illuminating to regard them as sharp discontinuities or state transitions in an otherwise continuously variable body? This notion would reinforce the view that the planet has an integrity. Whether it would be justified is unclear. However, as noted in Chapter 3, the lithosphere, asthenosphere, and mesosphere are essentially the same, the differences between them resulting mainly from temperature and viscosity changes. In the barysphere, the important core–mantle and inner-core–outer-core boundaries are also important physicochemical discontinuities that have existed for much of Earth history. They are zones of steep temperature, pressure, velocity, and density gradients. In this respect, they are analogous to baroclinic zones in the atmosphere that produce weather fronts. They play key roles in tectonic processes, acting as transition zones that modulate energy transmission from the hot inner core, and from the cold lithosphere. Again, the atmospheric layers, although on the face of it largely independent of one another, are known to interact – changes in the troposphere affect processes in the stratosphere, and vice versa. Even boundaries between spheres that appear essentially arbitrary, such as the sea–air and land–air interfaces, assume enormous importance as open borders in biogeochemical cycles. And these boundaries might not be quite so sharp on close inspection. As mentioned in Chapter 6, the pedosphere may be thought of as a continuum in the sediment-to-water ratio. This idea could be extended to the sediment-to-air ratio. Is dust in the air a part of the atmosphere, or is it a transient, solid aerosol extension of the pedosphere? These points might seem odd, but it is sometimes useful to reconsider basic assumptions that are accepted without question.

CAUSES OF ENVIRONMENTAL CHANGE

Much discussion in the previous chapters revolved around external and internal mechanisms of environmental change. Plainly, externality and internality may be defined only against a reference point. If the biosphere should be the reference point, then all living things and their interactions are internal factors, while all else (the non-living world) is external. This kind of assumption works well for individual spheres. Researchers are happy enough to set the limits of their 'sphere of interest' and to define internal and external factors according to these boundaries. However, an interesting and probably significant problem arises when the unitary nature of the ecosphere and its surrounding is deemed paramount. If everything is truly connected to everything else, albeit in varying degrees, all parts of the system are internal – there are no external factors. This may seem a strange idea, but the internal and external status of environmental factors makes a world of difference to their dynamics. This was demonstrated by Jonathan D. Phillips (1991) when considering human impacts on the environment. Humans may be defined as external factors that disturb or perturb ecosystems, or else as an integral part of ecosystems. When humans are regarded as external factors, ecosystems recover their stability after changes or perturbations. Conversely, when humans are regarded as internal factors, striving to maximize or minimize throughputs, ecosystems are intrinsically unstable, failing to regain their stability after a perturbation.

A difficult issue is the differential running rate of the terrestrial spheres. Manifestly, the terrestrial spheres, such as the atmosphere and lithosphere, run to different timetables. The atmosphere changes very quickly, the lithosphere very slowly. For this reason, it is common to regard some environmental factors as, effectively, independent. A good example of this thinking is a classic paper on 'Time, space, and causality in geomorphology'

(Schumm and Lichty 1965). This paper argued that the dependence or independence of variables influencing change in drainage basins and rivers varies with the time-scale being considered. So, over geological time-spans, relief is a dependent variable, but over shorter time-spans it is independent. The logic of this idea is seductive. It could be extended to change in all terrestrial spheres. It applies to investigations of biosphere–toposphere interactions (cf. J. D. Phillips 1995b). Vegetation changes almost continuously and rapidly, whilst landforms often change episodically and relatively slowly. The rate disparity creates a time-scale problem. This kind of argument leads to troublesome questions. Under what conditions may landforms and vegetation be treated independently? Is it possible for landform–vegetation systems to exist in a steady state?

Much work is being done on the 'scale problem'. However, none of this work recognizes a very challenging conceptual dilemma posed by the terrestrial spheres' having different running speeds. The dilemma is that the terrestrial spheres are always interacting among themselves and with their cosmic and geological environments. The interaction takes place second by second, year after year, aeon after aeon. Relief does not stop interacting with other variables over short time periods, and then somehow restart its influence over geological periods. No, the interaction is ever present. And out of this interaction emerges the functional whole that is the ecosphere. The interaction is simultaneous and instantaneous, in the sense that the forces involved are constantly interacting, even though they appear to produce long-term and short-term effects. Long-term effects might simply be short-term effects carried on over vast periods of time. Time-scales of operation, therefore, might be a delusion, the scale problem a chimera. This puzzling issue needs further exploration. An attractively simple solution rests in the presence of environmental thresholds, a topic to be explored in the next section. But I suspect that answer is not the whole story.

RATES OF ENVIRONMENTAL CHANGE

Rates of environmental change are surely one of the most debated issues in Earth and life science. The story of catastrophism versus gradualism assumes epic proportions. It is a well-rehearsed saga, though, like many stories in the history of Earth history, it has been recently retold by revisionists who show that the popular conception of 'the goody gradualists versus the baddy catastrophists' is a myth.

Catastrophes do occur in all the terrestrial spheres (and the cosmic ones, too). This book has mentioned the catastrophic collapse of megaliths stagnating at about 670 km depth, deglaciation, impact events, superfloods, giant landslides triggering giant tsunamis, lake bursts, supereruptions, and mass extinctions. Sometimes these events are signal. The possible late Archaean catastrophic inversion of the crust is an example (p. 84).

Acceptance of the more speculative catastrophes would topple esteemed beliefs. If, for instance, tillites should prove to be impact ejecta, and not signs of glaciation, then theories of cold palaeoclimates would need radically revising. But such speculations, valuable though they be, tend to overshadow a simple but significant point – big catastrophes, small catastrophes, and gradual changes all contribute to environmental change.

Some catastrophes involve slow changes accumulating until a critical point, or threshold, is reached, and then a sudden change occurs. This happens every time a kettle of water is boiled, or a geyser sends forth a waterspout. In the ecosphere, it would mean that quasi-steady-state systems resist stress to a breaking point and then flip smartly to a new quasi-steady state (cf. Gould 1984). Catastrophes of this kind are creeping catastrophes (cf. Moisseyev 1988). They involve continual, gradual changes leading, at thresholds, to discrete jumps (catastrophic change). Herbert R. Shaw (1994) took this kind of argument several steps further. The Earth, Sun, and Solar System, he averred, are coupled non-linear systems. Their interactions produce an innately steady-state and uniformitarian regime – the solar and planetary system is stable and has been since it first formed. But this uniformitarian regime emerges, not from a simple mechanical system near equilibrium, but from a dissipative system far from equilibrium. And, it breeds gradual change *and* catastrophes – gradual and catastrophic change in the living and non-living worlds are different expressions of the same non-linear processes. This idea paints a stunning new picture of Earth history, in which happenstance has no place. It depicts big and small catastrophes as part of a highly co-ordinated non-linear system, and not as the outcomes of rare and random events. Christopher Zeeman (1992), looking at the effects of random small variations and natural selection in biology, came to a similar conclusion. These Darwinian mechanisms are normally thought to produce gradual evolutionary change. Using a catastrophe model, Zeeman found that these ordinary processes imply several global discontinuities, including punctuated equilibria, speciation, and multiple speciation (radiations) at punctuation points.

Gradual changes and catastrophes both occur. This contention is bolstered by observational evidence and by theory. The interesting question then becomes, not whether catastrophic processes do cause change, but how much change they do cause in comparison with slow-acting processes. This question was faced by geomorphologists considering magnitude and frequency relationships in fluvial systems (Wolman and Miller 1960). Do low-magnitude events running over long periods of time do more work than isolated and short-lived events of high magnitude? In fluvial systems, it seems that events of middling magnitude, and thus also of middling frequency, are generally dominant in sediment transport. But most of this work considers the extreme end of 'normal' events – the 100-year flood, say.

It does not consider the potential work done by ultra-high-magnitude superfloods, such as those resulting from lake bursts, giant tsunamis, or asteroid impacts (cf. Gretener 1967). It is not certain how much work these superfloods do. If speculations about erosional features in continental lowlands and tillites being impact ejecta should be true, then they would be forces of change to be reckoned with, dwarfing the effects of normal erosion and deposition.

Despite all this fast and slow activity, there is mounting evidence that a deal of the Earth's surface is very old and has escaped appreciable change for tens of millions of years. An explanation is that some parts of the landscape are more susceptible to change than others (cf. Brunsden and Thornes 1979). Mobile, fast-responding parts (rivers, some soils, and beaches) change readily. They quickly adopt new configurations, and act as focal points for landscape change. Relatively immobile, slowly responding parts (plateaux and interfluves, some soils, and weathering features) lie far from susceptible parts. This differential landscape susceptibility would permit fast-changing 'soft-spots' to exist alongside 'stagnant' areas. But it does not explain why some areas are stagnant. Weathering should construct regolith and erosional processes should destroy it on all exposed surfaces, though the balance between constructive and destructive forces would vary in different environments. There is no obvious answer to this geomorphological conundrum. However, a recent theoretical study showed that landscape stability depends upon time-lags between pedogenic processes, which act normal to hillslope, and geomorphic processes, which act tangentially to hillslopes (Phillips 1995c). When there is no lag between debris production and its availability for removal, regolith thickness at a point along a hillside displays chaotic dynamics. On the other hand, when a time-lag is present, regolith thickness is stable and non-chaotic. The emergence of landscape stability at broad scales may therefore result from time-lags in different processes. Where regolith production is slow, and erosion even slower, stagnation might occur.

TRENDS IN ENVIRONMENTAL CHANGE

Like the rate debate, the state debate over directions of environmental change has a long pedigree. In the modern era, it was set alight by Charles Lyell, who was adamant that, in both the organic and inorganic worlds, change follows an approximate steady state and leads in no preferred direction. There is, he avouched, a mean condition that is roughly invariant, though cycles about this mean do occur. Geological climate change was the centrepiece in his argument for this perverse view of Earth history, which was not accepted by many of his contemporaries. The material in this book describes several directional changes in Earth history. Atmospheric gases, solar luminosity, sediment thickness, area, and type, biodiversity, and geodiversity all display secular trends. The Gaia hypothesis resurrects a steady-state view, whilst allowing progression. The overwhelming conclusion from all modern work on past environments is that the ecosphere, and each of its components, has evolved. Very few readers, if any, would disagree with that statement, but some might wonder about the appropriateness of the word 'evolution' when talking about non-biological systems. This matter requires clarification.

Julian Huxley (1953) discerned three lines of evolution – inorganic evolution (cosmic and planetary evolution), biological evolution (the evolution of living things), and psychological evolution (the evolution of humans and their cultures), for which the term societary evolution might be more fitting. But in what sense do inorganic systems, such as planets, soils, and landforms, evolve? The first step in answering this question is to define evolution. The word 'evolution' is emotive and equivocal, bearing many connotations and nuances of meaning. It literally means 'unfolding' or 'unrolling'. It is understood in two chief ways (Mayr 1970). First, evolution means the unfolding, or growth and development, of an individual organism. This process of ontogeny involves homeorhesis (Waddington 1957). Homeorhesis is a set of processes leading to the development of an individual organism, from egg to adult. It operates in conjunction with homeostasis, the processes that maintain an individual in a steady state. Second, in a grander sense, evolution means phylogenetic evolution – the derivation of all life forms from a single common ancestor. There is a crucial difference between these two ideas. Development (homeorhesis) produces a new organism that is almost identical to its progenitors (or identical in the case of asexual reproduction). Phylogenetic evolution creates organisms that have never before existed, and that may be more complex than their progenitors. A process of complexification occurs in both cases. With development, complexification leads to a familiar, pre-existing organism. With phylogenetic evolution, complexification leads to a novel organism, often at some higher level of organization. These then, briefly, are the biological meanings of 'evolution'. This information should help to decide whether the terms 'evolution' and 'evolve' are applicable to non-biological systems.

A developmental view of landforms, soils, and communities has been inherited from the late nineteenth and early twentieth centuries. The developmental paradigm may be, ultimately, the gift of Charles Darwin. It was handed down through William Morris Davis, Vasilii V. Dokuchaev, and Frederic E. Clements. According to these celebrated people, non-biological systems (including plant communities) follow, time after time, predetermined developmental (homeorhetic) pathways. In geomorphology, for instance, landscape development is, according to Davis, a process of reduction. It always has progressed, and always will do, through the same developmental sequence – youth, maturity, and old age – to end with an all but level plain. In pedology, the traditional view (soil-formation theory) is that soil forms or develops progressively under the influence of the environmental state factors. The developmental sequence continues until the soil is in equilibrium with prevailing environmental conditions. Once equilibrium is achieved, the soil will be 'mature' (e.g. a podzol or chernozem) and will change no more. In ecology, vegetation develops through successive seral stages until it attains a 'mature' or climax state (e.g. steppe grassland and deciduous forest) that is in balance with environmental, and especially climatic, conditions. Plainly, according to the developmental view, landforms, soils, and communities will always change in predictable ways, following the uplift or exposure of new land. The sequence of change envisaged is homeorhetic in character, and ends with a sort of permanent homeostasis (in the mature or climax stage).

Recent empirical work on environmental change undermines the tenets of the developmental view, for it shows that inconstancy, not constancy, of environmental conditions is the norm. In view of this fact, it is improbable that a developmental sequence of landforms, soils, or vegetation will ever run its full course under a constant environment. To complicate matters further, it has been found profitable to regard all ecospheric systems (including the climate, communities, landforms, and soils) as dissipative structures replete with non-linear relations and forced away from equilibrium states by driving variables. This non-linearity in systems removed from equilibrium may generate chaotic regimes. After having entered a chaotic regime, landforms, soil, and communities are driven by internal dynamics and thresholds into a series of fundamentally unpredictable states. The nature of these states is strongly dependent on the initial conditions. This is the antithesis of the developmental view, in which the initial state is considered unimportant.

Environmental inconstancy and non-linear dynamics lead, then, to a far more dynamic picture of environmental change than early generations of scientists could have imagined. The systems of the ecosphere are generally plastic in nature and respond to changes in their environment and to thresholds within themselves. The result is that climates, communities, landscapes, and soils evolve, rather than develop (cf. Huggett 1995). Their genesis involves continual creation and destruction at all scales, and may progress or retrogress depending on the environmental circumstances. They do not seem inevitably to pursue a preordained developmental path. Rather, they constantly evolve, responding to continual changes in their internal, cosmic, and geological environments.

This evolutionary view of environmental change is of the utmost significance. It means that, at any instant, ecospheric systems are unique and changing, and are greatly influenced by historical events (owing to the relevance of initial conditions). This is the basis of 'evolutionary geomorphology' and 'evolutionary pedology'. An evolutionary view makes it very difficult to predict change. Landforms, soils, and communities formed under the same environmental constraints are likely to be broadly similar, but they will invariably differ in detail. In short, an evolutionary view offers a new way of thinking about and studying environmental change. Its main conclusion is simple: environmental stability is fleeting, environmental change is perduring.

BIBLIOGRAPHY

Aaby, B. (1976) 'Cyclic variations in climate over the past 5,500 years reflected in raised bogs', *Nature* 263: 281–5.

Adushkin, V. V. and Nemchinov, I. V. (1994) 'Consequences of impacts of cosmic bodies on the surface of the Earth', pp. 721–78 in T. Gehrels (ed.), with the editorial assistance of M. S. Matthews and A. M. Schumann, *Hazards Due to Comets and Asteroids*, Tucson and London: University of Arizona Press.

Ager, D. V. (1973) *The Nature of the Stratigraphical Record*, 1st edn, London: Macmillan.

Ager, D. V. (1993a) *The Nature of the Stratigraphical Record*, 3rd edn, Chichester: John Wiley & Sons.

Ager, D. V. (1993b) *The New Catastrophism: The Importance of the Rare Event in Geological History*, Cambridge: Cambridge University Press.

Agterberg, F. P. and Banerjee, I. (1969) 'Stochastic model for the deposition of varves in glacial Lake Barlow–Ojibway, Ontario, Canada', *Canadian Journal of Earth Science* 6: 625–52.

Aharon, P. (1984) 'Implications of the coral-reef theory record from New Guinea concerning the astronomical theory of ice ages', pp. 379–89 in A. Berger, J. Imbrie, J. Hays, G. Kukla, and B. Saltzman (eds) *Milankovitch and Climate: Understanding the Response of Astronomical Forcing, Part I* (Proceedings of the NATO Advanced Research Workshop on Milankovitch and Climate, Palisades, New York, 1982; NATO ASI Series C, Mathematical and Physical Sciences, vol. 126), Dordrecht: D. Reidel.

Aldiss, B. (1982) *Helliconia Spring*, London: Jonathan Cape.

Aldiss, B. (1983) *Helliconia Summer*, London: Jonathan Cape.

Aldiss, B. (1985) *Helliconia Winter*, London: Jonathan Cape.

Algeo, T. J. and Seslavinsky, K. B. (1995) 'The Paleozoic world: continental flooding, hypsometry, and sealevel', *American Journal of Science* 295: 787–822.

Allard, H. A. (1948) 'Length of day in climates of past geological eras and its possible effects upon the changes in plant life', pp. 101–19 in A. E. Murneek and R. O. Whyte (eds) *Vernalization and Photoperiodism*, Waltham, Massachusetts: Chronica Botanica Co.

Alvarez, L. W., Alvarez, W., Asaro, F., and Michel, H. V. (1980) 'Extraterrestrial cause for the Cretaceous–Tertiary extinction', *Science* 208: 1095–108.

Amit, R., Gerson, R., and Yaalon, D. H. (1993) 'Stages and rate of the gravel shattering process by salts in desert reg soils', *Geoderma* 57: 295–324.

Ampe, C. and Langohr, R. (1993) 'Distribution and dynamics of shrub roots in recent coastal dune valley ecosystems of Belgium', *Geoderma* 56: 37–55.

Anderson, A. T. (1975) 'Some basaltic and andesitic gases', *Reviews of Geophysics and Space Physics* 13: 37–55.

Anderson, D. L. (1990) 'Planet Earth', pp. 65–76 in J. Kelly Beatty and Andrew

Chaikin (eds) *The New Solar System*, 3rd edn, Introduction by Carl Sagan, Cambridge: Cambridge University Press.

Anderson, R. Y. (1984) 'Orbital forcing of evaporite sedimentation', pp. 147–62 in A. Berger, J. Imbrie, J. Hays, G. Kukla, and B. Saltzman (eds) *Milankovitch and Climate: Understanding the Response of Astronomical Forcing, Part I* (Proceedings of the NATO Advanced Research Workshop on Milankovitch and Climate, Palisades, New York, 1982; NATO ASI Series C, Mathematical and Physical Sciences, vol. 126), Dordrecht: D. Reidel.

Anderson, R. Y. (1986) 'The varve microcosm: propagator of cyclic bedding', *Paleoceanography* 1: 373–82.

Anderson, R. Y. (1991) 'Solar variability captured in climate and high-resolution paleoclimatic records: a geological perspective', pp. 543–61 in C. P. Sonett, M. S. Giampapa, and M. S. Matthews (eds) *The Sun in Time*, Tucson, Arizona: University of Arizona Press.

Anderson, R. Y. (1992) 'Possible connection between surface winds, solar activity and the Earth's magnetic field', *Nature* 358: 51–3.

Anderson, R. Y. and Dean, W. E. (1988) 'Lacustrine varve formation through time', *Palaeogeography, Palaeoclimatology, Palaeoecology* 62: 215–35.

Anderson, R. Y. and Kirkland, D. W. (1960) 'Origin, varves, and cycles of Jurassic Todilto Formation, New Mexico', *Bulletin of the American Association of Petroleum Geologists* 44: 37–52.

Arkell, W. J. (1947) *The Geology of the Country around Weymouth, Swanage, Corfe & Lulworth (*Natural Environment Research Council, Institute of Geological Sciences, Memoirs of the Geological Survey of England and Wales), London: Her Majesty's Stationery Office.

Armstrong, R. L. (1969) 'Control of sea level relative to the continents', *Nature* 221: 1042–3.

Asimov, I. (1980) *A Choice of Catastrophes: The Disasters that Threaten Our World*, London: Hutchinson.

Atwater, B. F. (1987) 'Evidence for great Holocene earthquakes along the outer coast of Washington State', *Science* 236: 942–4.

Atwater, B. F., Stuiver, M., and Yamaguchi, D. K. (1991) 'Radiocarbon test of earthquake magnitude at the Cascadia subduction zone', *Nature* 353: 156–8.

Badgley, C. and Behrensmeyer, A. K. (1995a) 'Two long geological records of continental ecosystems', *Palaeogeography, Palaeoclimatology, Palaeoecology* 115: 1–11.

Badgley, C. and Behrensmeyer, A. K. (1995b) 'Preservational, paleoecological and evolutionary patterns in the Paleogene of Wyoming–Montana and the Neogene of Pakistan', *Palaeogeography, Palaeoecology, Palaeoclimatology* 115: 319–40.

Baer, A. J. (1981) 'Geothems, evolution of the lithosphere and plate tectonics', *Tectonophysics* 72: 203–27.

Bahcall, J. N. and Bahcall, S. (1985) 'The Sun's motion perpendicular to the galactic plane', *Nature* 316: 706–8.

Bailey, M. E., Clube, S. V. M., Hahn, G., Napier, W. M., and Valsecchi, G. B. (1994) 'Hazards due to giant comets: climate and short-term catastrophism', pp. 479–533 in T. Gehrels (ed.), with the editorial assistance of M. S. Matthews and A. M. Schumann, *Hazards Due to Comets and Asteroids*, Tucson and London: University of Arizona Press.

Bailey, M. E., Clube, S. V. M., and Napier, W. M. (1990) *The Origin of Comets*, Oxford: Pergamon Press.

Bailey, R. G. (1995) *Description of the Ecoregions of the United States*, 2nd edn, revised and enlarged (Miscellaneous Publication no. 1391), Washington, DC: United States Department of Agriculture, Forest Service.

Bailey, R. G. (1996) *Ecosystem Geography*. With a foreword by Jack Ward Thomas, Chief, USDA Forest Service, New York: Springer.

Baillie, M. G. L. (1994) 'Dendrochronology raises questions about the nature of the AD 536 dust-veil event', *The Holocene* 4: 212–17.

Bailon, S. and Rage, J.-C. (1992) 'Amphibiens et reptiles du Quaternaire: relations avec l'homme', *Mémoires de la Société Géologique de France* 160: 95–100.

Bain, D. C., Mellor, A., Robertson-Rintoul, M. S. E., and Buckland, S. T. (1993) 'Variations in weathering processes and rates with time in a chronosequence of soils from Glen Feshie, Scotland', *Geoderma* 57: 275–93.

Baker, A., Smart, P. L., Barnes, W. L., Edwards, R. L., and Farrant, A. (1995) 'The Hekla 3 volcanic eruption recorded in a Scottish speleothem?', *The Holocene* 5: 336–42.

Baker, A., Smart, P. L., and Edwards, R. L. (1995) 'Paleoclimate implications of mass spectrometric dating of a British flowstone', *Geology* 23: 309–12.

Baker, A., Smart, P. L., Edwards, R. L., and Richards, D. A. (1993) 'Annual growth banding in a cave stalagmite', *Nature* 364: 518–20.

Baker, A., Smart, P. L., and Ford, D. C. (1993) 'Northwestern European palaeoclimate as indicated by growth frequency variations of secondary calcite deposits', *Palaeogeography, Palaeoclimatology, Palaeoecology* 100: 291–301.

Baker, V. R. (1973) 'Paleohydrology and sedimentology of Lake Missoula flooding in eastern Washington', *Geological Society of America Special Paper 144*.

Baker, V. R. (1977) 'Stream channel response to floods with examples from Texas', *Bulletin of the Geological Society of America* 88: 1057–71.

Baker, V. R. (1978a) 'The Spokane Flood controversy', pp. 3–15 in V. R. Baker and D. Nummedal (eds) *The Channeled Scabland: A Guide to the Geomorphology of the Columbia Basin, Washington*, Washington, DC: National Aeronautics and Space Administration.

Baker, V. R. (1978b) 'The Spokane Flood controversy and Martian outflow channels', *Science* 202: 1249–56.

Baker, V. R. (1983) 'Large-scale fluvial palaeohydrology', pp. 453–78 in K. J. Gregory (ed.) *Background to Palaeohydrology*, Chichester: John Wiley & Sons.

Bakker, R. T. (1986) *The Dinosaur Heresies: A Revolutionary View of Dinosaurs*, Harlow, Essex: Longman.

Baliunas, S. and Jastrow, R. (1993) 'Evidence of the climate impact of solar variations', *Energy* 18: 1285–95.

Barber, K. E. (1981) *Peat Stratigraphy and Climatic Change: A Palaeoecological Test of the Theory of Cyclic Peat Bog Regeneration*, Rotterdam: A. A. Balkema.

Bardet, N. (1994) 'Extinction events among Mesozoic marine reptiles', *Historical Biology* 7: 313–24.

Barnola, J. M., Raynaud, D., Korotkevich, Y. S., and Lorius, C. (1987) 'Vostok ice core provides 160,000-year record of atmospheric CO_2', *Nature* 329: 408–14.

Barrell, J. (1917) 'Rhythms and the measurement of geologic time', *Bulletin of the Geological Society of America* 28: 745–904.

Barron, E. J. (1984) 'Climatic implications of the variable obliquity explanation of Cretaceous–Paleogene high-latitude floras', *Geology* 12: 595–8.

Barron, E. J. (1989) 'Studies of Cretaceous climate', pp. 149–57 in A. Berger, R. E. Dickinson, and J. W. Kidson (eds) *Understanding Climate Change* (International Union of Geodesy and Geophysics, vol. 7; Geophysical Monograph 52), Washington, DC: American Geophysical Union.

Barry, J. C. and Flynn, L. L. (1990) 'Key biostratigraphic events in the Siwalik sequence', pp. 557–71 in E. H. Lindsay, V. Fahlbusch, and P. Mein (eds) *European Neogene Mammal Chronology*, New York: Plenum Press.

Barry, J. C., Johnson, N. M., Raza, S. M., and Jacobs, L. L. (1985) 'Neogene mammalian faunal change in southern Asia: correlations with climatic, tectonic, and eustatic events', *Geology* 13: 637–40.

Barry, R. G. (1969) 'Long-term precipitation trends', pp. 513–23 in R. J. Chorley (ed.) *Water, Earth, and Man: A Synthesis of Hydrology, Geomorphology, and Socio-Economic Geography*, London: Methuen.

Barry, R. G. and Chorley, R. J. (1992) *Atmosphere, Weather and Climate*, 6th edn, London: Routledge.

Battiau-Queney, Y. (1984) 'The pre-glacial evolution of Wales', *Earth Surface Processes and Landforms* 9: 229–52.

Battiau-Queney, Y. (1987) 'Tertiary inheritance in the present landscape of the British Isles', pp. 979–89 in V. Gardiner (ed.) *International Geomorphology 1986, Part II* (Proceedings of the First International Conference on Geomorphology), Chichester: John Wiley & Sons.

Battiau-Queney, Y. (1991) 'Les marges passives', *Bulletin de l'Association de Géographes Français 1991–2*: 91–100.

Battiau-Queney, Y. (1996) 'A tentative classification of paleoweathering formations based on geomorphological criteria', *Geomorphology* 16: 87–102.

Baumgartner, A. and Reichel, E. (1975) *The World Water Balance*, Amsterdam and Oxford: Elsevier.

Beatty, J. K. and Chaikin, C. (eds) (1990) *The New Solar System*, 3rd edn, Introduction by Carl Sagan, Cambridge: Cambridge University Press.

Begét, J. E. and Hawkins, D. B. (1989) 'Influence of orbital parameters on Pleistocene loess deposition in central Alaska', *Nature* 337: 151–3.

Bell, M. and Walker, M. J. C. (1992) *Late Quaternary Environmental Change: Physical and Human Perspectives*, Harlow, Essex: Longman.

Belyy, V. F. (1994) 'Compensatory movements and contraction – the main factors of tectogenesis which accompany magmatism', *Geology of the Pacific Ocean* 8: 1085–99.

Benkö, F. (1985) *Geological and Cosmogonic Cycles as Reflected by the New Law of Universal Cyclicity*, Budapest: Akadémiai Kiadó.

Bennett, K. D. (1996) *Evolution and Ecology: The Pace of Life*, Cambridge: Cambridge University Press.

Bennett, M. R., Doyle, P., and Mather, A. E. (1996) 'Dropstones: their origin and significance', *Palaeogeography, Palaeoclimatology, Palaeoecology* 121: 331–9.

Benton, M. J. (1985) 'Mass extinction among non-marine tetrapods', *Nature* 316: 811–14.

Benton, M. J. (1990) 'The causes of the diversification of life', pp. 409–30 in P. D. Taylor and G. P. Larwood (eds) *Major Evolutionary Radiations* (Systematics Association, Special Volume 42), Oxford: Oxford University Press.

Benton, M. J. (1994) 'Palaeontological data and identifying mass extinctions', *Trends in Ecology and Evolution* 9: 181–5.

Berger, A. and Loutre, M. F. (1994) 'Astronomical forcing through geological time', pp. 15–24 in P. L. De Boer and D. G. Smith (eds) *Orbital Forcing and Cyclic Sequences* (Special Publication Number 19 of the International Association of Sedimentologists), Oxford: Blackwell Scientific Publications.

Berger, A., Loutre, M. F., and Laskar, J. (1992) 'Stability of the astronomical frequencies over the Earth's history for paleoclimate studies', *Science* 255: 560–6.

Bergeron, T. (1928) 'Über die dreidimensional verknüpfende Wetteranalyse', *Geofysiske Publikationer* 5, no. 6 (Oslo).

Berglund, B. E., Birks, H. J. B., Ralska-Jasiewiczowa, M., and Wright, H. E., Jr (eds) (1996) *Palaeoecological Events During the Last 15,000 Years: Regional Syntheses of Palaeoecological Studies in Lakes and Mires in Europe*, Chichester: John Wiley & Sons.

Berner, R. A. (1994) '3GEOCARB II: a revised model of atmospheric CO_2 over Phanerozoic time', *American Journal of Science* 294: 56–91.

Birchfield, G. E. and Ghil, M. (1993) 'Climate evolution in the Pliocene and Pleistocene from marine-sediment records and simulations: internal variability versus orbital forcing', *Journal of Geophysical Research* 98D6: 10,385–99.

Birks, H. H. (1993) 'The importance of plant macrofossils in late-glacial climatic reconstructions', *Quaternary Science Reviews* 12: 719–26.

Bishop, C. T. (1990) 'Historical variation of water levels in Lakes Erie and Michigan–Huron', *Journal of Great Lakes Research* 16: 406–25.

Bishop, P., Young, R. W., and McDougall, I. (1985) 'Stream profile change and long term landscape evolution: Early Miocene and modern rivers of the east Australian highland crest, central New South Wales, Australia', *Journal of Geology* 93: 455–74.

Bjerknes, J. (1969) 'Atmospheric teleconnections from the equatorial Pacific', *Monthly Weather Review* 97: 163–72.

Bloemendal, J. and DeMenocal, P. (1989) 'Evidence for a change in the periodicity of tropical climate cycles at 2.4 Myr from whole-core magnetic susceptibility measurements', *Nature* 342: 897–900.

Bloom, A. L. and Yonekura, N. (1985) 'Coastal terraces generated by sea-level change and tectonic uplift', pp. 139–54 in M. J. Woldenberg (ed.) *Models in Geomorphology* (The Binghamton Symposia in Geomorphology, International Series, no. 14), Boston: Allen & Unwin.

Bloom, A. L. and Yonekura, N. (1990) 'Graphic analysis of dislocated Quaternary shorelines', pp. 104–54 in Geophysics Study Commission (eds) *Sea-Level Change*, Washington, DC: National Academy Press.

Board on Global Change, Commission on Geosciences, Environment, and Resources, National Research Council (1994) *Solar Influences on Global Change*, Washington, DC: National Academy Press.

Bond, G. (C.), Broecker, W., Johnsen, S., McManus, J., Labeyrie, L., Jouzel, J., and Bonani, G. (1993) 'Correlations between climate records from North Atlantic sediments and Greenland ice', *Nature* 365: 143–7.

Bond, G. C., Devlin, W. J., Kominz, M. A., Beaven, J., and McManus, J. (1993) 'Evidence of astronomical forcing of Earth's climate in Cretaceous and Cambrian time', *Tectonophysics* 222: 295–315.

Boomer, I. (1993) 'Palaeoenvironmental indicators from late Holocene and contemporary Ostracoda of the Aral Sea', *Palaeogeography, Palaeoclimatology, Palaeoecology* 103: 141–53.

Bosák, P., Ford, D. C., Głazek, J., and Horáček, I. (eds) (1989) *Paleokarst: A Systematic and Regional Review* (Developments in Earth Surface Processes 1), Amsterdam: Elsevier.

Boston, P. J. and Thompson, S. L. (1991) 'Theoretical microbial and vegetational control of planetary environments', pp. 99–117 in S. H. Schneider and P. J. Boston (eds) *Scientists on Gaia*, Cambridge, Massachusetts and London: MIT Press.

Botkin, D. B. (1990) *Discordant Harmonies: A New Ecology for the Twenty-First Century*, New York: Oxford University Press.

Botkin, D. B. and Keller, E. A. (1995) *Environmental Science: Earth as a Living Planet*. New York: John Wiley & Sons.

Bourgeois, J., Hansen, T. A., Wiberg, P. L., and Kauffman, E. G. (1988) 'A tsunami deposit at the Cretaceous–Tertiary boundary in Texas', *Science* 214: 567–70.

Box, E. O. and Meentemeyer, V. (1991) 'Geographic modeling and modern ecology', pp. 773–804 in G. Esser and D. Overdieck (eds) *Modern Ecology: Basic and Applied Aspects*, Amsterdam: Elsevier.

Bradley, R. S. (1991) 'Pre-industrial climate: how has climate varied during the past 500 years?', in M. E. Schlesinger (ed.) *Greenhouse-Gas-Induced Climatic Change* (Developments in Atmospheric Science, vol. 19), Amsterdam: Elsevier.

Bradley, R. S. and Jones, P. D. (1992) *Climate since AD 1500*, London and New York: Routledge.

Brand, U. (1989) 'Global climatic changes during the Devonian–Mississippian: stable isotope biogeochemistry of brachiopods', *Palaeogeography, Palaeoclimatology, Palaeoecology (Global and Planetary Change Section)* 75: 311–29.

Brandt, D. S. and Elias, R. J. (1989) 'Temporal variations in tempestite thickness may be a geologic record of atmospheric CO_2', *Geology* 17: 951–2.

Brantley, S. L., Agustsdottir, A. M., and Rowe, G. L. (1993) 'Crater lakes reveal volcanic heat and volatile fluxes', *GSA–Today* 3: 173 and 176–8.

Brasier, M. D., Corfield, R. M., Derry, L. A., Rozanov, A. Yu., and Zhuravlev, A. Yu. (1994) 'Multiple $\delta^{13}C$ excursions spanning the Cambrian explosion to the Botomian crisis in Siberia', *Geology* 22: 455–8.

Bray, J. R. (1974) 'Volcanism and glaciation during the past 40 millennia', *Nature* 252: 679–80.

Bray, J. R. (1977) 'Volcanic dust veils and north Atlantic climatic change', *Nature* 268: 616–17.

Bremer, H. (1988) *Allgemeine Geomorphologie: Methodik – Grundvorstellungen – Ausblick auf den Landschaftshaushalt*, Berlin: Gebrüder Bornträger.

Bretz, J H. (1923) 'The Channeled Scabland of the Columbia Plateau', *Journal of Geology* 31: 617–77.

Bretz, J. H. (1978) 'Introduction', pp. 1–2 in V. R. Baker and D. Nummedal (eds) *The Channeled Scabland. A Guide to the Geomorphology of the Columbia Basin, Washington*, Washington, DC: National Aeronautics and Space Administration.

Brewer, R., Crook, A. W., and Speight, J. A. (1970) 'Proposal for soil–stratigraphic units in the Australian stratigraphic code', *Journal of the Geological Society of Australia* 17: 103–111.

Broecker, W. S. (1965) 'Isotope geochemistry and the Pleistocene climatic record', pp. 737–53 in H. E. Wright Jr and D. G. Frey (eds) *The Quaternary of the United States*, Princeton, New Jersey: Princeton University Press.

Broecker, W. S. (1995) 'Chaotic climate', *Scientific American* 273: 44–50.

Broecker, W. S. and Denton, G. H. (1990) 'What drives glacial cycles?', *Scientific American* 262: 42–50.

Broecker, W. S., Thurber, D. L., Goddard, J., Ku, T., Matthews, R. K., and Mesolella, K. J. (1968) 'Milankovitch hypothesis supported by precise dating of coral reefs and deep-sea sediments', *Science* 159: 1–4.

Brown, E. H. and Waters, R. S. (1974) 'Geomorphology in the United Kingdom since the First World War', pp. 3–9 in E. H. Brown and R. S. Waters (eds) *Progress in Geomorphology: Papers in Honour of David L. Linton* (Institute of British Geographers Special Publication no. 7), London: Institute of British Geographers.

Brown, G., Hawkesworth, C., and Wilson, C. (1992) *Understanding the Earth*, Cambridge: Cambridge University Press.

Brune, J. N. (1969) 'Surface waves and crustal structure', *American Geophysical Union Monograph* 13: 230–42.

Brunsden, D. and Kesel, R. H. (1973) 'The evolution of a Mississippi river bluff in historic time', *Journal of Geology* 81: 576–97.

Brunsden, D. and Thornes, J. B. (1979) 'Landscape sensitivity and change', *Transactions of the Institute of British Geographers* NS 4: 463–84.

Bryan, W. H. and Teakle, L. J. H. (1949) 'Pedogenic inertia – a concept in soil science', *Nature* 164: 969.

Büdel, J. (1957) 'Die "Doppelten Einebnungsflächen" in den feuchten Tropen', *Zeitschrift für Geomorphologie* NF 1: 201–28.

Büdel, J. (1982) *Climatic Geomorphology* (Translated by Lenore Fischer and Detlef Busche), Princeton, New Jersey: Princeton University Press.

Buggisch, W. (1991) 'The global Frasnian–Famennian "Kellwasser event"', *Geologische Rundschau* 80: 49–72.

Bull, W. B. (1992) *Geomorphic Responses to Climatic Change*, Oxford: Oxford University Press.

Buol, S. W., Hole, F. D., and McCracken, R. J. (1980) *Soil Genesis and Classification*, 2nd edn, Ames, Iowa: Iowa State University Press.

Burbank, D. W. and Beck, R. A. (1991) 'Rapid, long-term rates of denudation', *Geology* 19: 1169–72.

Butler, B. E. (1959) *Periodic Phenomena in Landscapes as a Basis for Soil Studies* (CSIRO Soil Publication 14), Canberra: CSIRO.

Butler, B. E. (1967) 'Soil periodicity in relation to landform development in south-eastern Australia', pp. 231–55 in J. Jennings and J. A. Marbut (eds) *Landform Studies from Australia and New Guinea*, Cambridge: Cambridge University Press.

Butler, B. E. (1982) 'A new system for soil studies', *Journal of Soil Science* 33: 581–95.

Butler, D. R. (1995) *Zoogeomorphology: Animals as Geomorphic Agents*, Cambridge: Cambridge University Press.

Butzer, K. W. (1962) 'Coastal geomorphology of Majorca', *Annals of the Association of American Geographers* 52: 191–212.

Butzer, K. W. (1975) 'Pleistocene littoral-sedimentary cycles of the Mediterranean basin: a Mallorquin view', pp. 25–71 in K. W. Butzer and G. L. L. Isaacs (eds) *After the Australopithecines*, The Hague: Mouton.

Butzer, K. W. (1976) *Geomorphology from the Earth*, New York: Harper & Row.

Caldeira, K. and Kasting, J. F. (1992) 'Possible irreversible glaciation of the early Earth caused by CO_2 clouds', *Nature* 359: 226–30.

Caldeira, K. and Rampino, M. R. (1991) 'The mid-Cretaceous super plume, carbon dioxide, and global warming', *Geophysical Research Letters* 18: 987–90.

Cameron, A. G. W. (1988) 'Origin of the Solar System', *Annual Review of Astronomy and Astrophysics* 26: 441–72.

Campbell, W. H., Blechman, J. B., and Bryson, R. A. (1983) 'Long-period tidal forcing of Indian monsoon rainfall: an hypothesis', *Journal of Climate and Applied Meteorology* 22: 289–96.

Carey, S. W. (1958) 'The tectonic approach to continental drift', pp. 177–355 in S. W. Carey (ed.) *Continental Drift – A Symposium*, Hobart, Tasmania: University of Tasmania.

Carey, S. W. (1963) 'The asymmetry of the Earth', *Australian Journal of Science* 25: 369–84, 479–88.

Carey, S. W. (1976) *The Expanding Earth*, Amsterdam: Elsevier.

Carey, S. W. (1988) *Theories of the Earth: A History of Dogma in the Earth Sciences*, Stanford, California: Stanford University Press.

Catt, J. A. (1989) 'Relict properties in soils of the central and north-west European temperate region', pp. 41–58 in A. Bronger and J. A. Catt (eds) *Paleopedology – Nature and Applications of Paleosols* (Catena Supplement 16), Cremlingen: Catena.

Cerling, T. E. (1992) 'Development of grasslands and savannahs in East Africa during the Neogene', *Palaeogeography, Palaeoclimatology, Palaeoecology* 97: 241–7.

Chadwick, O. A., Brimhall, G. H., and Hendricks, D. M. (1990) 'From a black to a gray box – a mass balance interpretation of pedogenesis', *Geomorphology* 3: 369–90.

Chahine, M. T. (1992a) 'GEWEX: the global energy and water cycle experiment', *Eos* 73: 9, 13–14.

Chahine, M. T. (1992b) 'The hydrological cycle and its influence on climate', *Nature* 359: 373–80.

Chan, M. A., Kvale, E. P., Archer, A. W., and Sonett, C. P. (1994) 'Oldest direct evidence of lunar–solar tidal forcing encoded in sedimentary rhythmites, Proterozoic Big Cottonwood Formation, central Utah', *Geology* 22: 791–4.

Chapman, C. R. and Morrison, D. (1994) 'Impacts on the Earth by asteroids and comets: assessing the hazard', *Nature* 367: 33–40.

Chesworth, W. (1982) 'Late Cenozoic geology and the second oldest profession', *Geoscience Canada* 9: 54–61.

Chesworth, W. (1992) 'Weathering systems', pp. 10–40 in I. P. Martini and W. Chesworth (eds) *Weathering, Soils and Paleosols* (Developments in Earth Surface Processes 2), Amsterdam: Elsevier.

Chorley, R. J. (1965) 'The application of quantitative methods to geomorphology', pp. 147–63 in R. J. Chorley and P. Haggett (eds) *Frontiers in Geographical Teaching: The Madingley Lectures for 1963*, London: Methuen.

Chyba, C. F. (1987) 'The cometary contribution to the oceans of primitive Earth', *Nature* 330: 632–5.

Chyba, C. F. (1990a) 'Impact delivery and erosion of planetary oceans in the early inner Solar System', *Nature* 343: 129–33.

Chyba, C. F. (1990b) 'Extraterrestrial amino acids and terrestrial life', *Nature* 348: 113–14.

Chyba, C. F. and McDonald, G. D. (1995) 'The origin of life in the Solar System: current issues', *Annual Review of Earth and Planetary Science* 23: 215–49.

Chyba, C. F. and Sagan, C. (1992) 'Endogenous production, exogenous delivery and impact-shock synthesis of organic molecules: an inventory for the origins of life', *Nature* 355: 125–32.

Chyba, C. F., Thomas, P. J., Brookshaw, L., and Sagan, C. (1990) 'Cometary delivery of organic molecules to the early Earth', *Science* 249: 366–73.

Claeys, P. and Casier, J. G. (1994) 'Microtektite-like impact glass associated with the Frasnian–Famennian boundary mass extinction', *Earth and Planetary Science Letters* 122: 303–15.

Claps, M. and Masetti, D. (1994) 'Milankovitch periodicities recorded in Cretaceous deep-sea sequences from the Southern Alps (Northern Italy)', pp. 99–107 in P. L. De Boer and D. G. Smith (eds) *Orbital Forcing and Cyclic Sequences* (Special Publication Number 19 of the International Association of Sedimentologists), Oxford: Blackwell Scientific Publications.

Claps, M., Masetti, D., Pedrielli, F., and Garavello, A. L. (1991) 'Analisi spettrale e cicli di Milankovitch in successioni Cretaciche del Sudalpino orientale', *Rivista Italiana di Paleontologia e Stratigrafia* 97: 153–74.

Clemens, S. C., Farrell, J. W., and Gromet, L. P. (1993) 'Synchronous changes in seawater strontium isotope composition and global climate', *Nature* 363: 607–10.

Clemens, S. (C.), Prell, W., Murray, D., Shimmield, G., and Weedon, G. (1991) 'Forcing mechanisms of the Indian Ocean monsoon', *Nature* 353: 720–5.

Clements, F. E. (1916) *Plant Succession: An Analysis of the Development of Vegetation* (Carnegie Institute of Washington, Publication no. 242), Washington, DC: Carnegie Institute of Washington.

Clements, F. E. and Shelford, V. E. (1939) *Bio-Ecology*, New York: John Wiley & Sons.

Clifford, N. J. (1993) 'Formation of riffle–pool sequences: field evidence for an autogenic process', *Sedimentary Geology* 85: 39–51.

Clifton, H. E. (ed.) (1988) *Sedimentologic Consequences of Convulsive Geologic Events* (Geological Society of America, Special Paper 229), Boulder, Colorado: Geological Society of America.

Clube, S. V. M. (1978) 'Does our Galaxy have a violent history?', *Vistas in Astronomy* 22: 77–118.

Clube, S. V. M. and Napier, W. M. (1982) 'The role of episodic bombardment in geophysics', *Earth and Planetary Science Letters* 57: 251–62.

Clube, S. V. M. and Napier, W. M. (1984) 'The microstructure of terrestrial catastrophism', *Monthly Notices of the Royal Astronomical Society* 211: 953–68.

Cocks, L. R. M. and Parker, A. (1981) 'The evolution of sedimentary environments', pp. 47–62 in L. R. M. Cocks (ed.) *The Evolving Earth*, Cambridge: Cambridge University Press; London: British Museum (Natural History).

Condie, K. C. (1980) 'Origin and development of the Earth's crust', *Precambrian Research* 11: 183–97.

Condie, K. C. (1986) 'Origin and early growth rate of continents', *Precambrian Research* 32: 261–78.

Condie, K. C. (1989a) *Plate Tectonics and Crustal Evolution*, 3rd edn, Oxford: Pergamon Press.

Condie, K. C. (1989b) 'Origin of the Earth's crust', *Palaeogeography, Palaeoclimatology, Palaeoecology (Global and Planetary Change Section)* 75: 57–81.

Condie, K. C. (1992) 'Introduction', pp. 1–6 in K. C. Condie (ed.) *Proterozoic Crustal Evolution* (Developments in Precambrian Geology, vol. 10), Amsterdam: Elsevier.

Condie, K. C. (ed.) (1994) *Archean Crustal Evolution* (Developments in Precambrian Geology, vol. 11), Amsterdam: Elsevier.

Coope, G. R. (1965) 'Fossil insect faunas from Late Quaternary deposits in Britain', *Advancement of Science. London* 21: 564–75.

Coope, G. R. (1994) 'The response of insect faunas to glacial–interglacial climatic fluctuations', *Philosophical Transactions of the Royal Society of London* 344B: 19–26.

Coplen, T. B., Winograd, I. J., Landwehr, J. M., and Riggs, A. C. (1994) '500,000-year stable carbon isotopic record from Devils Hole, Nevada', *Science* 263: 361–5.

Copper, P. (1986) 'Frasnian/Famennian mass extinction and cold-water oceans', *Geology* 14: 835–9.

Copper, P. (1994) 'Ancient reef ecosystem expansion and collapse', *Coral Reefs* 13: 3–11.

Courtillot, V. (1990) 'Deccan volcanism at the Cretaceous–Tertiary boundary: past climatic crises as a key to the future?', *Palaeogeography, Palaeoclimatology, Palaeoecology (Global and Planetary Change Section)* 189: 291–9.

Courtillot, V. and Besse, J. (1987) 'Magnetic field reversals, polar wander, and core–mantle coupling', *Science* 237: 1140–7.

Covey, C., Ghan, S. J., Walton, J. J., and Weissman, P. R. (1990) 'Global environmental effects of impact-generated aerosols; results from a general circulation model', pp. 263–70 in V. L. Sharpton and P. D. Ward (eds) *Global Catastrophes in Earth History: An Interdisciplinary Conference on Impacts, Volcanism, and Mass Mortality* (Geological Society of America, Special Paper 247), Boulder, Colorado: Geological Society of America.

Cowles, H. C. (1899) 'The ecological relations of the vegetation on the sand dunes of northeastern Minnesota', *Botanical Gazette* 27: 95–117, 167–202, 281–308, 361–91.

Cox, K. G. (1989) 'The role of mantle plumes in the development of continental drainage patterns', *Nature* 342: 873–6.

Cracraft, J. (1985) 'Biological diversification and its causes', *Annals of the Missouri Botanical Gardens* 72: 794–822.

Creemans, D. L., Brown, R. B., and Huddleston, J. H. (eds) (1994) *Whole Regolith Pedology* (Soil Science Society of America Special Publication 34), Madison, Wisconsin: Soil Science Society of America.

Crickmay, C. H. (1975) 'The hypothesis of unequal activity', pp. 103–9 in W. N. Melhorn and R. C. Flemal (eds) *Theories of Landform Development*, London: George Allen & Unwin.

Crocker, R. L. and Major, J. (1955) 'Soil development in relation to vegetation and surface age at Glacier Bay, Alaska', *Journal of Ecology* 43: 427–48.

Croll, J. (1875) *Climate and Time in Their Geological Relations: A Theory of Secular Changes of the Earth's Climate*, London: Daldy, Isbister & Co.

Cronin, J. R. (1989) 'Amino acids and bolide impacts', *Nature* 339: 423–4.

Crossley, D. (1993) 'The Earth's core', *Geoscience Canada* 29: 100–13.

Crough, S. T. (1983) 'Hotspot swells', *Annual Reviews of Earth and Planetary Science* 11: 165–93.

Crowell, J. C. (1978) 'Gondwana glaciation, cyclothems, continental positioning and climate change', *American Journal of Science* 278: 1345–72.

Currie, R. G. (1976) 'The spectrum of sea level from 4 to 40 years', *Geophysical Journal of the Royal Astronomical Society* 46: 513–20.

Currie, R. G. (1981) 'Evidence for 18.6-year signal in temperature and drought condition in North America since AD 1800', *Journal of Geophysical Research* 86: 11,055–64.

Currie, R. G. (1983) 'Detection of 18.6-year lunar nodal induced drought in the Patagonian Andes', *Geophysical Research Letters* 10: 1089–92.

Currie, R. G. (1984) 'On bistable phasing of 18.6-year induced flood in India', *Geophysical Research Letters* 11: 50–3.

Currie, R. G. (1987) 'On bistable phasing of 18.6-year induced drought and the flood in the Nile records since AD 650', *Journal of Climatology* 7: 373–89.

Currie, R. G. (1991a) 'Luni-solar 18.6-year signal in tree-rings from Argentina and Chile', *Pure and Applied Geophysics* 137: 281–300.

Currie, R. G. (1991b) 'Deterministic signals in tree-rings from North America', *International Journal of Climatology* 11: 861–76.

Currie, R. G. (1991c) 'Deterministic signals in tree-rings from the Corn Belt region', *Annales Geophysicae* 9: 565–70.

Currie, R. G. (1991d) 'Deterministic signals in tree-rings from Tasmania, New Zealand, and South Africa', *Annales Geophysicae* 9: 71–81.

Currie, R. G. (1992a) 'Deterministic signals in tree-rings from Europe', *Annales Geophysicae* 10: 241–53.

Currie, R. G. (1992b) 'Deterministic signals in height of sea level worldwide', pp. 403–21 in C. R. Smith *et al.* (eds) *Maximum Entropy Analysis and Bayesian Methods*, Seattle, 1991, Dordrecht: Kluwer Academic Publishers.

Currie, R. G. (1993a) 'Luni-solar 18.6- and solar cycle 10–11-year signals in USA air temperature records', *International Journal of Climatology* 13: 31–50.

Currie, R. G. (1993b) 'Luni-solar 18.6- and 10–11-year solar cycle signals in South African rainfall', *International Journal of Climatology* 13: 237–56.

Currie, R. G. (1994) 'Luni-solar 18.6- and 10–11-year solar cycle signals in H. H. Lamb's dust veil index', *International Journal of Climatology* 14: 215–26.

Currie, R. G. and Fairbridge, R. W. (1985) 'Periodic 18.6-year and cyclic 11-year induced drought and flood in northeastern China and some global implications', *Quaternary Science Reviews* 4: 109–34.

Currie, R. G. and O'Brien, D. P. (1990) 'Deterministic signals in USA precipitation records: Part I', *International Journal of Climatology* 10: 795–818.

Currie, R. G. and O'Brien, D. P. (1992) 'Deterministic signals in USA precipitation records: Part II', *International Journal of Climatology* 12: 281–304.

Currie, R. G., Wyatt, T., and O'Brien, D. P. (1993) 'Deterministic signals in European fish catches, wine harvests, and sea level, and further experiments', *International Journal of Climatology* 13: 665–87.

Dalrymple, J. B., Blong, R. J., and Conacher, A. J. (1968) 'A hypothetical nine-unit landsurface model', *Zeitschrift für Geomorphologie* NF 12: 60–76.

Daly, M., Behrends, P. R., Wilson, M. I., and Jacobs, L. F. (1992) 'Behavioural modulation of predation risk: moonlight avoidance and crepuscular compensation in a nocturnal desert rodent, *Dipodomys merriami*', *Animal Behaviour* 44: 1–9.

Dalziel, I. W. D. (1991) 'Pacific margins of Laurentia and East Antarctica–Australia as a conjugate rift pair: evidence and implications for an Eocambrian supercontinent', *Geology* 19: 598–601.

Damon, P. E. and Jirikowic, J. L. (1992) 'The Sun as a low-frequency harmonic oscillator', *Radiocarbon* 34: 199–205.

Dan, J. and Yaalon, D. H. (1968) 'Pedomorphic forms and pedomorphic surfaces', *Transactions of the Ninth International Congress of Soil Science, Adelaide* 4: 577–84.

Daniels, R. B. and Hammer, R. D. (1992) *Soil Geomorphology*, New York: John Wiley & Sons.

Darwin, C. R. (1881) *The Formation of Vegetable Mould through the Action of Worms, with Observations on Their Habits*, London: John Murray.

Davasse, B. and Galop, D. (1990) 'Le paysage forestier du haut Vicdessos (Ariège): l'évolution d'un milieu anthropise', *Revue Géographique des Pyrénées et du Sud Ouest* 61: 433–57.

Davies, G. F. (1977) 'Whole mantle convection and plate tectonics', *Geophysic Journal of the Royal Astronomical Society* 49: 459–86.

Davies, G. F. (1992) 'On the emergence of plate tectonics', *Geology* 20: 963–6.

Davis, M., Hut, P., and Muller, R. A. (1984) 'Extinction of species by periodic comet showers', *Nature* 308: 715–17.

Davis, W. M. (1889) 'The rivers and valleys of Pennsylvania', *National Geographic Magazine* 1: 183–253. (Also in *Geographical Essays.*)

Davis, W. M. (1899) 'The geographical cycle', *Geographical Journal* 14: 481–504. (Also in *Geographical Essays.*)

Davis, W. M. (1909) *Geographical Essays*, Boston, Massachusetts: Ginn.

Dawson, A. G. (1994) 'Geomorphological effects of tsunami run-up and backwash', *Geomorphology* 10: 83–94.

Dawson, A. G., Long, D., and Smith, D. E. (1988) 'The Storegga Slides: evidence from eastern Scotland for a possible tsunami', *Marine Geology* 82: 271–6.

Dawson, A. G., Long, D., Smith, D. E., Shi, S., and Foster, I. D. L. (1993) 'Tsunamis in the Norwegian Sea and North Sea caused by the Storegga submarine landslides', pp. 31–42 in S. Tinti (ed.) *Tsunamis of the World* (Fifteenth International Tsunami Symposium, 1991), Dordrecht: Kluwer Academic Publishers.

Dearnley, R. (1965) 'Orogenic fold-belts and continental drift', *Nature* 206: 1083–7.

Deepak, A. (ed.) (1983) *Atmospheric Effects and Potential Climatic Impact of the 1980 Eruptions of Mount St. Helens* (NASA Conference Publication 2240), Washington, DC: National Aeronautics and Space Administration.

Delcourt, H. R. and Delcourt, P. A. (1991) *Quaternary Ecology: A Paleoecological Perspective*, London: Chapman & Hall.

Delcourt, H. R. and Delcourt, P. A. (1994) 'Postglacial rise and decline of *Ostrya virginiana* (Mill.) K. Koch and *Carpinus caroliniana* Walt. in eastern North

America: predictable responses of forest species to cyclic changes in seasonality of climate', *Journal of Biogeography* 21: 137–50.

Delcourt, P. A. and Delcourt, H. R. (1987) *Long-term Forest Dynamics of the Temperate Zone: A Case Study of Late-Quaternary Forests in Eastern North America* (Ecological Studies 63), New York: Springer.

Dersch, M. and Stein, R. (1994) 'Late Cenozoic records of eolian quartz flux in the Sea of Japan (ODP Leg 128, Sites 798 and 799)', *Palaeogeography, Palaeoclimatology, Palaeoecology* 108: 523–35.

Desponts, M. and Payette, S. (1993) 'The Holocene dynamics of jack pine at its northern range limit in Quebec', *Journal of Ecology* 81: 719–27.

DeWeaver, E. and Nigam, S. (1995) 'Influence of mountain ranges on the mid-latitude atmospheric response to El Niño events', *Nature* 378: 706–8.

Dewey, H. (1916) 'On the origin of some river-gorges in Cornwall and Devon', *Quarterly Journal of the Geological Society, London* 72: 63–76.

Dewey, J. F. and Spall, H. (1975) 'Pre-Mesozoic plate tectonics', *Geology* 3: 422–4.

Dietz, R. S. (1947) 'Meteorite impact suggested by orientation of shatter cones at the Kentland, Indiana, disturbance', *Science* 105: 42–3.

Dietz, R. S. (1961) 'Astroblemes', *Scientific American* 205: 50–8.

DiMichele, W. A. and Nelson, W. J. (1989) 'Small-scale spatial heterogeneity in Pennsylvanian-age vegetation from the Roof Shale of the Springfield Coal (Illinois Basin)', *Palaios* 4: 276–80.

DiMichele, W. A. and Phillips, T. L. (1994) 'Paleobotanical and paleoecological constraints on models of peat formation in the Late Carboniferous of Euramerica', *Palaeogeography, Palaeoclimatology, Palaeoecology* 106: 39–90.

Donn, W. L. (1987) 'Terrestrial climate change from the Triassic to Recent', pp. 343–52 in M. R. Rampino, J. E. Sanders, W. S. Newman, and L. K. Königsson (eds) *Climate: History, Periodicity, and Predictability*, New York: Van Nostrand Reinhold.

Donovan, S. K. (1987) 'Mass extinctions. How sudden is sudden?', *Nature* 328: 109.

Douglas, I. (1987) 'Plate tectonics, palaeoenvironments and limestone geomorphology in west-central Britain', *Earth Surface Processes and Landforms* 12: 481–95.

Douglas, J. G. and Williams, G. E. (1982) 'Southern polar forests: the Early Cretaceous floras of Victoria and their paleoclimatic significance', *Palaeogeography, Palaeoclimatology, Palaeoecology* 39: 171–85.

Dowdeswell, J. A., Maslin, M. A., Andrews, J. T., and McCave, I. N. (1995) 'Iceberg production, debris rafting, and the extent and thickness of Heinrich layers (H-1, H-2) in North Atlantic sediments', *Geology* 23: 301–4.

Dowdeswell, J. A. and White, J. W. C. (1995) 'Greenland ice core records and rapid climate change', *Philosophical Transactions of the Royal Society of London* 352A: 359–71.

Drake, J. A. (1990) 'The mechanics of community assembly and succession', *Journal of Theoretical Biology* 147: 213-33.

Drury, W. H. and Nisbet, I. C. T. (1973) 'Succession', *Journal of the Arnold Arboretum* 54: 331–68.

Du, P.-R. (1994) '18.6 years seismic cycle and the preliminary exploration for its cause', *Acta Geophysica Sinica* 37: 362–9.

Duchaufour, P. (1982) *Pedology: Pedogenesis and Classification* (Translated by T. R. Paton), London: George Allen & Unwin.

Duff, B. A. and Langworthy, A. P. (1974) 'Orogenic zones in central Australia: intraplate tectonics?', *Nature* 249: 645–7.

Duffield, W. A. (1972) 'A naturally occurring model of global plate tectonics', *Journal of Geophysical Research* 77: 2543–55.

Dunn, G. E. (1940) 'Cyclogenesis in the tropical Atlantic', *Bulletin of the American Meteorological Society* 21: 215–29.
Dury, G. H. (1953) 'The shrinkage of the Warwickshire Itchen', *Proceedings of the Coventry Natural History and Science Society* 2: 208–14.
Dury, G. H. (1971) 'Relict deep weathering and duricrusting in relation to the palaeoenvironments of middle latitudes', *Geographical Journal* 137: 511–22.
Dutilleul, P. and Till, C. (1992) 'Evidence of periodicities related to climate and planetary behaviors in ring-width chronologies of Atlas cedar (*Cedrus atlantica*) in Morocco', *Canadian Journal of Forest Research* 22: 1469–82.
Eddy, J. A. (1977a) 'Anomalous solar radiation during the seventeenth century', *Science* 198: 824–9.
Eddy, J. A. (1977b) 'The case of the missing sunspots', *Scientific American* 236: 80–92.
Eddy, J. A. (1977c) 'Climate and the changing Sun', *Climatic Change* 1: 173–90.
Eddy, J. A. (1983) 'The Maunder minimum: a reappraisal', *Solar Physics* 89: 195–207.
Egyed, L. (1956a) 'Determination of changes in the dimensions of the Earth', *Nature* 178: 534.
Egyed, L. (1956b) 'The change of the Earth's dimensions determined from palaeogeographical data', *Geofisica Pura e Applicata* 33: 42–8.
Einsele, G., Ricken, W., and Seilacher, A. (eds) (1991a) *Cycles and Events in Stratigraphy*, Berlin: Springer.
Einsele, G., Ricken, W., and Seilacher, A. (1991b) 'Cycles and events in stratigraphy – basic concepts and terms', pp. 1–19 in G. Einsele, W. Ricken, and A. Seilacher (eds) *Cycles and Events in Stratigraphy*, Berlin: Springer.
El Sabh, M. I. and Murty, T. S. (1993) 'Sea level variations in eastern Canadian waterbodies', *Marine Geodesy* 16: 57–71.
Emiliani, C. (1982) 'Extinctive evolution', *Journal of Theoretical Biology* 97: 13–33.
Engebretson, D. C., Kelley, K. P., Cashman, H. J., and Richards, M. A. (1992) '180 million years of subduction', *GSA–Today* 2: 93–5 and 100.
Erhart, H. (1938) *Traité de pédologie*, 2 vols, Strasbourg: Institut Pédologique.
Erhart, H. (1956) *La Genèse des sols en tant que phénomène géologique*, Paris: Masson.
Evans, D. J. A. (ed.) (1994) *Cold Climate Landforms*, Chichester: John Wiley & Sons.
Fairbridge, R. W. (1978) 'Exo- and endogenetic geomagnetic modulation of climates on decadal to galactic scale', *Eos* 59: 269 (Abstract).
Fairbridge, R. W. (1984) 'Planetary periodicities and terrestrial climate stress', pp. 509–20 in N.-A. Mörner and W. Karlén (eds) *Climatic Changes on a Yearly to Millennial Basis*, Dordrecht, The Netherlands: D. Reidel.
Fairbridge, R. W. and Finkl, C. W., Jr (1980) 'Cratonic erosional unconformities and peneplains', *Journal of Geology* 88: 69–86.
Fairbridge, R. W. and Sanders, J. E. (1987) 'The Sun's orbit, AD 750–2050: basis for new perspectives on planetary dynamics and Earth–Moon linkage', pp. 446–541 in M. R. Rampino, J. E. Sanders, W. S. Newman, and L. K. Königsson (eds) *Climate: History, Periodicity, and Predictability*, New York: Van Nostrand Reinhold Company.
Fairbridge, R. W. and Shirley, J. H. (1987) 'Prolonged minima and the 179-yr cycle of the solar inertial motion', *Solar Physics* 110: 191–220.
Fanale, F. P. (1971) 'A case for catastrophic early degassing of the Earth', *Chemical Geology* 8: 79–105.
Farley, K. A. and Patterson, D. B. (1995) 'A 100-kyr periodicity in the flux of extraterrestrial ^{3}He to the sea floor', *Nature* 378: 600–3.
Faure, K., de Wit, M. J., and Willis, J. P. (1995) 'Late Permian global coal hiatus

linked to ^{13}C-depleted CO_2 flux into the atmosphere during the final consolidation of Pangaea', *Geology* 23: 507–10.

Ferrel, W. (1856) 'An essay on the winds and currents of the ocean', *Nashville Journal of Medicine and Surgery* 11: 287–301.

Fischer, A. G. (1964) 'Brackish ocean as the cause of the Permo-Triassic marine faunal crisis', pp. 566–77 in A. E. M. Nairn (ed.) *Problems in Palaeoclimatology*, London: Wiley Interscience.

Fischer, A. G. (1965) 'Fossils, early life, and atmospheric history', *Proceedings of the National Academy of Sciences* 53: 1205–15.

Fischer, A. G. (1972) 'Atmosphere and the evolution of life', *Main Currents in Modern Thought* 28, May–June: unpaginated.

Fischer, A. G. (1981) 'Climatic oscillations in the biosphere', pp. 103–31 in M. H. Nitecki (ed.) *Biotic Crises in Ecological and Evolutionary Time*, New York: Academic Press.

Fischer, A. G. (1984a) 'The two Phanerozoic supercycles', pp. 129–50 in W. A. Berggren and J. A. Van Couvering (eds) *Catastrophes and Earth History: The New Uniformitarianism*, Princeton, New Jersey: Princeton University Press.

Fischer, A. G. (1984b) 'Biological innovations and the sedimentary record', pp. 145–57 in H. D. Holland and A. F. Trendall (eds) *Patterns of Change in Earth Evolution* (Dahlem Konferenzen 1984), Berlin: Springer.

Fischer, A. G. (1986) 'Climatic rhythms recorded in strata', *Annual Review of Earth and Planetary Sciences* 14: 351–76.

Fischer, A. G. (1993) 'Cyclostratigraphy of Cretaceous chalk–marl sequences', *Geological Association of Canada Special Paper* 39: 283–95.

Fisher, R. F., Bourn, C. N., and Fisher, W. F. (1995) 'Opal phytoliths as an indicator of the floristics of prehistoric grasslands', *Geoderma* 68: 243–55.

Flannery, T. F., Rich, T. H., Turnbull, W. D., and Lundelius, E. L., Jr (1992) 'The Macropodoidea (Marsupialia) of the Early Pliocene Hamilton local fauna, Victoria, Australia', *Fieldiana: Geology*, New Series no. 25, Chicago, Illinois: Field Museum of Natural History.

Flynn, L. J. and Jacobs, L. L. (1982) 'Effects of changing environments on Siwalik rodent faunas of northern Pakistan', *Palaeogeography, Palaeoclimatology, Palaeoecology* 38: 129–38.

Foster, I. D. L., Albon, A. J., Bardell, K. M., Fletcher, J. L., Jardine, T. C., Mothers, R. J., Pritchard, M. A., and Turner, S. E. (1991) 'High energy coastal sedimentary deposits: an evaluation of depositional processes in southwest England', *Earth Surface Processes and Landforms* 16: 341–56.

Frakes, L. A., Francis, J. E., and Syktus, J. I. (1992) *Climate Modes of the Phanerozoic: The History of the Earth's Climate over the Past 600 Million Years*, Cambridge: Cambridge University Press.

Francis, P. (1993) *Volcanoes: A Planetary Perspective*, Oxford: Clarendon Press.

Fraser, G. S. (1989) *Clastic Depositional Sequences: Processes of Evolution and Principles of Interpretation*, Englewood Cliffs, New Jersey: Prentice-Hall.

Frey, H. (1980) 'Crustal evolution of the early Earth: the role of major impacts', *Precambrian Research* 10: 195–216.

Fricke, H. C., O'Neil, J. R., and Lynnerup, N. (1995) 'Oxygen isotope composition of human tooth enamel from medieval Greenland: linking climate and society', *Geology* 23: 869–72.

Friis-Christensen, E. and Lassen, K. (1991) 'Length of the solar cycle: an indicator of solar activity closely associated with climate', *Science* 254: 698–700.

Frye, J. C. (1959) 'Climate and Lester King's "Uniformitarian nature of hillslopes"', *Journal of Geology* 67: 111–13.

Fukao, Y., Maruyama, S., Obayashi, M., and Inoue, H. (1994) 'Geologic implication of the whole mantle P-wave tomography', *Journal of the Geological Society of Japan* 100: 4–23.

Gaffin, S. R. and O'Neill, B. C. (1994) 'Pulsations in seafloor spreading rates and transit time dynamics', *Geophysical Research Letters* 21: 1947–50.

Gallimore, R. G. and Kutzbach, J. E. (1989) 'Effects of soil moisture on the sensitivity of a climate model to Earth orbital forcing at 9000 yr BP', *Climatic Change* 14: 175–205.

Gallup, C. D., Edwards, R. L., and Johnson, R. G. (1994) 'The timing of high sea levels over the past 200,000 years', *Science* 263: 796–800.

Garrels, R. M. and Perry, E. A., Jr (1974) 'Cycling of carbon, sulfur, and oxygen through geological time', pp. 303–36 in E. D. Goldberg (ed.) *The Sea. Volume 5. Marine Chemistry*, New York: Wiley Interscience.

Gasse, F. and Fontes, J.-C. (1989) 'Palaeoenvironments and palaeohydrology of a tropical closed lake (Lake Asal, Djibouti) since 10,000 yr BP', *Palaeogeography, Palaeoclimatology, Palaeoecology* 69: 67–102.

Gastil, G. (1960) 'The distribution of mineral dates in time and space', *American Journal of Science* 258: 1–35.

Gates, D. M. (1993) *Climate Change and Its Biological Consequences*, Sunderland, Massachusetts: Sinauer Associates.

Gault, D. E. and Sonett, C. P. (1982) 'Laboratory simulation of pelagic asteroidal impact: atmospheric injection, benthic topography, and the surface wave radiation field', pp. 69–72 in L. T. Silver and P. H. Schultz (eds) *Geological Implications of Impacts of Large Asteroids and Comets on the Earth* (Geological Society of America Special Paper 190), Boulder, Colorado: Geological Society of America.

Gawthorpe, R. L., Hunt, D. W., and Underwood, J. R. (1994) *NERC Sequence Stratigraphy Workshop. Volume 1. Summary Course Notes*, Manchester: Department of Geology, University of Manchester.

Geldsetzer, H. H. J., Goodfellow, W. D., McLaren, D. J., and Orchard, M. J. (1987) 'Sulfur-isotope anomaly associated with the Frasnian–Famennian extinction, Medicine Lake, Alberta, Canada', *Geology* 15: 393–6.

Gentilli, J. (1948) 'Present day volcanicity and climatic change', *Geological Magazine* 85: 172–5.

Gérard, J.-C. (1989) 'Aeronomy and paleoclimate', pp. 139–48 in A. Berger, R. E. Dickinson, and J. W. Kidson (eds) *Understanding Climate Change* (International Union of Geodesy and Geophysics, vol. 7; Geophysical Monograph 52), Washington, DC: American Geophysical Union.

Gérard, J.-C., Hauglustaine, D. A., and François, L. M. (1992) 'The faint young Sun climatic paradox: a simulation with an interactive seasonal climate–sea ice model', *Palaeogeography, Palaeoclimatology, Palaeoecology* 97: 133–50.

Gerrard, A. J. (1992) *Soil Geomorphology: An Integration of Pedology and Geomorphology*, London: Chapman & Hall.

Gerson, R. and Grossman, S. (1987) 'Geomorphic activity on escarpments and associated fluvial systems in hot deserts', pp. 300–22 in M. R. Rampino, J. E. Sanders, W. S. Newman, and L. K. Königsson (eds) *Climate: History, Periodicity, and Predictability*, New York: Van Nostrand Reinhold.

Gilbert, G. K. (1877) *Geology of the Henry Mountains (Utah)* (United States Geographical and Geological Survey of the Rocky Mountains Region), Washington, DC: United States Government Printing Office.

Gilchrist, A. R., Kooi, H., and Beaumont, C. (1994) 'Post-Gondwana geomorphic evolution of southwestern Africa: implications for the controls of landscape development from observation and numerical experiments', *Journal of Geophysical Research* 99B: 12,211–28.

Gleissberg, W. (1955) 'The 80-year sunspot cycle', *Journal of the British Astronomical Association* 68: 148–52.

Gleissberg, W. (1965) 'The eighty-year cycle in auroral frequency numbers', *Journal of the British Astronomical Association* 75: 277–31.

Glenn, C. R. and Kelts, K. (1991) 'Sedimentary rhythms in lake deposits', pp. 188–221 in G. Einsele, W. Ricken, and A. Seilacher (eds) *Cycles and Events in Stratigraphy*, Berlin: Springer.

Glickson, A. Y. (1980) 'Precambrian sial–sima relations: evidence for Earth expansion', *Tectonophysics* 63: 193–234.

Godard, A. (1965) *Recherches de géomorphologie en Ecosse de nord-ouest* (Publications de la Faculté des Lettres de l'Université de Strasbourg, Fondation Baulig 1), Paris: Les Belles Lettres.

Gold, T. (1966) 'Long-term stability of the Earth', pp. 93–7 in B. G. Marsden and A. G. W. Cameron (eds) *The Earth–Moon System*, New York: Plenum Press.

Goldhammer, R. K., Oswald, E. J., and Dunn, P. A. (1994) 'High-frequency, glacio-eustatic cyclicity in the Middle Pennsylvanian of the Paradox Basin: an evaluation of Milankovitch forcing', pp. 243–83 in P. L. De Boer and D. G. Smith (eds) *Orbital Forcing and Cyclic Sequences* (Special Publication Number 19 of the International Association of Sedimentologists), Oxford: Blackwell Scientific Publications.

Gomez, B., Mertes, L. A. K., Phillips, J. D., Magilligan, F. J., and James, L. A. (1995) 'Sediment characteristics of an extreme flood: 1993 upper Mississippi River valley', *Geology* 23: 963–6.

Gordon, R. G. (1987) 'Polar wandering and paleomagnetism', *Annual Review of Earth and Planetary Sciences* 15: 567–93.

Gore, P. J. W. (1989) 'Toward a model for open- and closed-basin deposition in ancient lacustrine sequences: the Newark supergroup (Triassic–Jurassic), eastern North America', *Palaeogeography, Palaeoclimatology, Palaeoecology* 70: 29–51.

Gosse, J. C., Evenson, E. B., Klein, J., Lawn, B., and Middleton, R. (1995) 'Precise cosmogenic ^{10}Be measurements in western North America: support for a global Younger Dryas cooling event', *Geology* 23: 877–80.

Gossman, H. (1970) *Theorien zur Hangentwicklung in verschiedenen Klimazonen: mathematische Hangmodelle und ihre Beziehung zu den Abtragungsvorgängen* (Würzburger Geographische Arbeiten, vol. 31), Würzburg: Geographischen Instituts der Universität Würzburg.

Gossman, H. (1981) 'Fragen und Einsichten zum Einsatz von Hangmodellen in der geomorphologischen Analyse', *Geoökodynamik* 2: 205–18.

Goudie, A. S. (1985) 'Duricrusts and landforms', pp. 37–57 in K. S. Richards, R. R. Arnett, and S. Ellis (eds) *Geomorphology and Soils*, London: George Allen & Unwin.

Goudie, A. S. (1992) *Environmental Change*, 3rd edn, Oxford: Clarendon Press.

Goudie, A. S. (1995) *The Changing Earth: Rates of Geomorphological Processes*, Oxford: Blackwell.

Gould, S. J. (1984) 'Toward the vindication of punctuational change', pp. 9–34 in W. A. Berggren and J. A. van Couvering (eds) *Catastrophes in Earth History: The New Uniformitarianism*, Princeton, New Jersey: Princeton University Press.

Gradstein, F. M., Huang, Z.-H., Kristiansen, I. L., and Ogg, J. G. (1993) 'Optimum microfossil sequences and cyclic sediment patterns in Early Cretaceous pelagic strata', *Canadian Journal of Earth Sciences* 30: 391–411.

Graham, J. B., Dudley, R., Aguilar, N. M., and Gans, C. (1995) 'Implications of the late Palaeozoic oxygen pulse for physiology and evolution', *Nature* 375: 117–20.

Graham, R. W. (1979) 'Paleoclimates and late Pleistocene faunal provinces in North America', pp. 46–69 in R. L. Humphrey and D. J. Stanford (eds) *Pre-Llano Cultures of the Americas: Paradoxes and Possibilities*, Washington, DC: Anthropological Society of Washington.

Graham, R. W. (1992) 'Late Pleistocene faunal changes as a guide to understanding effects of greenhouse warming on the mammalian fauna of North America', pp. 76–87 in R. L. Peters and T. E. Lovejoy (eds) *Global Warming and Biological Diversity*, New Haven, Connecticut, and London: Yale University Press.

Graham, R. W. and Grimm, E. C. (1990) 'Effects of global climate change on the patterns of terrestrial biological communities', *Trends in Ecology and Evolution* 5: 289–92.

Graham, R. W. and Mead, J. I. (1987) 'Environmental fluctuations and evolution of mammalian faunas during the last deglaciation', pp. 371–402 in W. F. Ruddiman and H. E. Wright Jr (eds) *North America and Adjacent Ocean during the Last Deglaciation* (The Geology of North America, vol. K-3), Boulder, Colorado: Geological Society of America.

Grattan, J. and Charman, D. J. (1994) 'Non-climatic factors and the environmental impact of volcanic volatiles: implications of the Laki fissure eruption of AD 1783', *The Holocene* 4: 101–6.

Greeley, R. (1982) 'The Snake River Plain, Idaho: representative of a new category of volcanism', *Journal of Geophysical Research* 87: 2705–12.

Greenland Ice-Core Project (GRIP) Members (1993) 'Climate instability during the last interglacial period recorded in the GRIP ice core', *Nature* 364: 203–7.

Gregory, K. J., Starkel, L., and Baker, V. R. (eds) (1995) *Global Continental Palaeohydrology*, Chichester: John Wiley & Sons.

Gretener, P. E. (1967) 'The significance of the rare event in geology', *Bulletin of the American Association of Petroleum Geologists* 51: 2197–206.

Grieve, R. A. F. and Parmentier, E. M. (1984) 'Impact phenomena in the evolution of the Earth', *Proceedings of the 27th International Geological Congress, Moscow* (Utrecht: VNU Science) 19: 99–114.

Grieve, R. A. F. and Shoemaker, E. M. (1994) 'The record of past impacts on Earth', pp. 417–62 in T. Gehrels (ed.), with the editorial assistance of M. S. Matthews and A. M. Schumann, *Hazards Due to Comets and Asteroids*, Tucson and London: University of Arizona Press.

Grimm, E. C., Jacobson, G. L., Jr, Watts, W. A., Hansen, B. C. S., and Maasch, K. A. (1993) 'A 50,000-year record of climate oscillations from Florida and its temporal correlation with the Heinrich events', *Science* 261: 198–200.

Grootes, P. M., Stuiver, M., White, J. W. C., Johnsen, S. J., and Jouzel, J. (1993) 'Comparison of oxygen isotope records from the GISP2 and GRIP Greenland ice cores', *Nature* 366: 552–4.

Grove, J. M. (1988) *The Little Ice Age*, London and New York: Methuen.

Grove, J. M. and Battagel, A. (1989) 'The rains of December 1743 in western Norway', *Norsk Geografisk Tidsskrift* 43: 215–20.

Hack, J. T. (1960) 'Interpretation of erosional topography in humid temperate regions', *American Journal of Science* (Bradley Volume) 258-A: 80–97.

Hack, J. T. (1975) 'Dynamic equilibrium and landscape evolution', pp. 87–102 in W. N. Melhorn and R. C. Flemal (eds) *Theories of Landform Development*, London: George Allen & Unwin.

Hadley, G. (1735) 'Concerning the cause of the general tradewinds', *Philosophical Transactions, London* 29: 58–62.

Hagedorn, H. (1993) 'Beispiele aus der zentralen Sahara zu Grundvorstellungen in der klimatischen Geomorphologie', *Zeitschrift für Geomorphologie, Supplementband* 93: 1–11.

Halfman, J. D. and Johnson, T. C. (1988) 'High-resolution record of cyclic climatic change during the past 4 ka from Lake Turkana, Kenya', *Geology* 16: 496–500.

Halfman, J. D., Johnson, T. C., and Finney, B. P. (1994) 'New AMS dates, stratigraphic correlations and decadal climatic cycles for the past 4 ka at Lake Turkana, Kenya', *Palaeogeography, Palaeoclimatology, Palaeoecology* 111: 83–98.

Hall, A. M. (1985) 'Cenozoic weathering covers in Buchan, Scotland, and their significance', *Nature* 315: 392–5.

Hall, A. M. (1991) 'Pre-Quaternary landscape evolution in the Scottish Highlands', *Transactions of the Royal Society of Edinburgh: Earth Sciences* 82: 1–26.

Hall, V. A. (1994) 'Landscape development in northeast Ireland over the last half millennium', *Review of Palaeobotany and Palynology* 82: 75–82.

Hallam, A. (1963) 'Major epeirogenic and eustatic changes since the Cretaceous, and their possible relationship to crustal structure', *American Journal of Science* 261: 397–423.

Hallam, A. (1971) 'Re-evaluation of the palaeogeographical argument for an expanding Earth', *Nature* 232: 180–2.

Hallam, A. (1972) 'Continental drift and the fossil record', *Scientific American* 227 (November): 56–66.

Hallam, A. (1979) 'The end of the Cretaceous', *Nature* 281: 430–1.

Hallam, A. (1984) 'The unlikelihood of an expanding Earth', *Geological Magazine* 121: 653–5.

Hallet, B. (1990) 'Spatial self-organization in geomorphology: from periodic bedforms and patterned ground to scale-invariant topography', *Earth-Science Reviews* 29: 57–75.

Hameed, S. (1984) 'Fourier analysis of Nile flood level', *Geophysical Research Letters* 11: 843–5.

Hameed, S., Yeh, W.-M., Li, M.-T., Cess, R. D., and Wang, W.-C. (1983) 'An analysis of periodicities in the 1470 to 1974 Beijing precipitation record', *Geophysical Research Letters* 10: 436–9.

Hammond, E. H. (1954) 'Small-scale continental landform maps', *Annals of the Association of American Geographers* 44: 33–42.

Hang, K.-L. and Pimm, S. L. (1993) 'The assembly of ecological communities: a minimalist approach', *Journal of Animal Ecology* 62: 749–65.

Haq, B. U., Hardenbol, J., and Vail, P. R. (1987) 'Chronology of fluctuating sea levels since the Triassic', *Science* 235: 1156–67.

Haq, B. U. and Van Eysinga, F. W. B. (1987) *Geological Time Table*, Amsterdam: Elsevier.

Hare, F. K. (1996) 'Climatic variation and global change', pp. 482–507 in I. Douglas, R. J. Huggett, and M. E. Robinson (eds) *Companion Encyclopedia of Geography*, London: Routledge.

Harland, W. B. (1964) 'Critical evidence for a great infra-Cambrian glaciation', *Geologische Rundschau* 54: 45–61.

Harland, W. B., Armstrong, R. L., Cox, A. V., Craig, L. E., Smith, A. G., and Smith, D. G. (1990) *A Geologic Time Scale 1989*, Cambridge: Cambridge University Press.

Harland, W. B. and Bidgood, D. E. T. (1959) 'Palaeomagnetism in some Norwegian sparagmites and the late Pre-Cambrian ice age', *Nature* 184: 1860–2.

Hart, M. H. (1978) 'The evolution of the atmosphere of the Earth', *Icarus* 33: 23–39.

Hartmann, W. K. (1977) 'Cratering in the Solar System', *Scientific American* 236: 84–99.

Harvey, A. M. (1990) 'Factors influencing Quaternary alluvial fan development in southeast Spain', pp. 247–69 in A. H. Rachocki and M. Church (eds) *Alluvial Fans: A Field Approach*, Chichester: John Wiley & Sons.

Hauglustaine, D. A. and Gérard, J.-C. (1990) 'Possible composition and climatic changes due to past intense energetic particle precipitation', *Annales Geophysicae* 8: 87–96.

Hauglustaine, D. A. and Gérard, J.-C. (1992) 'A sensitivity study of the role of continental location and area on Paleozoic climate', *Palaeogeography, Palaeoclimatology, Palaeoecology (Global and Planetary Change Section)* 97: 311–23.

Hays, J. D. and Pitman III, W. C. (1973) 'Lithospheric plate motion, sea level changes and climatic and ecological consequences', *Nature* 246: 18–22.

Hedges, S. B., Parker, P. H., Sibley, C. G., and Kumar, S. (1996) 'Continental breakup and the ordinal diversification of birds and mammals', *Nature* 381: 226–9.

Heinrich, H. (1988) 'Origin and consequences of cyclic ice rafting in the northeast Atlantic Ocean during the past 130,000 years', *Quaternary Research* 29: 143–52.

Hendriks, E. M. L. (1923) 'The physiography of south-west Cornwall, the distribution of chalk flints, and the origin of the gravels of Crousa Common', *Geological Magazine* 60: 21–31.

Henkel, R. (1972) 'Evidence of an ultra-long cycle of solar activity', *Solar Physics* 25: 498–9.

Hide, R. and Dickey, J. O. (1991) 'Earth's variable rotation', *Science* 253: 629–37.

Higgins, C. G. (1975) 'Theories of landscape development: a perspective', pp. 1–28 in W. N. Melhorn and R. C. Flemal (eds) *Theories of Landform Development*, London: George Allen & Unwin.

Hildebrand, A. R., Penfield, G. T., Kring, D. A., Pilkington, M., Camargo, A. Z., Jacobsen, S. B., and Boynton, W. V. (1991) 'Chicxulub crater: a possible Cretaceous/Tertiary boundary impact crater on the Yucatán Peninsula, Mexico', *Geology* 19: 867–71.

Hilgen, F. J., Lourens, L. J., Berger, A., and Loutre, M. F. (1993) 'Evaluation of the astronomically calibrated time scale for the Late Pliocene and earliest Pleistocene', *Paleoceanography* 8: 549–65.

Hill, S. M., Ollier, C. D., and Joyce, E. B. (1995) 'Mesozoic deep weathering and erosion: an example from Wilsons Promontory, Australia', *Zeitschrift für Geomorphologie* NF 39: 331–9.

Hills, J. G., Nemchinov, I. V., Popov, S. P., and Teterev, A. V. (1994) 'Tsunami generated by small asteroid impacts', pp. 779–89 in T. Gehrels (ed.), with the editorial assistance of M. S. Matthews and A. M. Schumann, *Hazards Due to Comets and Asteroids*, Tucson and London: University of Arizona Press.

Hirano, M. (1975) 'Simulation of development process of interfluvial slopes with reference to graded form', *Journal of Geology* 83: 113–23.

Hirschboeck, K. K. (1987) 'Catastrophic flooding and atmospheric circulation anomalies', pp. 23–56 in L. Mayer and D. Nash (eds) *Catastrophic Flooding* (The Binghamton Symposia in Geomorphology, International Series, no. 18), Boston: Allen & Unwin.

Hirschboeck, K. K. (1991) 'Climate and floods', pp. 67–88 in R. W. Paulson, E. B. Chase, R. S. Roberts, and D. W. Moody (compilers) *National Water Summary 1988–89; Hydrologic Events and Floods and Droughts* (United States Geological Survey Water Supply Paper 2375), Reston, Virginia: United States Geological Survey.

Hole, F. D. (1961) 'A classification of pedoturbation and some other processes and factors of soil formation in relation to isotropism and anisotropism', *Soil Science* 91: 375–7.

Holland, H. D. (1984) *The Chemical Evolution of the Atmosphere and Oceans*, Princeton, New Jersey: Princeton University Press.

Holland, H. D., Feakes, C. R., and Zbinden, E. A. (1989) 'The Flin Flon paleosol and the composition of the atmosphere 1.8 by BP', *American Journal of Science* 289: 362–89.

Holliday, V. T. (1988) 'Mt. Blanco revisited: soil–geomorphic implications for the ages of the upper Cenozoic Blanco and Blackwater Draw Formations', *Geology* 16: 505–8.

Holliday, V. T. (1989) 'The Blackwater Draw Formation (Quaternary): a 1.4-plus-m.y. record of eolian sedimentation and soil formation in the Southern High Plains', *Bulletin of the Geological Society of America* 101: 1598–1607.

Holmes, A. (1965) *Principles of Physical Geology*, new and fully revised edn, London and Edinburgh: Thomas Nelson & Sons.

Holmes, J. A. (1992) 'Nonmarine ostracods as Quaternary palaeoenvironmental indicators', *Progress in Physical Geography* 16: 405–31.

Holser, W. T. (1977) 'Catastrophic chemical events in the history of the ocean', *Nature* 267: 403–8.

Holser, W. T. (1984) 'Gradual and abrupt shifts in ocean chemistry', pp. 123–43 in H. D. Holland and A. F. Trendall (eds) *Patterns of Change in Earth Evolution*, Berlin: Springer.

Horita, J., Friedman, T. J., Lazar, B., and Holland, H. D. (1991) 'The composition of Permian seawater', *Geochimica et Cosmochimica Acta* 55: 417–32.

Howard, A. D. (1988) 'Equilibrium models in geomorphology', pp. 49–72 in M. G. Anderson (ed.) *Modelling Geomorphological Systems*, Chichester: John Wiley & Sons.

Howard, A. D., Kochel, R. C., and Holt, H. E. (eds) (1988) *Sapping Features of the Colorado Plateau: A Comparative Planetary Geology Field Guide* (Scientific and Technical Information Division, National Aeronautics and Space Administration, NASA SP-491), Washington, DC: United States Government Printing Office.

Howard, J. L., Amos, D. F., and Daniels, W. L. (1993) 'Alluvial soil chronosequence in the inner Coastal Plain, central Virginia', *Quaternary Science* 39: 201–13.

Huggett, R. J. (1980) *Systems Analysis in Geography*, Oxford: Clarendon Press.

Huggett, R. J. (1985) *Earth Surface Systems* (Springer Series in Physical Environment, vol. 1), Heidelberg: Springer.

Huggett, R. J. (1988) 'Dissipative systems: implications for geomorphology', *Earth Surface Processes and Landforms* 13: 45–9.

Huggett, R. J. (1989a) 'Superwaves and superfloods: the bombardment hypothesis and geomorphology', *Earth Surface Processes and Landforms* 14: 433–42.

Huggett, R. J. (1989b) *Cataclysms and Earth History: The Development of Diluvialism*, Oxford: Clarendon Press.

Huggett, R. J. (1990) *Catastrophism: Systems of Earth History*, London: Edward Arnold.

Huggett, R. J. (1991) *Climate, Earth Processes and Earth History*, Heidelberg: Springer.

Huggett, R. J. (1993) *Modelling the Human Impact on Nature: Systems Analysis of Environmental Problems*, Oxford: Oxford University Press.

Huggett, R. J. (1994) 'Fluvialism or diluvialism? Changing views on superfloods and landscape change', *Progress in Physical Geography* 18: 335–42.

Huggett, R. J. (1995) *Geoecology: An Evolutionary Approach*, London: Routledge.

Hunt, B. G. (1979) 'The effects of past variations of the Earth's rotation rate on climate', *Nature* 281: 188–91.

Hunt, B. G. (1982) 'The impact of large variations of the Earth's obliquity on the climate', *Journal of the Meteorological Society of Japan* 60: 309–18.

Hunt, C. W. (1990) *Environment of Violence: Readings of Cataclysm Cast in Stone*, Calgary, Alberta: Polar Publishing.

Huntley, B. (1990) 'European vegetation history: palaeovegetation maps from pollen data – 13,000 yr BP to present', *Journal of Quaternary Science* 5: 103–22.

Huntley, B. (1992) 'Rates of change in the European palynological record of the last 13 000 years and their climatic interpretation', *Climate Dynamics* 6: 185–91.

Huntley, B. and Prentice, I. C. (1993) 'Holocene vegetation and climates of Europe', pp. 136–68 in H. E. Wright Jr, J. E. Kutzbach, T. Webb III, W. F. Ruddiman, F. A. Street-Perrott, and P. J. Bartlein (eds) *Global Climates since the Last Glacial Maximum*, Minneapolis and London: University of Minnesota Press.

Hurlbert, S. H. and Archibald, J. D. (1995) 'No statistical support for sudden (or gradual) extinction of dinosaurs', *Geology* 23: 881–4.

Hurt, P. R., Libby, L. M., Pandolphi, L. J., Levine, L. H., and Van Engel, W. A. (1979) 'Periodicities in blue crab population of Chesapeake Bay', *Climatic Change* 2: 75–8.

Hut, P., Alvarez, W., Elder, W. P., Hansen, T., Kauffman, E. G., Keller, G., Shoemaker, E. M., and Weissman, P. R. (1987) 'Comet showers as a cause of mass extinctions', *Nature* 329: 118–26.

Huxley, J. S. (1953) *Evolution in Action*, London: Chatto & Windus.

Hyvarinen, H. and Alhonen, P. (1994) 'Holocene lake-level changes in the Fennoscandian tree-line region, Western Finnish Lapland: diatom and cladoceran evidence', *The Holocene* 4: 251–8.

Imbrie, J. and Imbrie, K. P. (1986) *Ice Ages: Solving the Mystery*, Cambridge, Massachusetts, and London: Harvard University Press.

Ingram, B. L. and DePaolo, D. J. (1993) 'A 4300 year strontium isotope record of estuarine paleosalinity in San Francisco Bay, California', *Earth and Planetary Science Letters* 119: 103–19.

Ingrouille, M. (1995) *Historical Ecology of the British Flora*, London: Chapman & Hall.

Innanen, K. A., Patrick, A. T., and Duley, W. W. (1978) 'The interaction of the spiral density wave and the Sun's galactic orbit', *Astrophysics and Space Physics* 57: 511–15.

Isaac, K. P. (1981) 'Tertiary weathering profiles in plateau deposits of east Devon', *Proceedings of the Geologists' Association, London* 92: 159–68.

Isaac, K. P. (1983) 'Tertiary lateritic weathering in Devon, England, and the Palaeogene continental environment of south west England', *Proceedings of the Geologists' Association, London* 94: 105–14.

Jackson, S. T., Futyma, R. P., and Wilcox, D. A. (1988) 'A paleoecological test of a classical hydrosere in the Lake Michigan Dunes', *Ecology* 69: 928–36.

Jackson, S. T. and Givens, C. R. (1994) 'Late Wisconsinan vegetation and environment of the Tunica Hills region, Louisiana/Mississippi', *Quaternary Research* 41: 316–25.

Jacobson, R. B. and Oberg, K. A. (in press) *Geomorphic Changes of the Mississippi Floodplain at Miller City, Illinois, as a Result of the Flood of 1993* (United States Geological Survey Circular 1120-J), Washington, DC: United States Government Printing Office.

James, I. N. and James, P. M. (1989) 'Ultra-low-frequency variability in a simple atmospheric circulation model', *Nature* 342: 53–5.

Jannasch, H. W. and Mottl, M. J. (1985) 'Geomicrobiology of deep-sea hydrothermal vents', *Science* 229: 717–25.

Jansa, L. F. (1991) 'Processes affecting paleogeography, with examples from Tethys', *Palaeogeography, Palaeoclimatology, Palaeoecology* 87: 345–71.

Jarrett, R. D. and Malde, H. E. (1987) 'Paleodischarge of the late Pleistocene Bonneville Flood, Snake River, Idaho, computed from new evidence', *Bulletin of the Geological Society of America* 99: 127–34.

Jenkins, G. S. (1996) 'A sensitivity study of changes in Earth's rotation rate with an atmospheric general circulation model', *Global and Planetary Change* 11: 141–54.

Jenny, H. (1941) *Factors of Soil Formation: A System of Quantitative Pedology*, New York: McGraw-Hill.

Jenny, H. (1980) *The Soil Resource: Origin and Behaviour* (Ecological Studies, vol. 37), New York: Springer.

Jessen, O. (1936) *Reisen und Forschungen in Angola*, Berlin: D. Reimer.

Jimenez de Cisneros, C. and Vera, J. A. (1993) 'Milankovitch cyclicity in Purbeck peritidal limestones of the Prebetic (Berraisian, southern Spain)', *Sedimentology* 40: 513–37.

Johnsen, S. J., Clausen, H. B., Dansgaard, W., Fuher, K., Gundestrup, N., Hammer, C. U., Iversen, P., Jouzel, J., Stauffer, B., and Steffersen, J. P. (1992) 'Irregular glacial interstadials recorded in a new Greenland ice core', *Nature* 359: 311–13.

Johnson, D. L. (1990) 'Biomantle evolution and the redistribution of earth materials and artefacts', *Soil Science* 149: 84–102.

Johnson, D. L. (1993a) 'Dynamic denudation evolution of tropical, subtropical and temperate landscapes with three tiered soils: toward a general theory of landscape evolution', *Quaternary International* 17: 67–78.

Johnson, D. L. (1993b) 'Biomechanical processes and the Gaia paradigm in a unified pedo-geomorphic and pedo-archaeologic framework: dynamic denudation', pp. 41–67 in J. E. Foss, M. E. Timpson, and M. W. Morris (eds) *Proceedings of the First International Conference on Pedo-Archaeology* (University of Tennessee Agricultural Experimental Station, Special Paper 93-03), Knoxville, Tennessee: University of Tennessee Agricultural Experimental Station.

Johnson, D. L. (1994) 'Reassessment of early and modern soil horizon designation frameworks as associated pedogenetic processes: are midlatitude A E B–C horizons equivalent to tropical M S W horizons?', *Soil Science (Trends in Agricultural Science)* 2: 77–91.

Johnson, D. L. and Hole, F. D. (1994) 'Soil formation theory: a summary of its principal impacts on geography, geomorphology, soil-geomorphology, Quaternary geology and paleopedology', pp. 111–26 in R. Amundson (ed.) *Factors of Soil Formation: A Fiftieth Anniversary Retrospective* (Soil Science Society of America Special Publication 33), Madison, Wisconsin: Soil Science Society of America.

Jones, D. K. C. (1981) *Southeast and Southern England* (The Geomorphology of the British Isles), London and New York: Methuen.

Jong, W. J. (1976) 'Actualism in geology and in geography', *Catastrophist Geologist* 1: 32–42.

Jorda, M. and Roditis, J. C. (1993) 'Les épisodes de gel du Rhône depuis l'an mil. Périodisation, fréquence, interprétation paléoclimatique', *Méditerranée* 78: 19–30.

Jose, P. D. (1965) 'Sun's motion and sunspots', *Astronomical Journal* 70: 193–200.

Kajander, J. (1993) 'Methodological aspects on river cryophenology exemplified by a tricentennial break-up time series from Tornio', *Geophysica* 29: 73–95.

Kanari, S., Fuji, N., and Horie, S. (1984) 'The paleoclimatic constituents of paleotemperature in Lake Biwa', pp. 405–14 in A. Berger, J. Imbrie, J. Hays, G. Kukla, and B. Saltzman (eds) *Milankovitch and Climate: Understanding the Response of Astronomical Forcing, Part I* (Proceedings of the NATO Advanced Research Workshop on Milankovitch and Climate, Palisades, New York, 1982; NATO ASI Series C, Mathematical and Physical Sciences, vol. 126), Dordrecht: D. Reidel.

Kane, R. P. and Gobbi, D. (1992) 'Periodicities in the time series of annual minimum temperatures for the United States Gulf Coast region', *Pure and Applied Physics* 138: 323–33.

Kashiwaya, K., Atkinson, T. C., and Smart, P. L. (1991) 'Periodic variations in late Pleistocene speleothem abundance in Britain', *Quaternary Research* 35: 190–6.

Kashiwaya, K., Yamoto, A., and Fukuyama, K. (1987) 'Time variations of erosional force and grain size in Pleistocene lake sediments', *Quaternary Research* 28: 61–8.

Kasting, J. F. (1987) 'Theoretical constraints on oxygen and carbon dioxide concentrations in the Precambrian atmosphere', *Precambrian Research* 34: 205–29.

Kauffman, E. G. (1995) 'Global change leading to biodiversity crisis in a greenhouse world: the Cenomanian–Turonian (Cretaceous) mass extinction', pp. 47–71 in Board on Earth Sciences and Resources Commission on Geosciences, Environment, and Resources, National Research Council, *Effects of Past Global Change on Life*, Washington, DC: National Academy Press.

Kauffman, E. G. and Fagerstrom, J. A. (1993) 'The Phanerozoic evolution of reef diversity', pp. 365–404 in R. E. Ricklefs and D. Schulter (eds) *Species Diversity in Ecological Communities: Historical and Geographical Perspectives*, Chicago, Illinois: University of Chicago Press.

Kearey, P. and Vine, F. J. (1990) *Global Tectonics*, Oxford: Blackwell.

Keen, M. C. (1993) 'Ostracods as palaeoenvironmental indicators: examples from the Tertiary and Early Cretaceous', pp. 41–67 in D. G. Jenkins (ed.) *Applied Micropalaeontology*, Dordrecht: Kluwer Academic Publishers.

Kehew, A. E. and Teller, J. T. (1994) 'Glacial-lake spillway incision and deposition of a coarse-grained fan near Watrous, Saskatchewan', *Canadian Journal of Earth Sciences* 31: 544–53.

Keller, G. (1989) 'Extended periods of extinctions across the Cretaceous/Tertiary boundary in planktonic foraminifera of continental shelf sections: implications for impact and volcanic theories', *Bulletin of the Geological Society of America* 101: 1408–19.

Kelly, S. B. (1992) 'Milankovitch cyclicity recorded from Devonian non-marine sediments', *Terra Nova* 4: 578–84.

Kennedy, K. A. (1994) 'Early-Holocene geochemical evolution of saline Medicine Lake, South Dakota', *Journal of Paleolimnology* 10: 69–84.

Kennett, J. P. and Stott, L. D. (1995) 'Terminal Paleocene mass extinction in the deep sea: association with global warming', pp. 94–107 in Board on Earth Sciences and Resources Commission on Geosciences, Environment, and Resources, National Research Council, *Effects of Past Global Change on Life*, Washington, DC: National Academy Press.

Kent, R. (1991) 'Lithospheric uplift in eastern Gondwana: evidence for a long-lived mantle plume system?', *Geology* 19: 19–23.

Kerr, R. A. (1981) 'Mount St Helens and a climate quandary', *Science* 211: 371–4.

Kieft, T. L. (1994) 'Grazing and plant canopy effects on semiarid soil microbial biomass and respiration', *Biology and Fertility of Soils* 18: 155–62.

King, L. C. (1953) 'Canons of landscape evolution', *Bulletin of the Geological Society of America* 64: 721–52.

King, L. C. (1957) 'The uniformitarian nature of hillslopes', *Transactions of the Edinburgh Geological Society* 17: 81–102.

King, L. C. (1967) *The Morphology of the Earth*, 2nd edn, Edinburgh: Oliver & Boyd.

King, L. C. (1983) *Wandering Continents and Spreading Sea Floors on an Expanding Earth*, Chichester: John Wiley & Sons.

Kirchner, J. W. (1991) 'The Gaia hypotheses: are they testable? Are they useful?', pp. 38–46 in S. H. Schneider and P. J. Boston (eds) *Scientists on Gaia*, Cambridge, Massachusetts, and London: MIT Press.

Kirkby, M. J. (1990) 'The landscape viewed through models', *Zeitschrift für Geomorphologie* NF 79: 63–81.

Kirkby, M. J. (1993) 'Long term interactions between networks and hillslopes', pp. 255–93 in K. Beven and M. J. Kirkby (eds) *Channel Network Hydrology*, Chichester: John Wiley & Sons.

Kirschvink, J. L. (1992) 'Late Proterozoic low-latitude global glaciation: the snowball Earth', pp. 51–2 in J. W. Schopf and C. Klein (eds) *The Proterozoic Biosphere: A Multidisciplinary Study*, Cambridge: Cambridge University Press.

Kitchell, J. A. and Carr, T. R. (1985) 'Nonequilibrium model of diversification: faunal turnover dynamics', pp. 277–310 in J. W. Valentine (ed.) *Phanerozoic Diversity Patterns*, Princeton, New Jersey: Princeton University Press.

Knoll, A. H. (1984) 'Patterns of extinction in the fossil record of vascular plants', pp. 21–68 in M. H. Nitecki (ed.) *Extinctions*, Chicago, Illinois: University of Chicago Press.

Knoll, A. H. and Holland, H. D. (1995) 'Oxygen and Proterozoic evolution: an update', pp. 21–33 in Board on Earth Sciences and Resources Commission on Geosciences, Environment, and Resources, National Research Council, *Effects of Past Global Change on Life*, Washington, DC: National Academy Press.

Kocharov, G. E. *et al.* (1992) 'Variation of radiocarbon content in tree rings during the Maunder Minimum of solar activity', *Radiocarbon* 34: 213–17.

Kohls, S. J., Van Kessel, C., Baker, D. D., Grigal, D. F., and Lawrence, D. B. (1994) 'Assessment of N_2 fixation and N cycling by *Dryas* along a chronosequence within the forelands of the Athabasca Glacier, Canada', *Soil Biology and Biochemistry* 26: 623–32.

Koizumi, I. (1994) 'Spectral analysis of the diatom palaeotemperature records at DSDP sites 579 and 580 near the subarctic front in the western North Pacific', *Palaeogeography, Palaeoclimatology, Palaeoecology* 108: 475–85.

Koltermann, C. E. and Gorelick, S. M. (1992) 'Paleoclimatic signature in terrestrial flood deposits', *Science*, 256: 1775–82.

Kondrat'ev, K. Ya. (1993) 'Water cycle and feedbacks in global climate change', *Russian Meteorology and Hydrology* 3: 1–8.

Kozlenko, V. G. and Shen, E. L. (1993) 'Expansion of the Earth: theoretical construction and factual data', *Geophysical Journal* 12: 177–85.

Krapez, B. (1993) 'Sequence stratigraphy of the Archaean supracrustal belts of the Pilbara Block, Western Australia', *Precambrian Research* 60: 1–45.

Kristan-Tollmann, E. and Tollmann, A. (1992) 'Der Sintflut-Impakt. The Flood impact', *Mitteilungen der Österreichische geologischen Gesellschaft* 84: 1–63.

Kröner, A. (1977) 'Precambrian mobile belts of southern and eastern Africa: ancient sutures or sites of ensialic mobility? A case for crustal evolution towards plate tectonics', *Tectonophysics* 40: 101–5.

Kröner, A. (1985) 'Evolution of the Archean continental crust', *Annual Review of Earth and Planetary Sciences* 13: 49–74.

Kuhn, W. R., Walker, J. C. G., and Marshall, H. G. (1989) 'The effect on the Earth's surface temperature from variations in rotation rate, continent formation, solar luminosity, and carbon dioxide', *Journal of Geophysical Research* 94D: 11,129–36.

Kuhry, P. (1994) 'The role of fire in the development of *Sphagnum*-dominated peatlands in western boreal Canada', *Journal of Ecology* 82: 899–910.

Kukla, G. J. (1968) 'Comment', *Current Anthropology* 9: 37–9.

Kumazawa, M. and Maruyama, S. (1994) 'Whole Earth tectonics', *Journal of the Geological Society of Japan* 100: 81–102.

Kumazawa, M. and Mizutani, H. (1981) 'A presence of a 2 AE period in the meteorite flux on the Earth and Moon', pp. 313–21 in *Proceedings of the*

Fourteenth Institute of Space and Astronautic Sciences, Japan, Lunar and Planetary Science Symposium, Tokyo: Institute of Space and Astronautic Sciences.

Kumazawa, M., Yoshida, S., Ito, T., and Yoshioka, H. (1994) 'Archaean–Proterozoic boundary interpreted as a catastrophic collapse of the stable density stratification in the core', *Journal of the Geological Society of Japan* 100: 50–9.

Kurtén, B. (1969) 'Continental drift and evolution', *Scientific American* 220 (March): 54–64.

Kutzbach, J. E. and Gallimore, R. G. (1989) 'Pangaean climates: megamonsoons of the megacontinent', *Journal of Geophysical Research* 94D: 3341–57.

Kutzbach, J. E. and Guetter, P. J. (1986) 'The influence of changing orbital parameters and surface boundary conditions on climate simulations for the past 18,000 years', *Journal of the Atmospheric Sciences* 43: 1726–59.

Kutzbach, J. E., Guetter, P. J., Behling, P. J., and Selin, R. (1993) 'Simulated climatic changes: results of the COHMAP climate-model experiments', pp. 24–93 in H. E. Wright Jr, J. E. Kutzbach, T. Webb III, W. F. Ruddiman, F. A. Street-Perrott, and P. J. Bartlein (eds) *Global Climates since the Last Glacial Maximum*, Minneapolis and London: University of Minnesota Press.

Kutzbach, J. E., Guetter, P. J., Ruddiman, W. F., and Prell, W. L. (1989) 'Sensitivity of climate to late Cenozoic uplift in southern Asia and the American west: numerical experiments', *Journal of Geophysical Research* 94D: 18,393–407.

Kutzbach, J. E. and Otto-Bliesner, B. L. (1982) 'The sensitivity of the African–Asian monsoonal climate to orbital parameter changes for 9000 years BP in a low-resolution general circulation model', *Journal of the Atmospheric Sciences* 39: 1177–88.

Kutzbach, J. E. and Street-Perrott, F. A. (1985) 'Milankovitch forcing of fluctuations in the level of tropical lakes from 18 to 0 kyr BP', *Nature* 317: 130–4.

Kutzbach, J. E., and Webb III, T. (1991) 'Late Quaternary climatic and vegetational change in eastern North America: concepts, models, and data', pp. 175–271 in L. C. K. Shane and E. J. Cushing (eds) *Quaternary Landscapes*, Minneapolis: University of Minnesota Press.

Kutzbach, J. E. and Webb III, T. (1993) 'Conceptual basis for understanding late-Quaternary climates', pp. 5–11 in H. E. Wright Jr, J. E. Kutzbach, T. Webb III, W. F. Ruddiman, F. A. Street-Perrott, and P. J. Bartlein (eds) *Global Climates since the Last Glacial Maximum*, Minneapolis and London: University of Minnesota Press.

Kvet, R. (1993) 'Valleys and river benches from the viewpoint of morphotectonics', *GeoJournal* 30: 403–8.

Labandeira, C. and Sepkoski, J. J., Jr (1993) 'Insect diversity in the fossil record', *Science* 261: 310–15.

Lamb, H. H. (1995) *Climate, History and the Modern World*, 2nd edn, London: Routledge.

Lambeck, K. (1980) *The Earth's Rotation: Geophysical Causes and Consequences*, Cambridge: Cambridge University Press.

Lambeck, K. (1990) 'Late Pleistocene, Holocene and present sea-levels: constraints on future change', *Palaeogeography, Palaeoclimatology, Palaeoecology (Global and Planetary Change Section)* 89: 205–17.

Lancaster, N. (1995) *Geomorphology of Desert Dunes*, London: Routledge.

Landsberg, H. E. (1980) 'Variable solar emissions, the "Maunder Minimum" and climatic temperature fluctuations', *Archiv für Meteorologie, Geophysik und Bioclimatologie* 28B: 181–91.

Landscheidt, T. (1983) 'Solar oscillations and climatic change', pp. 293–398 in B. M. McCormac (ed.) *Weather and Climatic Response to Solar Variations*, Boulder, Colorado: Colorado Associated University Press.

Lang, K. R. (1995) *Sun, Earth and Sky*, Berlin: Springer.
Larson, R. L. (1991) 'Latest pulse of the Earth: evidence for a mid-Cretaceous superplume', *Geology* 19: 547–50.
Larson, R. L. and Pitman III, W. C. (1972) 'World-wide correlation of Mesozoic magnetic anomalies, and its implications', *Bulletin of the Geological Society of America* 83: 3645–62.
Laskar, J., Joutel, F., and Robutel, P. (1993) 'Stabilization of the Earth's obliquity by the Moon', *Nature* 361: 615–17.
Lay, T. (1994) 'The fate of descending slabs', *Annual Review of Earth and Planetary Science* 22: 33–61.
Leeder, M. R. (1991) 'Denudation, vertical crustal movements and sedimentary basin infill', *Geologische Rundschau* 80: 441–58.
Lees, B. G., Stanner, J., Price, D. M., and Yanchou, L. (1995) 'Thermoluminescence dating of dune podzols at Cape Arnhem, northern Australia', *Marine Geology* 129: 63–75.
Legrand, J. P., Le Goff, M., Mazaudier, C., and Schröder, W. (1992) 'Solar and auroral activities during the seventeenth century', pp. 40–76c in W. Schröder and J. P. Legrand (eds) *Solar–Terrestrial Variability and Global Change* (Selected Papers from the Symposia of the Interdivisional Commission on History of the IAGA during the IUGG/IAGA Assembly, held in Vienna, 1991), Bremen-Roennebeck, Germany: Interdivisional Commission on History of the International Association of Geomagnetism and Aeronomy (IAGA).
Leigh, D. S. (1996) 'Soil chronosequence of Brasstown Creek, Blue Ridge Mountains, USA', *Catena* 26: 99–114.
Leigh, D. S. and Feeney, T. P. (1995) 'Paleochannels indicating wet climate and lack of response to lower sea level, southeast Georgia', *Geology* 23: 687–90.
Lenardic, A. and Kaula, W. M. (1994) 'Tectonic plates, D″ thermal structure, and the nature of mantle plumes', *Journal of Geophysical Research* 99(B8): 15,697–708.
Lessios, H. A. (1991) 'Presence and absence of monthly reproductive rhythms among eight Caribbean echinoids off the coast of Panama', *Journal of Experimental Marine Biology and Ecology* 153: 27–47.
Lézine, A.-M. (1988a) 'New pollen data from the Sahel, Senegal', *Review of Palaeobotany and Palynology* 55: 141–54.
Lézine, A.-M. (1988b) 'Les variations de la couverture forestière mésophile d'Afrique occidentale au cours de l'Holocène', *Comptes rendus de l'Académie des Sciences, Paris, Series II* 307: 439–45.
Lézine, A.-M. (1989) 'Late Quaternary vegetation and climate of the Sahel', *Quaternary Research* 32: 317–34.
Lézine, A.-M. and Casanova, J. (1989) 'Pollen and hydrological evidence for the interpretation of past climates in tropical West Africa during the Holocene', *Quaternary Science Reviews* 8: 45–55.
Lézine, A.-M. and Casanova, J. (1991) 'Correlated oceanic and continental records demonstrate past climate and hydrology of North Africa (0–140 ka)', *Geology* 19: 307–10.
Lézine, A.-M. and Vergnaud-Grazzini, C. (1993) 'Evidence of forest expansion in West Africa since 22 000 BP: a pollen record from the eastern tropical Atlantic', *Quaternary Science Reviews* 12: 203–10.
Lidmar-Bergström, K. (1989) 'Exhumed Cretaceous landforms in south Sweden', *Zeitschrift für Geomorphologie, Supplementband* 72: 21–40.
Lidmar-Bergström, K. (1993) 'Denudation surfaces and tectonics in the southernmost part of the Baltic Shield', *Precambrian Research* 64: 337–45.

Lidmar-Bergström, K. (1995) 'Relief and saprolites through time on the Baltic Shield', *Geomorphology* 12: 45–61.

Lidmar-Bergström, K. (1996) 'Long term morphotectonic evolution in Sweden', *Geomorphology* 16: 33–59.

Linton, D. L. (1955) 'The problem of tors', *Geographical Journal* 121: 289–91.

Liu, H.-S. (1992) 'Frequency variations of the Earth's obliquity and the 100-kyr ice-age cycles', *Nature* 358: 397–9.

Liu, H.-S. (1995) 'A new view on the driving mechanisms of Milankovitch glaciation cycles', *Earth and Planetary Science Letters* 131: 17–26.

Livermore, R. A., Vine, F. J., and Smith, A. G. (1984) 'Plate motions and the geomagnetic field. II. Jurassic to Tertiary', *Geophysical Journal of the Royal Astronomical Society* 79: 939–61.

Locker, S. and Martini, E. (1989) 'Phytoliths at DSDP Site 591 in the southwest Pacific and the aridification of Australia', *Geologische Rundschau* 78: 1165–72.

Loper, D. E. and McCartney, K. (1990) 'On impacts as a cause of geomagnetic field reversals or flood basalts', pp. 19–25 in V. L. Sharpton and P. D. Ward (eds) *Global Catastrophes in Earth History; An Interdisciplinary Conference on Impacts, Volcanism, and Mass Mortality* (Geological Society of America Special Paper 247), Boulder, Colorado: Geological Society of America.

Loper, D. E., McCartney, K., and Buzyna, G. (1988) 'A model of correlated episodicity in magnetic-field reversals, climate, and mass extinctions', *Journal of Geology* 96: 1–15.

Love, S. G. and Brownlee, D. E. (1993) 'A direct measurement of the terrestrial mass accretion rate of cosmic dust', *Science* 262: 50–3.

Lovelock, J. E. (1979) *Gaia: A New Look at Life on Earth*, Oxford and New York: Oxford University Press.

Lovelock, J. E. (1988) *The Ages of Gaia: A Biography of Our Living Earth*, Oxford: Oxford University Press.

Lovelock, J. E. (1991) 'Geophysiology – the science of Gaia', pp. 3–10 in S. H. Schneider and P. J. Boston (eds) *Scientists on Gaia*, Cambridge, Massachusetts: MIT Press.

Lowe, J. J., Coope, G. R., Sheldrik, C., Harness, D. D., and Walker, M. J. C. (1995) 'Direct comparison of UK temperatures and Greenland snow accumulation rates 15,000–12,000 yr ago', *Journal of Quaternary Science* 10: 175–80.

Lowe, J. J. and Walker, M. J. C. (1997) *Reconstructing Quaternary Environments* (2nd edn), Harlow, Essex: Longman.

Ludwig, K. R., Muhs, D. R., Simmons, K. R., Halley, R. B., and Shinn, E. A. (1996) 'Sea-level records at ~80 ka from tectonically stable platforms: Florida and Bermuda', *Geology* 24: 211–14.

Lundelius, E. L., Jr (1991) 'The Avenue local fauna, late Pleistocene vertebrates from terrace deposits at Austin, Texas', *Annales Zoologici Fennici* 28: 329–40.

Lundelius, E. L., Jr, Graham, R. W., Anderson, E., Guilday, J., Holman, J. A., Steadman, D., and Webb, S. D. (1983) 'Terrestrial vertebrate faunas', pp. 311–53 in S. C. Porter (ed.) *Late-Quaternary Environments of the United States. Volume 1. The Late Pleistocene*, London: Longman.

McBean, G. A. (1991) 'Global climate, energy and water cycle', pp. 177–85 in J. Jäger and H. L. Ferguson (eds) *Climate Change* (Proceedings of the 2nd World Climate Conference, Geneva, 1990), Cambridge: Cambridge University Press.

McCrea, W. H. (1981) 'Long time-scale fluctuations in the evolution of the Earth', *Proceedings of the Royal Society of London* 375A: 1–41.

MacDonald, G. J. F. (1964) 'Tidal friction', *Reviews of Geophysics* 2: 467–541.

McFarlane, M. J. (1976) *Laterite and Landscape*, London: Academic Press.

McGowran, B. (1990) 'Fifty million years ago', *American Scientist* 78: 30–9.
Mackenzie, F. T. (1975) 'Sedimentary cycling and the evolution of sea water', pp. 309–64 in J. P. Riley and G. Skirrow (eds) *Chemical Oceanography*, vol. 1, London: Academic Press.
McLaren, D. J. (1970) 'Presidential address: Time, life, and boundaries', *Journal of Paleontology* 44: 801–15.
McLaren, D. J. (1988) 'Detection and significance of mass killings', pp. 1–7 in N. J. McMillan, A. F. Embry, and D. J. Glass (eds) *Devonian of the World. Volume III: Paleontology, Paleoecology and Biostratigraphy* (Proceedings of the Second International Symposium on the Devonian System, Calgary, Canada), Calgary, Canada: Canadian Society of Petroleum Geologists.
McLean, D. M. (1981) 'A test of terminal Mesozoic "catastrophe"', *Earth and Planetary Science Letters* 53: 845–54.
McLean, D. M. (1985) 'Deccan traps mantle degassing in the terminal Cretaceous marine extinctions', *Cretaceous Research* 6: 235–59.
McWilliams, M. O. and Kröner, A. (1981) 'Paleomagnetism and tectonic evolution of the Pan-African Damara Belt, southern Africa', *Journal of Geophysical Research* 86: 5147–62.
Malde, H. E. (1968) *The Catastrophic Late Pleistocene Bonneville Flood in the Snake River Plain, Idaho* (United States Geological Survey Professional Paper 596), Washington, DC: United States Government Printing Office.
Margulis, L. and Hinkle, G. (1991) 'The biota and Gaia: 150 years support for environmental sciences', pp. 11–18 in S. H. Schneider and P. J. Boston (eds) *Scientists on Gaia*, Cambridge, Massachusetts and London: MIT Press.
Marsden, B. G. and Steel, D. I. (1994) 'Warning times and impact probabilities of long-period comets', pp. 221–39 in T. Gehrels (ed.), with the editorial assistance of M. S. Matthews and A. M. Schumann, *Hazards Due to Comets and Asteroids*, Tucson and London: University of Arizona Press.
Martin, R. E. (1995) 'Cyclic and secular variation in microfossil biomineralization: clues to the biogeochemical evolution of Phanerozoic oceans', *Global and Planetary Change* 11: 1–23.
Maruyama, S. (1994) 'Plume tectonics', *Journal of the Geological Society of Japan* 100: 24–49.
Maruyama, S., Kumazawa, M., and Kawakami, S. (1994) 'Towards a new paradigm on the Earth's dynamics', *Journal of the Geological Society of Japan* 100: 1–3.
Mason, I. M., Guzkowska, M. A. J., Rapley, C. G., and Street-Perrott, F. A. (1994) 'The response of lake levels and areas to climatic change', *Climatic Change* 27: 161–97.
Masurel, H. (1989) 'Ostracods as palaeoenvironmental indicators in the Lower Carboniferous Yoredale Series of northern England', *Journal of Micropalaeontology* 8: 157–82.
Matese, J. J. and Whitmire, D. P. (1986) 'Planet X and the origins of the shower and steady state flux of short-period comets', *Icarus* 65: 37–50.
Mayr, E. (1970) *Population, Species, and Evolution* (An abridgement of *Animal Species and Evolution*), Cambridge, Massachusetts and London: The Belknap Press of Harvard University Press.
Mellor, A. (1987) 'A pedogenic investigation of some soil chronosequences on neoglacial moraine ridges, southern Norway: examination of soil chemical data using principal components analysis', *Catena* 14: 369–81.
Meybeck, M. (1979) 'Concentrations des eaux fluviales en éléments majeurs et apports en solution aux océans', *Revue de Géologie Dynamique et de Géographie Physique* 21: 215–46.

Miall, A. D. (1995) 'Whither stratigraphy?' *Sedimentary Geology* 100: 5–20.

Milani, A. and Nobili, A. M. (1992) 'An example of stable chaos in the Solar System', *Nature* 357: 569–71.

Milankovitch, M. (1930) 'Mathematische Klimalehre und astronomische Theorie der Klimaschwankungen', pp. 1–176 in W. Köppen and R. Geiger (eds) *Handbuch der Klimatologie, I(A)*, Berlin: Gebrüder Bornträger.

Millero, F. J. (1974) 'Seawater as a multicomponent electrolyte solution', pp. 3–80 in E. D. Goldberg (ed.) *The Sea. Volume 5. Marine Chemistry*, New York: John Wiley & Sons.

Mills, H. H. (1990) 'Thickness and character of regolith on mountain slopes in the vicinity of Mountain Lake, Virginia, as indicated by seismic refraction, and implications for hillslope evolution', *Geomorphology* 3: 143–57.

Milly, P. C. D. and Dunne, K. A. (1994) 'Sensitivity of the global water cycle to the water-holding capacity of land', *Journal of Climate* 7: 506–26.

Milsom, T. P., Rochard, J. B. A., and Poole, S. J. (1990) 'Activity patterns of lapwings *Vanellus vanellus* in relation to the lunar cycle', *Ornis Scandinavica* 21: 147–56.

Mitchell, G. F. (1980) 'The search for Tertiary Ireland', *Journal of Earth Science* 3: 13–33.

Miyashiro, A. (1981) 'Precambrian orogenies', pp. 213–37 in A. Miyashiro, K. Aki, and A. M. C. Şengör, *Orogeny*, Chichester: John Wiley & Sons.

Moisseyev, N. M. (1988) 'The ecological imperative', pp. 199–203 in D. C. Pitt (ed.) *The Future of the Environment: The Social Dimensions of Conservation and Ecological Alternatives*, London and New York: Routledge.

Molfino, B., Heusser, L. H., and Woillard, G. M. (1984) 'Frequency components of a Grand Pile pollen record: evidence of precessional orbital forcing', pp. 392–404 in A. Berger, J. Imbrie, J. Hays, G. Kukla, and B. Saltzman (eds) *Milankovitch and Climate: Understanding the Response of Astronomical Forcing, Part I* (Proceedings of the NATO Advanced Research Workshop on Milankovitch and Climate, Palisades, New York, 1982; NATO ASI Series C, Mathematical and Physical Sciences, vol. 126), Dordrecht: D. Reidel.

Monin, A. S. (1986) *An Introduction to the Theory of Climate*, Dordrecht: D. Reidel.

Moorbath, S. (1977) 'The oldest rocks and the growth of continents', *Scientific American* 236: 92–104.

Moore, G. W. and Moore, J. G. (1984) 'Deposit from a giant wave on the island of Lanai, Hawaii', *Science* 226: 1312–15.

Moore, J. G., Bryan, W. B., and Ludwig, K. R. (1994) 'Chaotic deposition by a giant wave, Molokai, Hawaii', *Bulletin of the Geological Society of America* 106: 962–7.

Moore, J. G., Clague, D. A., Holcomb, R. T., Lipman, P. W., Normark, W. R., and Torresan, M. E. (1989) 'Prodigious submarine landslides on the Hawaiian ridge', *Journal of Geophysical Research* 94B: 17,465–84.

Moore, J. G., Normark, W. R., and Holcomb, R. T. (1994) 'Giant Hawaiian landslides', *Annual Review of Earth and Planetary Sciences* 22: 119–44.

Moores, E. M. (1993) 'Neoproterozoic oceanic crustal thinning, emergence of continents, and origin of the Phanerozoic ecosystem: a model', *Geology* 21: 5–8.

Mörner, N.-A. (1980) 'The northwest European "sea-level laboratory" and regional Holocene eustasy', *Palaeogeography, Palaeoclimatology, Palaeoecology* 29: 281–300.

Mörner, N.-A. (1987) 'Models of global sea-level changes', pp. 332–55 in M. J. Tooley and I. Shennan (eds) *Sea-Level Changes*, Oxford: Basil Blackwell.

Mörner, N.-A. (1993a) 'Global change: the last millennia', *Global and Planetary Change* 7: 211–17.

Mörner, N.-A. (1993b) 'Global change: the high-amplitude changes 13–10 ka ago – novel aspects', *Global and Planetary Change* 7: 243–50.

Mörner, N.-A. (1994) 'Internal response to orbital forcing and external cyclic sedimentary sequences', pp. 25–33 in P. L. De Boer and D. G. Smith (eds) *Orbital Forcing and Cyclic Sequences* (Special Publication Number 19 of the International Association of Sedimentologists), Oxford: Blackwell Scientific Publications.

Mosalam Shaltout, M. A., Tadros, M. T. Y., and Mesiha, S. L. (1992) 'Power spectra analysis for world-wide and North Africa historical earthquakes data in relation with sunspots periodicities', pp. 163–74 in W. Schröder and J.-P. Legrand (eds) *Solar–Terrestrial Variability and Global Change*, Bremen–Roennebeck, Germany: Interdivisional Commission on History of the International Association of Geomagnetism and Aeronomy (IAGA).

Muhs, D. R. (1982) 'The influence of topography on the spatial variability of soils in Mediterranean climates', pp. 269–84 in C. E. Thorn (ed.) *Space and Time in Geomorphology*, London: George Allen & Unwin.

Muhs, D. R. (1984) 'Intrinsic thresholds in soil systems', *Physical Geography* 5: 99–110.

Muller, R. A. and MacDonald, G. J. (1995) 'Glacial cycles and orbital inclination', *Nature* 377: 107–8.

Muller, R. A. and Morris, D. E. (1986) 'Geomagnetic reversals from impacts on Earth', *Geophysical Research Letters* 13: 1177–80.

Nahon, D. B. (1991) 'Self-organization in chemical lateritic weathering', *Geoderma* 51: 5–13.

Nakai, K., Yanagisawa, Y., Sato, T., Niimura, Y., and Gashagaza, M. M. (1990) 'Lunar synchronization of spawning in cichlid fishes of the tribe Lamprologini in Lake Tanganyika', *Journal of Fish Biology* 37: 589–98.

Namias, J. (1950) 'The index cycle and its role in the general circulation', *Journal of Meteorology* 17: 130–9.

Nance, R. D., Worsley, T. R., and Moody, J. B. (1988) 'The supercontinent cycle', *Scientific American* 259: 44–51.

Napier, W. M. and Clube, S. V. M. (1979) 'A theory of terrestrial catastrophism', *Nature* 282: 455–9.

Nash, D. B. (1981) 'Fault: a FORTRAN program for modelling the degradation of active normal fault scarps', *Computers & Geosciences* 7: 249–66.

Nesme-Ribes, E. and Mangeney, A. (1992) 'On a plausible mechanism linking the Maunder Minimum to the Little Ice Age', *Radiocarbon* 34: 263–70.

Newell, R. E. (1970) 'Stratospheric temperature change from the Mount Agung volcanic eruption of 1963', *Journal of the Atmospheric Sciences* 27: 977–8.

Newell, R. E. (1981) 'Further studies of the atmospheric temperature change from the Mount Agung volcanic eruption of 1963', *Journal of Volcanology and Geothermal Research* 11: 61–6.

Newsom, H. E. and Taylor, S. R. (1989) 'Geochemical implications of the formation of the Moon by a single giant impact', *Nature* 338: 29–34.

Nieuwenhuyse, A., Jongmans, A. G., and Van Breemen, N. (1994) 'Mineralogy of a Holocene chronosequence on andesitic beach sediments in Costa Rica', *Journal of the Soil Science Society of America* 58: 485–94.

Nikiforoff, C. C. (1935) 'Weathering and soil formation', *Transactions of the Third International Congress of Soil Science, Oxford, England, 1935* 1: 324–6.

Nikiforoff, C. C. (1959) 'Reappraisal of the soil', *Science* 129: 186–96.

Niklas, K. J., Tiffney, B. H., and Knoll, A. H. (1983) 'Patterns in vascular land plant diversification', *Nature* 303: 614–16.

Nininger, H. H. (1942) 'Cataclysm and evolution', *Popular Astronomy* 50: 270–2.

O'Brien, D. P. and Currie, R. G. (1993) 'Observations of the 18.6-year cycle of air pressure and a theoretical model to explain certain aspects of the signal', *Climate Dynamics* 8: 287–98.

O'Hara, S. L. (1993) 'Historical evidence of fluctuations in the level of Lake Pátzcuaro, Michoacán, Mexico over the last 600 years', *Geographical Journal* 159: 51–62.

O'Keefe, J. D. and Ahrens, T. J. (1989) 'Impact production of CO_2 by the K/T extinction bolide and the resultant heating of the Earth', *Nature* 338: 247–9.

Oberbeck, V. R., Marshall, J. R., and Aggarwal, H. (1993) 'Impacts, tillites, and the break up of Gondwanaland', *Journal of Geology* 101: 1–19.

Odum, H. T. (1971) *Environment, Power, and Society*, New York: John Wiley & Sons.

Officer, C. B., Hallam, A., Drake, C. L., and Devine, J. D. (1987) 'Late Cretaceous and paroxysmal Cretaceous/Tertiary extinctions', *Nature* 326: 143–9.

Ollier, C. D. (1959) 'A two-cycle theory of tropical pedology', *Journal of Soil Science* 10: 137–48.

Ollier, C. D. (1960) 'The inselbergs of Uganda', *Zeitschrift für Geomorphologie* NF 4: 470–87.

Ollier, C. D. (1967) 'Landform description without stage names', *Australian Geographical Studies* 5: 73–80.

Ollier, C. D. (1981) *Tectonics and Landforms* (Geomorphology Texts 6), London and New York: Longman.

Ollier, C. D. (1988) 'Deep weathering, groundwater and climate', *Geografiska Annaler* 70A: 285–90.

Ollier, C. D. (1990) 'Morphotectonics of the Lake Albert Rift Valley and its significance for continental margins', *Journal of Geodynamics* 11: 343–55.

Ollier, C. D. (1991) *Ancient Landforms*, London: Belhaven Press.

Ollier, C. D. (1992) 'Global change and long-term geomorphology', *Terra Nova* 4: 312–19.

Ollier, C. D. (1995) 'Tectonics and landscape evolution in southeast Australia', *Geomorphology* 12: 37–44.

Ollier, C. D. (1996) 'Planet Earth', pp. 15–43 in I. Douglas, R. J. Huggett, and M. E. Robinson (eds) *Companion Encyclopedia of Geography*, London: Routledge.

Ollier, C. D. and Pain, C. F. (1994) 'Landscape evolution and tectonics in southeastern Australia', *AGSO Journal of Australian Geology and Geophysics* 15: 335–45.

Ollier, C. D. and Pain, C. (1996) *Regolith, Soils and Landforms*, Chichester: John Wiley & Sons.

Olsen, P. E. (1984) 'Periodicity of lake-level cycles in the Late Triassic Lockatong Formation of the Newark Basin (Newark Supergroup, New Jersey and Pennsylvania)', pp. 129–46 in A. Berger, J. Imbrie, J. Hays, G. Kukla, and B. Saltzman (eds) *Milankovitch and Climate: Understanding the Response of Astronomical Forcing, Part I* (Proceedings of the NATO Advanced Research Workshop on Milankovitch and Climate, Palisades, New York, 1982; NATO ASI Series C, Mathematical and Physical Sciences, vol. 126), Dordrecht: D. Reidel.

Olsen, P. E. (1986) 'A 40-million-year lake record of early Mesozoic orbital climatic forcing', *Science* 234: 842–8.

Olsen, P. E., Kent, D. V., Cornet, B., Witte, W. K., and Schische, R. W. (1996) 'High-resolution stratigraphy of the Newark rift basin (early Mesozoic, eastern North America)', *Bulletin of the Geological Society of America* 108: 40–77.

Olson, J. S. (1958) 'Rates of succession and soil changes on Southern Lake, Michigan, sand dunes', *Botanical Gazette* 119: 125–70.

Öpik, E. J. (1958a) 'Climate and the changing Sun', *Scientific American* 198: 85–92.

Öpik, E. J. (1958b) 'Solar variability and palaeoclimatic changes', *Irish Astronomical Journal* 5: 97–109.

Orme, A. J. and Orme, A. R. (1991) 'Relict barrier beaches as paleoenvironmental indicators in the Californian desert', *Physical Geography* 12: 334–46.

Osterkamp, T. E., Zhang, T., and Romanovsky, V. E. (1994) 'Evidence for a cyclic variation of permafrost temperatures in northern Alaska', *Permafrost and Periglacial Processes* 5: 137–44.

Owen, H. G. (1976) 'Continental displacement and expansion of the Earth during the Mesozoic and Cenozoic', *Philosophical Transactions of the Royal Society of London* 281A: 223–91.

Owen, H. G. (1981) 'Constant dimensions or an expanding Earth?', pp. 172–92 in L. R. M. Cocks (ed.) *The Evolving Earth*, Cambridge: Cambridge University Press and London: British Museum (Natural History).

Pain, C. F. and Ollier, C. D. (1995) 'Inversion of relief – a component of landscape evolution', *Geomorphology* 12: 151–65.

Palais, J. M. and Sigurdsson, H. (1989) 'Petrologic evidence of volatile emissions from major historic and pre-historic volcanic eruptions', pp. 31–53 in A. Berger, R. E. Dickinson, and J. W. Kidson (eds) *Understanding Climate Change* (International Union of Geodesy and Geophysics, vol. 7; Geophysical Monograph 52), Washington, DC: American Geophysical Union.

Pannella, G. (1972) 'Paleontological evidence on the Earth's rotational history since early Precambrian', *Astrophysics and Space Physics* 16: 212–37.

Parenago, P. P. (1952) 'On the gravitational potential of the Galaxy', *Astronomical Journal* 29: 245–87.

Park, R. G. (1988) *Geological Structures and Moving Plates*, London: Blackie Academic and Professional. An Imprint of Chapman & Hall.

Parkes, R. J., Cragg, B. A., Bale, S. J., Getliff, J. M., Goodman, K., Rochelle, P. A., Fry, J. C., Weightman, A. J., and Harvey, S. M. (1994) 'Deep bacterial biosphere in Pacific Ocean sediments', *Nature* 371: 410–13.

Parrish, J. T. (1990) 'Paleoceanographic and paleoclimatic setting of the Miocene phosphogenic episode', pp. 223–40 in W. C. Burnett and S. R. Riggs (eds) *Phosphate Deposits of the World. Volume 3. Neogene to Modern Phosphorites*, Cambridge: Cambridge University Press.

Parrish, J. T. (1993) 'Climate of the supercontinent Pangea', *Journal of Geology* 101: 215–33.

Parrish, J. T., Ziegler, A. M., and Scotese, C. R. (1982) 'Rainfall patterns and the distribution of coals and evaporites in the Mesozoic and Cenozoic', *Palaeogeography, Palaeoclimatology, Palaeoecology* 40: 67–101.

Partridge, J. and Baker, V. R. (1987) 'Paleoflood hydrology of the Salt River, Arizona', *Earth Surface Processes and Landforms* 12: 109–25.

Partridge, T. C. (1993) 'The evidence for Cainozoic aridification in southern Africa', *Quaternary International* 17: 105–10.

Paton, T. R., Humphreys, G. S., and Mitchell, P. B. (1995) *Soils: A New Global View*, London: UCL Press.

Patrick, S. T., Timberlid, J. A., and Stevenson, A. C. (1990) 'The significance of land-use and land-management change in the acidification of lakes in Scotland and Norway: an assessment utilizing documentary sources and pollen analysis', *Philosophical Transactions of the Royal Society of London* 327B: 363–7.

Paul, C. R. C. and Mitchell, S. F. (1994) 'Is famine a common factor in marine mass extinctions?', *Geology* 22: 679–82.

Payette, S. (1988) 'Late-Holocene development of subarctic ombrotrophic peatlands: allogenic and autogenic succession', *Ecology* 69: 516–31.

Payette, S. (1993) 'The range limit of boreal tree species in Quebec–Labrador: an ecological and palaeoecological interpretation', *Review of Palaeobotany and Palynology* 79: 7–30.

Pedersen, K. (1993) 'The deep subterranean biosphere', *Earth-Science Reviews* 34: 243–60.

Pellatt, M. G. and Mathewes, R. W. (1994) 'Paleoecology of postglacial tree line fluctuations on the Queen Charlotte Islands, Canada', *Ecoscience* 1: 71–81.

Penck, W. (1924) *Die morphologische Analyse, ein Kapitel der physikalischen Geologie*, Stuttgart: Engelhorn.

Penck, W. (1953) *Morphological Analysis of Landforms* (Translated and edited by H. Czech and K. C. Boswell), London: Macmillan.

Penvenne, L. J. (1995) 'Turning up the heat', *New Scientist* 148 (no. 2008): 26–30.

Peper, T. and Cloetingh, S. (1995) 'Autocyclic perturbations of orbitally forced signals in the sedimentary record', *Geology* 23: 937–40.

Pfister, C., Zhongwei, Y., and Schule, H. (1994) 'Climatic variations in western Europe and China, AD 1645–1715: a preliminary continental-scale comparison of documentary evidence', *The Holocene* 4: 206–11.

Phillips, B. A. M. and Fralick, P. W. (1994) 'Interpretation of the sedimentology and morphology of perched glaciolacustrine deltas on the flanks of the Lake Superior Basin, Thunder Bay, Ontario', *Journal of Great Lakes Research* 20: 390–406.

Phillips, J. D. (1991) 'The human role in Earth surface systems: some theoretical considerations', *Geographical Analysis* 23: 316–31.

Phillips, J. D. (1992) 'Nonlinear dynamical systems in geomorphology: revolution or evolution?', *Geomorphology* 5: 219–29.

Phillips, J. D. (1995a) 'Self-organization and landscape evolution', *Progress in Physical Geography* 19: 309–21.

Phillips, J. D. (1995b) 'Biogeomorphology and landscape evolution: the problem of scale', *Geomorphology* 13: 337–47.

Phillips. J. D. (1995c) 'Time lags and emergent stability in morphogenic/pedogenic system models', *Ecological Modelling* 78: 267–76.

Piccolo, M. C., Neill, C., and Cerri, C. C. (1994) 'Net nitrogen mineralization and nitrification along a tropical forest-to-pasture chronosequence', *Plant and Soil* 162: 61–70.

Pickerill, R. K. and Brenchley, P. J. (1991) 'Benthic macrofossils as paleoenvironmental indicators in marine siliciclastic facies', *Geoscience Canada* 18: 119–38.

Pimm, S. L. (1991) *Balance of Nature? Ecological Issues in the Conservation of Species and Communities*, Chicago, Illinois: University of Chicago Press.

Piper, J. D. A. (1987) *Palaeomagnetism and the Continental Crust*, Milton Keynes: Open University Press.

Piper, J. D. A. (1990) 'Implications of some recent sedimentological studies to the history of the Earth–Moon system', pp. 227–33 in P. Brosche and J. Sündermann (eds) *Earth's Rotation from Eons to Days* (Proceedings of a Workshop held at the Centre for Interdisciplinary research (ZiF) of the University of Bielefeld, FRG, 26–30 September 1988), Berlin: Springer.

Pitman III, W. C. (1978) 'Relationships between eustasy and stratigraphic sequences of passive margins', *Bulletin of the Geological Society of America* 89: 1389–403.

Pitman III, W. C. and Golovchenko, X. (1991) 'The effect of sea level changes on the morphology of mountain belts', *Journal of Geophysical Research* 96B4: 6879–91.

Posamentier, H. W. and James, D. P. (1993) 'An overview of sequence-stratigraphic concepts: uses and abuses', pp. 3–18 in H. W. Posamentier, C. P. Summerhayes,

B. U. Haq, and G. P. Allen (eds) *Sequence Stratigraphy and Facies Associations* (Special Publication Number 18 of the International Association of Sedimentologists), Oxford: Blackwell Scientific Publications.

Prell, W. L. and Kutzbach, J. E. (1987) 'Monsoon variability over the past 150,000 years', *Journal of Geophysical Research* 92D: 8411–25.

Pye, K., Winspear, N. R., and Zhou, L. P. (1995) 'Thermoluminescence ages of loess and associated sediments in central Nebraska, USA', *Palaeogeography, Palaeoclimatology, Palaeoecology* 118: 73–87.

Quade, J. and Cerling, T. E. (1995) 'Expansion of C_4 grasses in the Late Miocene of northern Pakistan: evidence from stable isotopes in paleosols', *Palaeogeography, Palaeoclimatology, Palaeoecology* 115: 91–116.

Queiroga, H., Costlow, J. D., and Moreira, M. H. (1994) 'Larval abundance patterns of *Carcinus maenas* (Decapoda, Brachyura) in Cana de Mira (Ria de Aveiro, Portugal)', *Marine Ecology Progress Series* 111: 63–72.

Quinn, F. H. and Sellinger, C. E. (1990) 'Lake Michigan record levels of 1838: a present perspective', *Journal of Great Lakes Research* 6: 133–8.

Rabinowitz, D., Bowell, E., Shoemaker, E., and Muinonen, K. (1994) 'The population of Earth-crossing asteroids', pp. 285–312 in T. Gehrels (ed.), with the editorial assistance of M. S. Matthews and A. M. Schumann, *Hazards Due to Comets and Asteroids*, Tucson and London: University of Arizona Press.

Rampino, M. R. (1989) 'Dinosaurs, comets and volcanoes', *New Scientist* 121: 54–8.

Rampino, M. R. (1994) 'Tillites, diamictites, and ballistic ejecta of large impacts', *Journal of Geology* 102: 439–56.

Rampino, M. R. and Caldeira, K. (1993) 'Major episodes of geologic change: correlations, time structure and possible causes', *Earth and Planetary Science Letters* 114: 215–27.

Rampino, M. R. and Haggerty, B. M. (1994) 'Extraterrestrial impacts and mass extinctions of life', pp. 827–57 in T. Gehrels (ed.), with the editorial assistance of M. S. Matthews and A. M. Schumann, *Hazards Due to Comets and Asteroids*, Tucson and London: University of Arizona Press.

Rampino, M. R. and Self, S. (1992) 'Volcanic winter and accelerated glaciation following the Toba super-eruption', *Nature* 359: 50–2.

Rampino, M. R. and Self, S. (1993) 'Climate–volcanism feedback and the Toba eruption of approximately 74,000 years ago', *Quaternary Research* 40: 269–80.

Rampino, M. R., Self, S., and Stothers, R. B. (1988) 'Volcanic winters', *Annual Review of Earth and Planetary Sciences* 16: 73–99.

Rampino, M. R. and Stothers, R. B. (1984a) 'Terrestrial mass extinctions, cometary impacts and the Sun's motion perpendicular to the galactic plane', *Nature* 308: 709–12.

Rampino, M. R. and Stothers, R. B. (1984b) 'Geological rhythms and cometary impacts', *Science* 226: 1427–31.

Rampino, M. R., Stothers, R. B., and Self, S. (1985) 'Climatic effects of volcanic eruptions', *Nature* 313: 272.

Ran, E. H. T., Bohncke, S. J. P., Van Huissteden, K. J., and Vandenberghe, J. (1990) 'Evidence of episodic permafrost conditions during the Weichselian Middle Pleniglacial in the Hengelo Basin (the Netherlands)', *Geologie en Mijnbouw* 69: 207–18.

Rasmussen, E. M., Wang, X., and Ropelewski, C. F. (1990) 'The biennial component of ENSO variability', *Journal of Marine Systems* 1: 71–96.

Raup, D. M. (1972) 'Taxonomic diversity during the Phanerozoic', *Science* 177: 1065–77.

Raup, D. M. (1985) 'Magnetic reversals and mass extinction', *Nature* 314: 341–3.

Raup, D. M. (1993) 'Extinction from a paleontological perspective', *European Review* 1: 207–16.
Raup, D. M. and Sepkoski, J. J. (1982) 'Mass extinctions in the marine fossil record', *Science* 215: 1501–3.
Raymo, M. E. and Ruddiman, W. F. (1992) 'Tectonic forcing of late Cenozoic climate', *Nature* 359: 117–22.
Reid, G. C., McAfee, J. R., and Crutzen, P. J. (1978) 'Effects of intense stratospheric ionisation events', *Nature* 275: 489–92.
Renne, P. R. and Basu, A. R. (1991) 'Rapid eruption of the Siberian Traps flood basalts at the Permo-Triassic boundary', *Science* 253: 176–9.
Retallack, G. J. (1986a) 'Reappraisal of a 2200 Ma-old paleosol near Waterval Onder, South Africa', *Precambrian Research* 32: 195–232.
Retallack, G. J. (1986b) 'The fossil record of soils', pp. 1–57 in V. P. Wright (ed.) *Palaeosols: Their Recognition and Interpretation*, Oxford: Blackwell Scientific.
Retallack, G. J. (1990) *Soils of the Past: An Introduction to Paleopedology*, Boston: Unwin Hyman.
Retallack, G. J. (1992a) 'Paleozoic paleosols', pp. 543–64 in I. P. Martini and W. Chesworth (eds) *Weathering, Soils and Paleosols* (Developments in Earth Surface Processes 2), Amsterdam: Elsevier.
Retallack, G. J. (1992b) 'What to call early plant formations on land', *Palaios* 7: 508–20.
Retallack, G. J. and Krinsley, D. H. (1993) 'Metamorphic alteration of a Precambrian (2.2 Ga) paleosol from South Africa revealed by backscattered electron imaging', *Precambrian Research* 63: 27–41.
Retallack, G. J., Veevers, J. J., and Morante, R. (1996) 'Global coal gap between Permian–Triassic extinction and Middle Triassic recovery of peat-forming plants', *Bulletin of the Geological Society of America* 108: 195–207.
Rhodes II, R. S. (1984) 'Paleoecological and regional paleoclimatic implications of the Farmdalian Craigmile and Woodfordian Waubonsie mammalian local faunas, southwestern Iowa', *Illinois State Museum Report of Investigations* 40: 1–51.
Ricard, Y., Spada, G., and Sabadini, R. (1993) 'Polar wandering of a dynamic Earth', *Geophysical Journal International* 113: 284–98.
Rich, J. E., Johnson, G. L., Jones, J. E., and Campsie, J. (1986) 'A significant correlation between fluctuations in seafloor spreading rates and evolutionary pulsations', *Paleoceanography* 1: 85–95.
Riehl, H. (1954) *Tropical Meteorology*, New York and London: McGraw-Hill.
Rind, D. and Overpeck, J. (1993) 'Hypothesized causes of decade-to-century-scale climate variability: climate model results', *Quaternary Science Reviews* 12: 357–74.
Ripepe, M., Roberts, L. T., and Fischer, A. G. (1991) 'ENSO and sunspot cycles in varved Eocene oil shales from image analysis', *Journal of Sedimentary Petrology* 61: 1155–63.
Rivenæs, J. C. (1992) 'Application of a dual-lithology, depth dependent diffusion equation in stratigraphic simulation', *Basin Research* 4: 133–46.
Roberts, N. (1989) *The Holocene: An Environmental History*, Oxford: Blackwell.
Robertson, D. R. (1992) 'Patterns of lunar settlement and early recruitment in Caribbean reef fishes at Panama', *Marine Biology* 114: 527–37.
Robinson, J, M. (1990) 'Lignin, land plants and fungi: biological evolution affecting Phanerozoic oxygen balance', *Geology* 15: 607–10.
Robinson, J. M. (1991) 'Phanerozoic atmospheric reconstructions: a terrestrial perspective', *Palaeogeography, Palaeoclimatology, Palaeoecology (Global and Planetary Change Section)* 97: 51–62.

Robinson, P. L. (1973) 'Paleoclimatology and continental drift', pp. 451–76 in D. H. Tarling and S. K. Runcorn (eds) *Implications of Continental Drift to the Earth Sciences, Volume 2*, New York: Academic Press.

Robock, A. (1991) 'The volcanic contribution to climate change of the past 100 years', pp. 429–43 in M. E. Schlesinger (ed.) *Greenhouse-Gas-Induced Climatic Change: A Critical Appraisal of Simulations and Observations*, Amsterdam: Elsevier.

Robock, A. and Mass, C. (1982) 'The Mount St. Helens volcanic eruption of 18 May 1980: large short-term surface temperature effects', *Science* 216: 628–30.

Rochester, M. G. (1973) 'The Earth's rotation', *Transactions of the American Geophysical Union* 54: 769–80.

Roddy, D. J., Schuster, S. H., Rosenblatt, M., Grant, L. B., Hassig, P. J., and Kreyenhagen, K. N. (1987) 'Computer simulations of large asteroid impacts into oceanic and continental sites – preliminary results on atmospheric, cratering and ejecta dynamics', *International Journal of Impact Engineering* 5: 525–41.

Rona, P. (1988) 'Hydrothermal mineralization at ocean ridges', *Canadian Mineralogist* 26: 431–65.

Ronov, A. B. (1964) 'Common tendencies in the chemical evolution of the Earth's crust, ocean and atmosphere', *Geochemistry International* 1: 713–37.

Ropelewski, C. F. and Halpert, M. S. (1987) 'Global and regional scale precipitation patterns associated with the El Niño/Southern Oscillation', *Monthly Weather Review* 115: 1606–26.

Rose, W. I. and Chesner, C. A. (1990) 'Worldwide dispersal of ash and gases from earth's largest known eruption: Toba, Sumatra, 75 ka', *Palaeogeography, Palaeoclimatology, Palaeoecology (Global and Planetary Change Section)* 89: 269–75.

Rosenblatt, P., Pinet, P. C., and Thouvenot, E. (1994) 'Comparative hypsometric analysis of Earth and Venus', *Geophysical Research Letters* 21: 465–8.

Rosenzweig, M. L. (1995) *Species Diversity in Space and Time*, Cambridge: Cambridge University Press.

Rossby, C.-G. (1941) 'The scientific basis of modern meteorology', pp. 599–655 in *Climate and Man* (United States Department of Agriculture, Yearbook of Agriculture for 1941), Washington, DC: United States Government Printing Office.

Rossignol-Strick, M. and Planchais, N. (1989) 'Climate patterns revealed by pollen and oxygen isotope records of a Tyrrhenian Sea core', *Nature* 342: 413–16.

Rubey, W. W. (1951) 'Geologic history of sea water', *Bulletin of the Geological Society of America* 62: 1111–47.

Rubincam, D. P. (1993) 'The obliquity of Mars and "climate friction"', *Journal of Geophysical Research* 98E: 10,827–32.

Ruddiman, W. F. and Kutzbach, J. E. (1989) 'Forcing of late Cenozoic Northern Hemisphere climate by plateau uplift in southern Asia and the American west', *Journal of Geophysical Research* 94D: 18,409–27.

Ruddiman, W. F. and Kutzbach, J. E. (1991) 'Plateau uplift and climatic change', *Scientific American* 264: 42–50.

Ruddiman, W. F., Prell, W. L., and Raymo, M. E. (1989) 'Late Cenozoic uplift in southern Asia and the American west: rationale for general circulation modeling experiments', *Journal of Geophysical Research* 94D: 18,379–91.

Ruderman, M. A. (1974) 'Possible consequences of nearby supernova explosions for atmospheric ozone and terrestrial life', *Science* 184: 1079–81.

Rudnick, R. L. (1995) 'Making continental crust', *Nature* 378: 571–8.

Rudnick, R. L. and Fountain, D. M. (1995) 'Nature and composition of the continental crust: a lower crustal perspective', *Reviews of Geophysics* 33: 267–309.

Ruhe, R. V. (1960) 'Elements of the soil landscape', *Transactions of the Seventh International Congress of Soil Science, Madison* 4: 165–70.

Ruhe, R. V. (1975) 'Climatic geomorphology and fully developed slopes', *Catena* 2: 309–20.

Runcorn, S. K. (1982) 'Primeval displacements of the lunar pole', *Physics of the Earth and Planetary Interiors* 29: 135–47.

Runcorn, S. K. (1987) 'The Moon's ancient magnetism', *Scientific American* 257: 34–43.

Russel, K. L. (1968) 'Ocean ridges and eustatic changes in sea level', *Nature* 261: 680–2.

Russell, D. A. (1979) 'The enigma of the extinction of the dinosaurs', *Annual Review of Earth and Planetary Sciences* 7: 163–82.

Sabadini, R. and Yuen, D. A. (1989) 'Mantle stratification and long-term polar wander', *Nature* 339: 373–5.

Sabadini, R., Yuen, D. A., and Boschi, E. (1983) 'Dynamic effects from mantle phase transitions on true polar wander during ice ages', *Nature* 303: 694–6.

Salisbury, E. J. (1925) 'Note on the edaphic succession in some dune soils with special reference to the time factor', *Journal of Ecology* 13: 322–8.

Saltzman, B. (1987) 'Carbon dioxide and the $\delta^{18}O$ record of late-Quaternary climatic change: a global model', *Climate Dynamics* 1: 77–85.

Saltzman, B., Hansen, A. R., and Maasch, K. A. (1984) 'The late Quaternary glaciations as the response of a three-component feedback system to Earth-orbital forcing', *Journal of the Atmospheric Sciences* 41: 3380–9.

Saltzman, B. and Maasch, K. A. (1988) 'Carbon cycle instability as a cause of the late Pleistocene ice age oscillations: modeling the asymmetric response', *Global Biogeochemical Cycles* 2: 177–85.

Saltzman, B. and Maasch, K. A. (1990) 'A first-order global model of late Cenozoic climatic change', *Transactions of the Royal Society of Edinburgh: Earth Sciences* 81: 315–25.

Saltzman, B. and Sutera, A. (1984) 'A model of the internal feedback system involved in late Quaternary climatic variations', *Journal of the Atmospheric Sciences* 41: 736–45.

Saltzman, B. and Sutera, A. (1987) 'The mid-Quaternary climatic transition as a free response of a three-variable dynamical model', *Journal of the Atmospheric Sciences* 44: 236–41.

Saltzman, B. and Verbitsky, M. Ya. (1993) 'Multiple instabilities and modes of glacial rhythmicity in the Plio-Pleistocene: a general theory of late Cenozoic climatic change', *Climate Dynamics* 9: 1–15.

Sanderson, M. (1989) 'Water levels in the Great Lakes – past, present, future', *Ontario Geography* 33: 1–21.

Saunders, I. and Young, A. (1983) 'Rates of surface processes on slopes, slope retreat and denudation', *Earth Surface Processes and Landforms* 8: 473–501.

Savigear, R. A. G. (1952) 'Some observations on slope development in South Wales', *Transactions of the Institute of British Geographers* 18: 31–52.

Savigear, R. A. G. (1956) 'Technique and terminology in the investigations of slope forms', pp. 66–75 in *Premier Rapport de la Commission pour l'Etude des Versants*, Amsterdam: Union Géographique Internationale.

Savigear, R. A. G. (1965) 'A technique of morphological mapping', *Annals of the Association of American Geographers* 55: 514–38.

Schaake, J. C., Jr (1994) 'Science strategy of the GEWEX Continental-scale International Project (GCIP)', *Advances in Water Resources* 17: 117–27.

Scheidegger, A. E. (1961) 'Mathematical models of slope development', *Bulletin of the Geological Society of America* 72: 37–49.

Scheidegger, A. E. (1979) 'The principle of antagonism in the Earth's evolution', *Tectonophysics* 55: T7–T10.

Scheidegger, A. E. (1983) 'Instability principle in geomorphic equilibrium', *Zeitschrift für Geomorphologie* NF 27: 1–19.

Scheidegger, A. E. (1986) 'The catena principle in geomorphology', *Zeitschrift für Geomorphologie* NF 30: 257–73.

Scheidegger, A. E. (1991) *Theoretical Geomorphology*, 3rd, completely revised edn, Berlin: Springer.

Scheidegger, A. E. (1994) 'Hazards: singularities in geomorphic systems', *Geomorphology* 10: 19–25.

Scheidegger, A. E. and Hantke, R. (1994) 'On the genesis of river gorges', *Transactions of the Japanese Geomorphological Union* 15: 91–110.

Schidlowski, M. (1988) 'A 3,800-million-year isotopic record of life from carbon in sedimentary rocks', *Nature* 333: 313–18.

Schindewolf, O. H. (1963) 'Neokatastrophismus?', *Zeitschrift der Deutschen Geologischen Gesellschaft* 114: 430–45.

Schmidt, P. W., Williams, G. E., and Embleton, B. J. J. (1991) 'Low palaeolatitude of late Proterozoic glaciation: early timing of remanence in haematite of the Elatina Formation, South Australia', *Earth and Planetary Science Letters* 105: 355–67.

Schneider, S. H. and Boston, P. J. (eds) (1991) *Scientists on Gaia*, Cambridge, Massachusetts and London: MIT Press.

Schultz, J. (1995) *The Ecozones of the World: The Ecological Divisions of the Geosphere*, Hamburg: Springer.

Schultz, P. H. (1985) 'Polar wandering on Mars', *Scientific American* 253: 82–90.

Schumm, S. A. (1979) 'Geomorphic thresholds: the concept and its applications', *Transactions of the Institute of British Geographers* NS 4: 485–515.

Schumm, S. A. and Lichty, R. W. (1965) 'Time, space, and causality in geomorphology', *American Journal of Science* 263: 110–19.

Sclater, J. G., Jaupart, C., and Galson, D. (1980) 'The heat flow through oceanic and continental crust and the heat loss of the Earth', *Reviews of Geophysics and Space Physics* 18: 269–311.

Seilacher, A. (1991) 'Events and their signature – an overview', pp. 222–6 in G. Einsele, W. Ricken, and A. Seilacher (eds) *Cycles and Events in Stratigraphy*, Berlin: Springer.

Self, S., Rampino, M. R., and Carr, M. J. (1989) 'A reappraisal of the 1835 eruption of Cosiguina and its atmospheric impact', *Bulletin of Volcanology* 52: 57–65.

Sepkoski, J. J., Jr (1989) 'Periodicity in extinction and the problem of catastrophism in the history of life', *Journal of the Geological Society, London* 146: 7–19.

Sepkoski, J. J., Jr (1993) 'Ten years in the library: new data confirm paleontological patterns', *Paleobiology* 19: 43–51.

Sepkoski, J. J., Jr, Bambach, R. K., and Droser, M. L. (1991) 'Secular changes in Phanerozoic event bedding and the biological overprint', pp. 298–312 in G. Einsele, W. Ricken, and A. Seilacher (eds) *Cycles and Events in Stratigraphy*, Berlin: Springer.

Sepkoski, J. J., Jr, Bambach, R. K., Raup, D. M., and Valentine, J. W. (1981) 'Phanerozoic marine diversity and the fossil record', *Nature* 293: 435–7.

Shackleton, R. M. (1973) 'Correlation of structures across Precambrian orogenic belts in Africa', pp. 1091–5 in D. H. Tarling and S. K. Runcorn (eds) *Implications of Continental Drift to the Earth Sciences, Volume 2*, London: Academic Press.

Shaw, H. R. (1994) *Craters, Cosmos, Chronicles: A New Theory of Earth*, Stanford, California: Stanford University Press.

Shaw, H. R. and Moore, J. G. (1988) 'Magmatic heat and the El Niño cycle', *Eos* 69: 1552, 1564–5.
Sheenan, P. M. (1991) 'Patterns of synecology during the Phanerozoic', pp. 103–18 in E. C. Dudley (ed.) *The Unity of Evolutionary Biology, Volume 1*, Portland, Oregon: Dioscorides Press.
Sheenan, P. M., Fastovsky, D. E., Hoffman, D. E., Berghaus, R. G., and Gabriel, D. L. (1991) 'Sudden extinction of the dinosaurs: latest Cretaceous, upper Great Plains, U.S.A.', *Science* 254: 835–9.
Sheenan, P. M. and Russell, D. A. (1994) 'Faunal change following the Cretaceous–Tertiary impact: using paleontological data to assess the hazard of impacts', pp. 879–93 in T. Gehrels (ed.), with the editorial assistance of M. S. Matthews and A. M. Schumann, *Hazards Due to Comets and Asteroids*, Tucson and London: University of Arizona Press.
Shelford, V. E. (1911) 'Ecological succession. II. Pond fishes', *Biological Bulletin* 21: 127–51.
Shelford, V. E. (1913) *Animal Communities in Temperate America as Illustrated in the Chicago Region*, Chicago: University of Chicago Press.
Shepard, F. P. (1963) *Submarine Geology*, 2nd edn, New York: Harper & Row.
Sheridan, R. E. (1987) 'Pulsation tectonics as the control of long-term stratigraphic cycles', *Paleoceanography* 2: 97–118.
Shields, O. (1979) 'Evidence for the initial opening of the Pacific Ocean in the Jurassic', *Palaeogeography, Palaeoclimatology, Palaeoecology* 26: 181–200.
Shklovskii, I. S. (1968) *Supernovae* (Translated from the original manuscript by Literaturprojekt Innsbruck, Austria), New York: John Wiley & Sons.
Shoemaker, E. M., Weissman, P. R., and Shoemaker, C. S. (1994) 'The flux of periodic comets near Earth', pp. 313–35 in T. Gehrels (ed.), with the editorial assistance of M. S. Matthews and A. M. Schumann, *Hazards Due to Comets and Asteroids*, Tucson and London: University of Arizona Press.
Shoemaker, E. M., Wolfe, R. F., and Shoemaker, C. S. (1990) 'Asteroid and comet flux in the neighborhood of Earth', pp. 155–70 in V. L. Sharpton and P. D. Ward (eds) *Global Catastrophes in Earth History: An Interdisciplinary Conference on Impacts, Volcanism, and Mass Mortality* (Geological Society of America Special Paper 247), Boulder, Colorado: Geological Society of America.
Shopov, Y. Y., Ford, D. C., and Schwartz, H. P. (1994) 'Luminescent microbanding in speleothems: high-resolution chronology and paleoclimate', *Geology* 22: 407–10.
Shotyk, W. (1992) 'Organic soils', pp. 203–24 in I. P. Martini and W. Chesworth (eds) *Weathering, Soils and Paleosols* (Developments in Earth Surface Processes 2), Amsterdam: Elsevier.
Signor, P. W. (1994) 'Biodiversity in geological time', *American Zoologist* 34: 23–32.
Sigurdsson, H. (1990) 'Assessment of the atmospheric impact of volcanic eruptions', pp. 99–110 in V. L. Sharpton and P. D. Ward (eds) *Global Catastrophes in Earth History: An Interdisciplinary Conference on Impacts, Volcanism, and Mass Mortality* (Geological Society of America Special Paper 247), Boulder, Colorado: Geological Society of America.
Simonson, R. W. (1959) 'Outline of a generalized theory of soil genesis', *Proceedings of the Soil Science Society of America* 23: 152–6.
Simonson, R. W. (1978) 'A multiple-process model of soil genesis', pp. 1–25 in W. C. Mahaney (ed.) *Quaternary Soils*, Norwich: Geo Abstracts.
Simonson, R. W. (1995) 'Airborne dust and its significance to soils', *Geoderma* 65: 1–43.
Simpson, G. G. (1953) *The Major Features of Evolution*, New York: Columbia University Press.

Singer, A. (1980) 'The paleoclimatic interpretation of clay minerals in soils and weathering profiles', *Earth-Science Reviews* 15: 303–26.

Skjemstad, J. O., Fitzpatrick, R. W., Zarcinas, B. A., and Thompson, C. H. (1992) 'Genesis of podzols on coastal dunes in southern Queensland. II. Geochemistry and forms of elements as deduced from various soils extraction procedures', *Australian Journal of Soil Research* 30: 615–44.

Smit, J. (1994) 'Extinctions at the Cretaceous–Tertiary boundary: the link to the Chicxulub impact', pp. 859–78 in T. Gehrels (ed.), with the editorial assistance of M. S. Matthews and A. M. Schumann, *Hazards Due to Comets and Asteroids*, Tucson and London: University of Arizona Press.

Smith, A. G., Smith, D. G., and Funnell, B. M. (1994) *Atlas of Mesozoic and Cenozoic Coastlines*, Cambridge: Cambridge University Press.

Smith, D. G. and Fisher, T. G. (1993) 'Glacial Lake Agassiz: the northwestern outlet and paleoflood', *Geology* 21: 9–12.

Soil Survey Staff (1975) *Soil Taxonomy: A Basic System of Soil Classification for Making and Interpreting Soil Surveys* (US Department of Agriculture, Agricultural Handbook 436), Washington, DC: US Government Printing Office.

Sonett, C. P. (1991) 'Is radiocarbon a "tracer" for long period solar variability?', *Journal of Geomagnetism and Geoelectricity* 43, Supplement: 803–10.

Sonett, C. P., Williams, C. R., and Mörner, N.-A. (1992) 'The Fourier spectrum of Swedish riverine varves: evidence of sub-arctic quasi-biennial (QBO) oscillations', *Palaeogeography, Palaeoclimatology, Palaeoecology* 98: 57–65.

Sonett, C. P. and Williams, G. E. (1985) 'Solar periodicities expressed in varves from glacial Skilak Lake, southern Alaska', *Journal of Geophysical Research* 90A: 12,019–26.

Southward, A. J. (1991) 'Forty years of changes in species composition and population density of barnacles on a rocky shore near Plymouth', *Journal of the Marine Biological Association of the United Kingdom* 71: 495–513.

Southward, A. J., Butler, E. I., and Pennycuick, L. (1975) 'Recent cyclic changes of climate and in abundance of marine life', *Nature* 253: 714–17.

Spada, G., Ricard, Y., and Sabadini, R. (1992) 'Excitation of true polar wander by subduction', *Nature* 360: 452–4.

Sparks, B. W. (1957) 'The tjaele gravel near Thriplow, Cambridgeshire', *Geological Magazine* 94: 194–200.

Sparks, B. W. (1964) 'The distribution of non-marine Mollusca in the last interglacial in south-east England', *Proceedings of the Malacological Society, London* 34: 302–15.

Spedicato, E. (1990) *Apollo Objects, Atlantis and the Deluge: A Catastrophical Scenario for the End of the Last Glaciation,* Istituto Universitario di Bergamo, Quaderni del Dipartimento di Matematica, Statistica, Informatica e Applicazioni, Anno 1990 N. 22.

Spicer, R. A. (1993) 'Palaeoecology, past climate systems, and C_3/C_4 photosynthesis', *Chemosphere* 27: 947–78.

Sponholz, B., Baumhauer, R., and Felix-Henningsen, P. (1993) 'Fulgurites in the southern central Sahara, Republic of Niger and their palaeoenvironmental significance', *The Holocene* 3: 97–104.

Sprovieri, R. (1992) 'Mediterranean Pliocene biochronology: a high resolution record based on quantitative planktonic foraminifera distribution', *Rivista Italiana di Paleontologia e Stratigrafia* 98: 61–100.

Stanley, S. M. (1988a) 'Paleozoic mass extinctions: shared patterns suggest global cooling as a common cause', *American Journal of Science* 288: 334–52.

Stanley, S. M. (1988b) 'Climatic cooling and mass extinction of Paleozoic reef communities', *Palaios* 3: 228–32.

Stanley, S. M. (1990a) 'Adaptive radiation and macroevolution', *Systematics Association Special Volume 42*: 1–16.
Stanley, S. M. (1990b) 'Delayed recovery and the spacing of major extinctions', *Paleobiology* 16: 401–14.
Stanley, S. M. (1992) 'An ecological theory for the origin of *Homo*', *Paleobiology* 18: 237–57.
Stanley, S. M. (1995) 'Climatic forcing and the origin of the human genus', pp. 233–43 in Board on Earth Sciences and Resources Commission on Geosciences, Environment, and Resources, National Research Council, *Effects of Past Global Change on Life*, Washington, DC: National Academy Press.
Stanley, S. M. and Yang, X. (1994) 'A double mass extinction at the end of the Paleozoic era', *Science* 266: 1340–4.
Starkel, L. (1987) 'Long-term and short-term rhythmicity in terrestrial landforms and deposits', pp. 323–32 in M. R. Rampino, J. E. Sanders, W. S. Newman, and L. K. Königsson (eds) *Climate: History, Periodicity, and Predictability*, New York: Van Nostrand Reinhold.
Steel, D. I. (1991) 'Our asteroid-pelted planet', *Nature* 354: 265–7.
Steel, D. I., Asher, D. J., Napier, W. M., and Clube, S. V. M. (1994) 'Are impacts correlated in time?', pp. 463–77 in T. Gehrels (ed.), with the editorial assistance of M. S. Matthews and A. M. Schumann, *Hazards Due to Comets and Asteroids*, Tucson and London: University of Arizona Press.
Steiner, J. (1967) 'The sequence of geological events and the dynamics of the Milky Way Galaxy', *Journal of the Geological Society of Australia* 14: 99–131.
Steiner, J. (1973) 'Possible galactic causes for synchronous sedimentation sequences of the North American and eastern European cratons', *Geology* 1: 89–92.
Steiner, J. (1978) 'Lead isotope events of the Canadian Shield, ad hoc solar galactic orbits and glaciations', *Precambrian Research* 6: 269–74.
Steiner, J. (1979) 'Regularities of the revised Phanerozoic time scale and the Precambrian time scale', *Geologische Rundschau* 68: 825–31.
Steiner, J. and Grillmair, E. (1973) 'Possible galactic causes for periodic and episodic glaciations', *Bulletin of the Geological Society of America* 84: 1003–18.
Stephens, D. W. (1990) 'Changes in lake levels, salinity and the biological community of Great Salt Lake (Utah, USA)', *Hydrobiologia* 197: 139–46.
Stevens, C. H. (1977) 'Was development of brackish oceans a factor in Permian extinctions?', *Bulletin of the Geological Society of America* 88: 133–8.
Stewart, C. A. and Rampino, M. R. (1992) 'Time dependent thermal convection in the Earth's mantle: theory and observation', *Transactions of the American Geophysical Union* 43: 303.
Stoddart, D. R. (1969) 'Climatic geomorphology: review and re-assessment', *Progress in Physical Geography* 1: 160–222.
Stoltz, J. F., Botkin, D. B., and Dastoor, M. N. (1989) 'The integral biosphere', pp. 31–49 in M. B. Rambler, L. Margulis, and R. Fester (eds) *Global Ecology: Towards a Science of the Biosphere*, San Diego, California: Academic Press.
Stothers, R. B. and Rampino, M. R. (1990) 'Periodicity in flood basalts, mass extinctions, and impacts; a statistical view and a model', pp. 9–18 in V. L. Sharpton and P. D. Ward (eds) *Global Catastrophes in Earth History: An Interdisciplinary Conference on Impacts, Volcanism, and Mass Mortality* (Geological Society of America Special Paper 247), Boulder, Colorado: Geological Society of America.
Strahler, A. H. and Strahler, A. N. (1994) *Introducing Physical Geography*, New York: John Wiley & Sons.
Street-Perrott, F. A. and Perrott, R. A. (1990) 'Abrupt climate fluctuations in the tropics: the influence of Atlantic Ocean circulation', *Nature* 343: 607–12.

Strobel, J., Cannon, R., Kendall, C. G. St C., Biswas, G., and Bedzek, J. (1989) 'Interactive (SEDPAK) simulation of clastic and carbonate sediments in shelf to basin settings', *Computers & Geosciences* 15: 1279–90.

Suess, H. E. and Linick, T. W. (1990) 'The ^{14}C record in bristlecone pine wood of the past 8000 years based on the dendrochronology of the late C. W. Ferguson', *Philosophical Transactions of the Royal Society, London* 330A: 403–12.

Summerfield, M. A. (1983) 'Silcrete as a palaeoclimatic indicator: evidence from southern Africa', *Palaeogeography, Palaeoclimatology, Palaeoecology* 41: 65–79.

Summerfield, M. A. (1984) 'Plate tectonics and landscape development on the African continent', pp. 27–51 in M. Morisawa and J. T. Hack (eds) *Tectonic Geomorphology* (The Symposia on Geomorphology, International Series no. 15), Boston, Massachusetts: George Allen & Unwin.

Summerfield, M. A. (1987) 'Neotectonics and landform genesis', *Progress in Physical Geography* 11: 384–97.

Summerfield, M. A. (1988) 'Global tectonic and landform development', *Progress in Physical Geography* 12: 389–404.

Summerfield, M. A. (1991a) *Global Geomorphology: An Introduction to the Study of Landforms*, Harlow, Essex: Longman.

Summerfield, M. A. (1991b) 'Sub-aerial denudation of passive margins: regional elevation versus local relief models', *Earth and Planetary Science Letters* 102: 460–9.

Summerfield, M. A. and Thomas, M. F. (1987) 'Long-term landform development: editorial comment', pp. 927–33 in V. Gardiner (ed.) *International Geomorphology 1986, Part II* (Proceedings of the First International Conference on Geomorphology), Chichester: John Wiley & Sons.

Sutton, J. (1967) 'The extension of the geological record into the Pre-Cambrian', *Proceedings of the Geologists' Association, London* 78: 493–534.

Swinburne, N. (1993) 'It came from outer space', *New Scientist* 137 (1861): 28-32.

Tagliaferri, E., Spalding, R., Jacobs, C., Worden, S. P., and Erlich, A. (1994) 'Detection of meteoroid impacts by optical sensors in Earth orbit', pp 199–220 in T. Gehrels (ed.), with the editorial assistance of M. S. Matthews and A. M. Schumann, *Hazards Due to Comets and Asteroids*, Tucson and London: University of Arizona Press.

Tallis, J. H. (1994) 'Pool-and-hummock patterning in a southern Pennine blanket mire. II. The formation and erosion of the pool systems', *Journal of Ecology* 82: 789–803.

Tandon, S. K. and Gibling, M. R. (1994) 'Calcrete and coal in Late Carboniferous cyclothems of Nova Scotia, Canada: climate and sea-level changes linked', *Geology* 22: 755–8.

Tansley, A. G. (1939) *The British Islands and Their Vegetation*, Cambridge: Cambridge University Press.

Tappan, H. (1982) 'Extinction or survival: selectivity and causes of Phanerozoic crises', pp. 265–76 in L. T. Silver and P. H. Schultz (eds) *Geological Implications of Impacts of Large Asteroids and Comets on the Earth* (Geological Society of America Special Paper 190), Boulder, Colorado: Geological Society of America.

Tappan, H. (1986) 'Phytoplankton: below the salt at the global table', *Journal of Paleontology* 60: 545–54.

Tardy, Y., N'Kounkou, R., and Probst, J.-L. (1989) 'The global water cycle and continental erosion during Phanerozoic time (570 my)', *American Journal of Science* 289: 455–83.

Taylor, S. R. and McLennan, S. M. (1996) 'The evolution of continental crust', *Scientific American* 274: 60–5.

Terry, K. D. and Tucker, W. H. (1968) 'Biological effects of supernovae', *Science* 159: 421–3.

Tetzlaff, D. M. and Harbaugh, J. W. (1990) *Simulating Clastic Sedimentation*, New York: Van Nostrand Reinhold.

Thierstein, H. and Berger, W. H. (1978) 'Injection events in ocean history', *Nature* 276: 461–6.

Thiry, M. and Simon-Coinçon, R. (1996) 'Tertiary paleoweathering and silcretes in the southern Paris Basin', *Catena* 26: 1–26.

Thomas, M. F. (1965) 'Some aspects of the geomorphology of tors and domes in Nigeria', *Zeitschrift für Geomorphologie* NF 9: 63–81.

Thomas, M. F. (1978) 'Denudation in the tropics and the interpretation of the tropical legacy in higher latitudes – a view of the British experience', pp. 185–202 in C. Embleton (ed.) *Geomorphology: Present Problems, Future Prospects*, Oxford: Oxford University Press.

Thomas, M. F. (1989a) 'The role of etch processes in landform development. I. Etching concepts and their application', *Zeitschrift für Geomorphologie* NF 33: 129–42.

Thomas, M. F. (1989b) 'The role of etch processes in landform development. II. Etching and the formation of relief', *Zeitschrift für Geomorphologie* NF 33: 257–74.

Thomas, M. F. (1994) *Geomorphology in the Tropics: A Study of Weathering and Denudation in Low Latitudes*, Chichester: John Wiley & Sons.

Thomas, M. F. (1995) 'Models for landform development on passive margins. Some implications for relief development in glaciated areas', *Geomorphology* 12: 3–15.

Thomas, M. F. and Thorp, M. B. (1985) 'Environmental change and episodic etchplanation in the humid tropics of Sierra Leone: the Koidu etchplain', pp. 239–67 in I. Douglas and T. Spencer (eds) *Environmental Change and Tropical Geomorphology*, London: George Allen & Unwin.

Thompson, A. B. (1991) 'Petrology of a dynamic Earth's mantle', *Ecologae Geologicae Helvetiae* 84: 285–96.

Thompson, C. H. (1992) 'Genesis of podzols on coastal dunes in southern Queensland. I. Field relationships and profile morphology', *Australian Journal of Soil Research* 30: 593–613.

Thompson, W. F. (1990) 'Climate related landscapes in world mountains: criteria and maps', *Zeitschrift für Geomorphologie Supplementband* 78.

Thorn, C. E. (1988) *An Introduction to Theoretical Geomorphology*, Boston, Massachusetts: Unwin Hyman.

Thornbury, W. D. (1954) *Principles of Geomorphology*, New York: John Wiley & Sons; London: Chapman & Hall.

Tickell, O. (1996) 'Healing the planet', *Independent On Sunday*, 4 August 1996, pp. 40–1.

Tiedemann, R., Sarnthein, M., and Shackleton, N. J. (1994) 'Astronomic timescale for the Pliocene Atlantic $\delta^{18}O$ and dust flux records of Ocean Drilling Program Site 659', *Paleoceanography* 9: 619–38.

Tipping, R. (1994) 'Fluvial chronology and valley floor evolution of the upper Bowmont Valley, Borders Region, Scotland', *Earth Surface Processes and Landforms* 19: 641–57.

Titley, S. R. (1993) 'Relationship of stratabound ores with tectonic cycles of the Phanerozoic and Proterozoic', *Precambrian Research* 61: 295–322.

Toon, O. B., Zahnle, K., Turco, R. P., and Covey, C. (1994) 'Environmental perturbations caused by asteroid impacts', pp. 791–826 in T. Gehrels (ed.), with the editorial assistance of M. S. Matthews and A. M. Schumann, *Hazards Due to Comets and Asteroids*, Tucson and London: University of Arizona Press.

Torbett, M. V. (1989) 'Solar system and galactic influences on the stability of the Earth', *Palaeogeography, Palaeoclimatology, Palaeoecology (Global and Planetary Change Section)* 75: 3–33.

Trefil, J. S. and Raup, D. M. (1990) 'Crater taphonomy and bombardment rates in the Phanerozoic', *Journal of Geology* 98: 385–98.

Tricart, J. and Cailleux, A. (1972) *Introduction to Climatic Geomorphology* (Translated from the French by Conrad J. Kiewiet de Jonge), London: Longman.

Trimble, S. W. (1995) 'Catchment sediment budgets and change', pp. 201–15 in A. Gurnell and G. Petts (eds) *Changing River Channels*, Chichester: John Wiley & Sons.

Trupin, A. and Wahr, J. (1990) 'Spectroscopic analysis of global tide gauge sea level data', *Geophysical Journal International* 100: 441–53.

Tubia, J. M. (1994) 'The Ronda peridotites (Los Reales nappe): an example of the relationship between lithospheric thickening by oblique tectonics and late extensional deformation within the Betic Cordillera (Spain)', *Tectonophysics* 238: 381–98.

Tucker, M. E. (1992) 'The Precambrian–Cambrian boundary: seawater chemistry, ocean circulation and nutrient supply in metazoan evolution, extinction and biomineralization', *Journal of the Geological Society of London* 149: 655–68.

Tucker, M. E. and Benton, M. J. (1982) 'Triassic environments, climates and reptile evolution', *Palaeogeography, Palaeoclimatology, Palaeoecology* 40: 361–79.

Twidale, C. R. (1976) 'On the survival of paleoforms', *American Journal of Science* 276: 77–95.

Twidale, C. R. (1991) 'A model of landscape evolution involving increased and increasing relief amplitude', *Zeitschrift für Geomorphologie* NF 35: 85–109.

Twidale, C. R. (1994) 'Gondwanan (Late Jurassic and Cretaceous) palaeosurfaces of the Australian craton', *Palaeogeography, Palaeoclimatology, Palaeoecology* 112: 157–86.

Twidale, C. R., Bourne, J. A., and Smith, D. M. (1974) 'Reinforcement and stabilisation mechanisms in landform development', *Revue de Géomorphologie Dynamique* 28: 81–95.

Twidale, C. R. and Campbell, E. M. (1995) 'Pre-Quaternary landforms in the low latitude context: the example of Australia', *Geomorphology* 12: 17–35.

Twidale, C. R. and Lageat, Y. (1994) 'Climatic geomorphology: a critique', *Progress in Physical Geography* 18: 319–34.

Useinova, I. (1989) 'The astounding continental factor', *Geographical Magazine* 61: 24–6.

Vail, P. R., Audemard, F., Bowman, S. A., Eisner, P. N., and Perez-Cruz, G. (1991) 'The stratigraphic signatures of tectonics, eustasy and sedimentology – an overview', pp. 617–59 in G. Einsele, W. Ricken, and A. Seilacher (eds) *Cycles and Events in Stratigraphy*, Berlin: Springer.

Vail, P. R., Mitchum, R. M., Jr, and Thompson III, S. (1977) 'Seismic stratigraphy and global changes of sea level. Part 4: Global cycles of relative changes of sea levels', pp. 83–97 in C. E. Payton (ed.) *Seismic Stratigraphy: Applications to Hydrocarbon Exploration* (American Association of Petroleum Geologists Memoir 26), Tulsa, Oklahoma: American Association of Petroleum Geologists.

Valentine, J. W. (1989) 'Phanerozoic marine faunas and the stability of the Earth system', *Palaeogeography, Palaeoclimatology, Palaeoecology (Global and Planetary Change Section)* 75: 137–55.

Valentine, J. W., Foin, T. C., and Peart, D. (1978) 'A provincial model of Phanerozoic marine diversity', *Paleobiology* 4: 55–66.

Valeton, I. (1994) 'Element concentration and formation of ore deposits by weathering', *Catena* 21: 99–129.

Van der Voo, R. (1994) 'True polar wander during the middle Paleozoic?', *Earth and Planetary Science Letters* 122: 239–43.

van Houten, F. W. (1964) 'Cyclic lacustrine sedimentation, Upper Triassic Lockatong Formation, central New Jersey and adjacent Pennsylvania', *Kansas State Geological Survey Bulletin* 169: 497–531.

Van Valkenburgh, B. and Janis, C. M. (1993) 'Historical diversity patterns in North American large herbivores and carnivores', pp. 330–40 in R. E. Ricklefs and D. Schluter (eds) *Species Diversity in Ecological Communities: Historical and Geographical Perspectives*, Chicago, Illinois: University of Chicago Press.

Van Wagoner, J. C., Mitchum, R. M., Campion, K. M., and Rahmanian, V. D. (1990) *Siliciclastic Sequence Stratigraphy in Well Logs, Cores, and Outcrops: Concepts for High-Resolution Correlation of Time and Facies* (AAPG Methods in Exploration Series, no. 7), Tulsa, Oklahoma: American Association of Petroleum Geologists.

Varnes, D. J. (1978) 'Slope movement and types and processes', pp. 11–33 in R. L. Schuster and R. J. Krizek (eds) *Landslides: Analysis and Control* (Transportation Research Board Special Report 176), Washington, DC: National Academy of Sciences.

Vasavada, A. R., Milavec, T. J., and Paige, D. A. (1993) 'Microcraters of Mars: evidence for past climate variations', *Journal of Geophysical Research* 98E: 3469–76.

Veevers, J. J. (1990) 'Tectonic–climatic supercycle in the billion-year plate-tectonic eon: Permian Pangean icehouse alternates with Cretaceous dispersed-continents greenhouse', *Sedimentary Geology* 68: 1–16.

Veizer, J. and Jansen, S. L. (1979) 'Basement and sedimentary recycling and continental evolution', *Journal of Geology* 87: 341–70.

Verardo, D. J. and McIntyre, A. (1994) 'Production and destruction: control of biogenous sedimentation in the tropical Atlantic 0–300 000 years BP', *Paleoceanography* 9: 63–86.

Vita-Finzi, C. (1973) *Recent Earth History*, London and Basingstoke: Macmillan.

Vogt, P. R. (1979) 'Global magmatic episodes: new evidence and implications for steady-state mid-ocean ridge', *Geology* 7: 93–8.

Vrba, E. S., Denton, G. H., and Prentice, M. L. (1989) 'Climatic influences on early hominid behaviour', *Ossa* 14: 127–56.

Waddington, C. H. (1957) *The Strategy of the Genes: A Discussion of Some Aspects of Theoretical Biology*, London: Macmillan.

Walker, D. (1970) 'Direction and rate in some British post-glacial hydroseres', pp. 117–39 in D. Walker and R. G. West (eds) *Studies in the Vegetational History of the British Isles*, Cambridge: Cambridge University Press.

Walter, H. (1985) *Vegetation of the Earth and Ecological Systems of the Geo-Biosphere*, 3rd revised and enlarged edn (Translated from the 5th revised German edn by O. Muise), Berlin: Springer.

Waltham, D. (1992) 'Mathematical modelling of sedimentary basin processes', *Marine and Petroleum Geology* 9: 265–73.

Wang, K., Geldsetzer, H. H. J., and Krouse, H. R. (1994) 'Permian–Triassic extinction: organic $\delta^{13}C$ evidence from British Columbia, Canada', *Geology* 22: 580–4.

Wang, Y., Evans, M. E., Rutter, N., and Ding, Z. (1990) 'Magnetic susceptibility of Chinese loess and its bearing on paleoclimate', *Geophysical Research Letters* 17: 2449–51.

Ward, W. R. (1973) 'Large-scale variations in the obliquity of Mars', *Science* 181: 260–2.

Watkins, J. R. (1967) 'The relationship between climate and the development of landforms in Cainozoic rocks of Queensland', *Journal of the Geological Society of Australia* 14: 153–68.

Watson, J. P. (1961) 'Some observations on soil horizons and insect activity in granite soils', *Proceedings of the First Federal Science Congress, 1960, Salisbury, Southern Rhodesia* 1: 271–6.

Watts, W. A. and Hansen, B. C. S. (1994) 'Pre-Holocene and Holocene pollen records of vegetation history from the Florida Peninsula and their climatic implications', *Palaeogeography, Palaeoclimatology, Palaeoecology* 109: 163–76.

Wayland, E. J. (1933) 'Peneplains and some other erosional platforms', *Annual Report and Bulletin, Protectorate of Uganda Geological Survey, Department of Mines, Notes* 1: 77–9.

Webb, N. R. and Thomas, J. A. (1994) 'Conserving insect habitats in heathland biotopes: a question of scale', pp. 129–51 in P. J. Edwards, R. M. May, and N. R. Webb (eds) *Large-Scale Ecology and Conservation Biology* (35th Symposium of the British Ecological Society with the Society for Conservation Biology, University of Southampton, 1993), Oxford: Blackwell Scientific Publications.

Webb, S. D. (1983) 'The rise and fall of the Late Miocene ungulate fauna in North America', pp. 267–306 in M. H. Nitecki (ed.) *Coevolution*, Chicago and London: University of Chicago Press.

Webb, S. D. and Opdyke, N. D. (1995) 'Global climatic influence on Cenozoic land mammal faunas', pp. 184–208 in Board on Earth Sciences and Resources Commission on Geosciences, Environment, and Resources, National Research Council, *Effects of Past Global Change on Life*, Washington, DC: National Academy Press.

Webb III, T., Ruddiman, W. F., Street-Perrott, F. A., Markgraf, V., Kutzbach, J. E., Bartlein, P. J., Wright, H. E., Jr, and Prell, W. L. (1993) 'Climatic changes during the past 18,000 years: regional syntheses, mechanisms, and causes', pp. 514–35 in H. E. Wright Jr, J. E. Kutzbach, T. Webb III, W. F. Ruddiman, F. A. Street-Perrott, and P. J. Bartlein (eds) *Global Climates since the Last Glacial Maximum*, Minneapolis and London: University of Minnesota Press.

Weijermars, R. (1986) 'Slow but not fast global expansion may explain the surface dichotomy of the Earth', *Physics of the Earth and Planetary Interiors* 43: 67–89.

Weischet, W. (1991) *Einführung in die Allgemeine Klimatologie*, 5th edn, Stuttgart: Teubner.

Weissman, P. R. (1984) 'Cometary showers and unseen solar companions', *Nature* 312: 380–1.

Wells, J. W. (1963) 'Coral growth and geochronometry', *Nature* 197: 948–50.

Werner, B. T. and Fink, T. M. (1994) 'Beach cusps as self-organized patterns', *Science* 260: 968–71.

Westbroek, P. (1991) *Life as a Geologial Force: Dynamics of the Earth*, New York and London: W. W. Norton.

Wetherill, G. W. (1985) 'Occurrence of giant impacts during the growth of the terrestrial planets', *Science* 228: 877–9.

Wetherill, G. W. (1990) 'Formation of the Earth', *Annual Review of Earth and Planetary Sciences* 18: 205–56.

Wetzel, A. (1991) 'Stratification in black shales: depositional models and timing – an overview', pp. 508–23 in G. Einsele, W. Ricken, and A. Seilacher (eds) *Cycles and Events in Stratigraphy*, Berlin: Springer.

Wexler, H. (1952) 'Volcanoes and climate', *Scientific American* 186: 74–80.

Whitmire, D. P. and Jackson, A. A. (1984) 'Are periodic mass extinctions driven by a solar companion?', *Nature* 308: 713–15.

Whitmire, D. P. and Matese, J. J. (1985) 'Periodic comet showers and Planet X', *Nature* 313: 36.

Whitney, J. W. and Harrington, C. D. (1993) 'Relict colluvial boulder deposits as paleoclimatic indicators in the Yucca Mountain region, southern Nevada', *Bulletin of the Geological Society of America* 105: 1008–18.

Whyte, M. A. (1977) 'Turning points in Phanerozoic history', *Nature* 267: 679–82.

Wigley, T. M. L. and Kelly, P. M. (1990) 'Holocene climatic change, ^{14}C wiggles and variations in solar irradiance', *Philosophical Transactions of the Royal Society of London* 330A: 547–60.

Wilby, R. (ed.) (1996) *Contemporary Hydrology*, Chichester: John Wiley & Sons.

Willett, H. C. (1949) 'Solar variability as a factor in the fluctuations of climate during geological time', *Geografiska Annaler* 31: 295–315.

Willett, H. C. (1980) 'Solar prediction of climatic change', *Physical Geography* 1: 95–117.

Williams, G. E. (1975a) 'Possible relation between periodic glaciation and the flexure of the Galaxy', *Earth and Planetary Science Letters* 26: 361–9.

Williams, G. E. (1975b) 'Late Precambrian glacial climate and the Earth's obliquity', *Geological Magazine* 112: 441–4.

Williams, G. E. (1981) 'Editor's comments on papers 21 through 27', pp. 216–23 in G. E. Williams (ed.) *Megacycles: Long-term Episodicity in Earth and Planetary History* (Benchmark Papers in Geology, vol. 57), Stroudsberg, Pennsylvania: Dowden, Hutchinson & Ross.

Williams, G. E. (1993) 'History of the Earth's obliquity', *Earth-Science Reviews* 34: 1–45.

Williams, G. E., Schmidt, P. W., and Embleton, B. J. J. (1995) 'Comment on "The Neoproterozoic (1000–540 Ma) glacial intervals: no more snowball earth?" by Joseph G. Meert and Rob van der Voo', *Earth and Planetary Science Letters* 131: 115–22.

Williams, M. A. J. (1968) 'Termites and soil development near Brocks Creek, northern Australia', *Australian Journal of Soil Science* 31: 153–4.

Williams, M. A. J., Dunkerly, D. L., De Dekker, P., Kershaw, A. P., and Stokes, T. (1993) *Quaternary Environments*, London: Edward Arnold.

Williams, M. E. (1994) 'Catastrophic versus noncatastrophic extinction of the dinosaurs: testing, falsifiability, and the burden of proof', *Journal of Paleontology* 68: 183–90.

Willis, K. J., Sümegi, P., Braun, M., and Tóth, A. (1995) 'The late Quaternary environmental history of Bátorliget, N.E. Hungary', *Palaeogeography, Palaeoclimatology, Palaeoecology* 118: 25–47.

Wilson, J. T. (1963) 'Evidence from islands on the spreading of the ocean floor', *Nature* 197: 536–8.

Wilson, J. T. (1968) 'Static or mobile Earth: the current scientific revolution', *Proceedings of the American Philosophical Society* 112: 309–20.

Windley, B. F. (1993) 'Uniformitarianism today: plate tectonics is the key to the past', *Journal of the Geological Society, London* 150: 7–19.

Windley, B. F. (1995) *The Evolving Continents*, 3rd edn, Chichester: John Wiley & Sons.

Wise, D. U. (1973) 'Freeboard of continents through time', *Memoirs of the Geological Society of America* 132: 87–100.

Wolfe, J. A. (1978) 'A paleobotanical interpretation of Tertiary climates in the Northern Hemisphere', *American Scientist* 66: 694–703.

Wolfe, J. A. (1980) 'Tertiary climates and floristic relationships at high latitudes in the Northern Hemisphere', *Palaeogeography, Palaeoclimatology, Palaeoecology* 30: 313–23.

Wolfe, J. A. (1993) 'A method of obtaining climatic parameters from leaf assemblages', *United States Geological Survey Bulletin* 2040: 1–71.

Wolfe, J. A. (1995) 'Paleoclimatic estimates from Tertiary leaf assemblages', *Annual Review of Earth and Planetary Science* 23: 119–42.

Wolfe, J. A. and Poore, R. Z. (1982) 'Tertiary marine and nonmarine climatic trends', pp. 154–8 in W. H. Berger and J. C. Crowell (eds) *Climate in Earth History* (Studies in Geophysics), Washington, DC: National Academy of Sciences.

Wolman, M. G. and Miller, J. P. (1960) 'Magnitude and frequency of forces in geomorphic processes', *Journal of Geology* 68: 54–74.

Womack, W. R. and Schumm, S. A. (1977) 'Terraces of Douglas Creek, northwestern Colorado: an example of episodic erosion', *Geology* 5: 72–6.

Wood, A. (1942) 'The development of hillside slopes', *Proceedings of the Geologists' Association, London* 53: 128–39.

Wood, F. J. (1985) *Tidal Dynamics: Coastal Flooding and Cycles of Gravitational Force*, Dordrecht: D. Reidel.

Worsley, T. R. and Nance, R. D. (1989) 'Carbon redox and climate control through Earth history: a speculative reconstruction', *Palaeogeography, Palaeoclimatology, Palaeoecology (Global and Planetary Change Section)* 75: 259–82.

Worsley, T. R., Nance, R. D., and Moody, J. B. (1984) 'Global tectonics and eustasy for the past 2 billion years', *Marine Geology* 58: 373–400.

Worster, D. (1994) *Nature's Economy: A History of Ecological Ideas,* 2nd edn, Cambridge: Cambridge University Press.

Wright, H. E., Jr, Kutzbach, J. E., Webb III, T., Ruddiman, W. F., Street-Perrott, F. A., and Bartlein, P. J. (eds) (1993) *Global Climates since the Last Glacial Maximum*, Minneapolis and London: University of Minnesota Press.

Wright, V. P. (1990) 'Equatorial aridity and climatic oscillations during the Early Carboniferous, southern Britain', *Journal of the Geological Society, London* 147: 359–63.

Wyers, S. C., Barnes, H. S., and Smith, S. R. (1991) 'Spawning of hermatypic corals in Bermuda: a pilot study', *Hydrobiologia* 216–17: 109–16.

Wyllie, P. J. (1971) *The Dynamic Earth: Textbook in Geosciences*, New York, John Wiley & Sons.

Wynne-Edwards, H. R. (1976) 'Proterozoic ensialic orogenesis: the millipede model of ductile plate tectonics', *American Journal of Science* 276: 927–53.

Wyrwoll, K.-H. and McConchie, D. (1986) 'Accelerated plate motion and rates of volcanicity as controls on Archaean climates', *Climatic Change* 8: 257–65.

Xu, Q.-Q. (1979) 'On the causes of ice ages', *Scientia Geologica Sinica* 7: 252–63. (In Chinese with English summary.)

Xu, Q.-Q. (1980) 'Climatic variation and the obliquity', *Vertebrata PalAsiatica* 18: 334–43. (In Chinese with English summary.)

Yaalon, D. H. (1971) 'Soil-forming processes in time and space', pp. 29–39 in D. H. Yaalon (ed.) *Paleopedology – Origin, Nature and Dating of Paleosols*, Jerusalem: International Society of Soil Science and Israel Universities Press.

Yabushita, S. (1994) 'Are periodicities in crater formations and mass extinctions related?', *Earth, Moon, and Planets* 64: 207–16.

Yapp, C. Y. and Poths, H. (1992) 'Ancient atmospheric CO_2 pressures inferred from natural goethites', *Nature* 355: 342–4.

Yeh, T.-C. and Fu, C.-B. (1985) 'Climatic change – a global and interdisciplinary theme', pp. 127–45 in T. F. Malone and J. G. Roederer (eds) *Global Change*, Cambridge: Published on behalf of the ICSU Press by Cambridge University Press.

Yemane, K. (1993) 'Contribution of Late Permian palaeogeography in maintaining a temperate climate in Gondwana', *Nature* 361: 51–4.

Young, A. (1974) 'The rate of slope retreat', pp. 65–78 in E. H. Brown and R. S.

Waters (eds) *Progress in Geomorphology: Papers in Honour of David L. Linton* (Institute of British Geographers Special Publication no. 7), London: Institute of British Geographers.

Young, R. W. (1983) 'The tempo of geomorphological change: evidence from southeastern Australia', *Journal of Geology* 91: 221–30.

Young, R. W. and Bryant, E. A. (1992) 'Catastrophic wave erosion on the southeastern coast of Australia: impact of the Lanai tsunamis ca. 105 ka?', *Geology* 20: 199–202.

Yu, Z.-W. (1994) 'Comparison of paleoclimatic variation periodicity between loess and ocean over the past 2.5 Ma', *Acta Geophysica Sinica* 37: 128–31.

Zeeman, C. (1992) 'Evolution and catastrophe theory', pp. 83–101 in J. Bourriau (ed.) *Understanding Catastrophe*, Cambridge: Cambridge University Press.

Zhang De'er (1994) 'Evidence for the existence of the Medieval Warm Period in China', *Climatic Change* 26: 289–97.

Zhao, M. and Bada, J. L. (1989) 'Extraterrestrial amino acids in Cretaceous/Tertiary boundary sediments at Stevns Klint, Denmark', *Nature* 339: 463–5.

Ziegler, A. M. *et al.* (1993) 'Early Mesozoic phytogeography and climate', *Philosophical Transactions of the Royal Society of London* 341B: 297–305.

INDEX